Molecular Relaxation in Liquids

Molecular Relaxation in Liquids

BIMAN BAGCHI

OXFORD
UNIVERSITY PRESS

Oxford University Press, Inc., publishes works that further
Oxford University's objective of excellence
in research, scholarship, and education.

Oxford   New York
Auckland   Cape Town   Dar es Salaam   Hong Kong   Karachi
Kuala Lumpur   Madrid   Melbourne   Mexico City   Nairobi
New Delhi   Shanghai   Taipei   Toronto

With offices in
Argentina   Austria   Brazil   Chile   Czech Republic   France   Greece
Guatemala   Hungary   Italy   Japan   Poland   Portugal   Singapore
South Korea   Switzerland   Thailand   Turkey   Ukraine   Vietnam

Copyright © 2012 by Oxford University Press

Published by Oxford University Press, Inc.
198 Madison Avenue, New York, New York 10016
www.oup.com

Oxford is a registered trademark of Oxford University Press

All rights reserved. No part of this publication may be reproduced,
stored in a retrieval system, or transmitted, in any form or by any means,
electronic, mechanical, photocopying, recording, or otherwise,
without the prior permission of Oxford University Press.

Library of Congress Cataloging-in-Publication Data

Bagchi, B. (Biman)
  Molecular relaxation in liquids / Biman Bagchi.
     p. cm. Includes bibliographical references and index.
  ISBN 978-0-19-986332-7 (hardcover : alk. paper) 1. Relaxation phenomena.
2. Liquids.   I. Title.
QD543.B118 2012
538'.3–dc23   2011031643

---

1 3 5 7 9 8 6 4 2

Printed in the United States of America
on acid-free paper

Dedicated to the loving memory of my wife, Suhrita Bagchi.

CONTENTS

Preface xiii
Acknowledgments xv
Foreword xvii

1 Basic Concepts 3
   1.1 Introduction 3
   1.2 Response Functions and Fluctuations 4
   1.3 Time-Correlation Functions 6
   1.4 Linear Response Theory 6
   1.5 Fluctuation-Dissipation Theorem 8
   1.6 Diffusion, Friction, and Viscosity 8
   1.7 Summary 10

2 Phenomenological Description of Relaxation in Liquids 12
   2.1 Introduction 12
   2.2 Langevin Equation 13
   2.3 Fokker–Planck Equation 14
   2.4 Smoluchowski Equation 15
   2.5 Master Equations 16
   2.6 The Special Case of Harmonic Potential 16
   2.7 Summary 17

3 Density and Momentum Relaxation in Liquids 19
   3.1 Introduction 19
   3.2 Hydrodynamics at Large Length Scales 20
      3.2.1 Rayleigh–Brillouin Spectrum 22
   3.3 Hydrodynamic Relations between Self-Diffusion Coefficient and Viscosity 24
   3.4 Slow Dynamics at Large Wave Numbers: de Gennes Narrowing 25
   3.5 Extended Hydrodynamics: Dynamics at Intermediate Length Scales 27
   3.6 Mode-Coupling Theory 29
   3.7 Summary 30

4 Relationship between Theory and Experiment 32
    4.1 Introduction 32
    4.2 Dynamic Light Scattering: Probe of Density Fluctuation at Long Length Scales 34
    4.3 Magnetic Resonance Experiments: Probe of Single-Particle Dynamics 36
    4.4 Kerr Relaxation 38
    4.5 Dielectric Relaxation 38
    4.6 Fluorescence Depolarization 39
    4.7 Solvation Dynamics (Time-Dependent Fluorescence Stokes Shift) 40
    4.8 Neutron Scattering: Coherent and Incoherent 41
    4.9 Raman Line-Shape Measurements 43
    4.10 Coherent Anti-Stokes Raman Scattering (CARS) 45
    4.11 Echo Techniques 45
    4.12 Ultrafast Chemical Reactions 47
    4.13 Fluorescence Quenching 47
    4.14 Two-Dimensional Infrared (2D-IR) Spectroscopy 48
    4.15 Single-Molecule Spectroscopy 49
    4.16 Summary 49

5 Orientational and Dielectric Relaxation 51
    5.1 Introduction 51
    5.2 Equilibrium and Time-Dependent Orientational Correlation Functions 55
    5.3 Relationship with Experimental Observables 57
    5.4 Molecular Hydrodynamic Description of Orientational Motion 57
        5.4.1 The Equations of Motion 58
        5.4.2 Limiting Situations 59
    5.5 Markovian Theory of Collective Orientational Relaxation: Berne Treatment 59
        5.5.1 Generalized Smoluchowski Equation Description 60
        5.5.2 Solution by Spherical Harmonic Expansion 62
        5.5.3 Relaxation of Longitudinal and Transverse Components 64
        5.5.4 Molecular Theory of Dielectric Relaxation 64
        5.5.5 Hidden Role of Translational Motion in Orientational Relaxation 65
        5.5.6 Orientational de Gennes Narrowing at Intermediate Wave Numbers 66
        5.5.7 Reduction to the Continuum Limit 67
    5.6 Memory Effects in Orientational Relaxation 68
    5.7 Relationship between Macroscopic and Microscopic Orientational Relaxations 70
    5.8 The Special Case of Orientational Relaxation of Water 72
    5.9 Lattice Models of Orientational Relaxation 74
    5.10 Nonassociated Liquids 75
    5.11 Summary 76

6 Solvation Dynamics in Dipolar Liquid 78
  6.1 Introduction 78
  6.2 Physical Concepts and Measurement 79
    6.2.1 Measuring Ultrafast, Sub-100 fs Decay 83
  6.3 Phenomenological Theories: Continuum-Model Descriptions 86
    6.3.1 Homogeneous Dielectric Models 86
    6.3.2 Inhomogeneous Dielectric Models 89
    6.3.3 Dynamic Exchange Model 91
  6.4 Experimental Results: A Chronological Overview 93
    6.4.1 Discovery of Multiexponential Solvation Dynamics: Phase-I (1980–1990) 93
    6.4.2 Discovery of Subpicosecond Ultrafast Solvation Dynamics: Phase-II (1990–2000) 94
    6.4.3 Solvation Dynamics in Complex Systems: Phase-III (2000–) 95
  6.5 Microscopic Theories 97
    6.5.1 Molecular Hydrodynamics Description 97
    6.5.2 Polarization and Dielectric Relaxation of Pure Liquid 98
      6.5.2.1 Effects of Translational Diffusion in Solvation Dynamics 98
  6.6 Simple Idealized Models 100
    6.6.1 Overdamped Solvation: Brownian Dipolar Lattice 101
    6.6.2 Underdamped Solvation: Stockmayer Liquid 102
  6.7 Solvation Dynamics in Water, Acetonitrile, and Methanol Revisited 102
    6.7.1 The Sub-100 fs Ultrafast Component: Microscopic Origin 104
  6.8 Effects of Solvation on Chemical Processes in the Solution Phase 106
    6.8.1 Limiting Ionic Conductivity of Electrolyte Solutions: Control of a Slow Phenomenon by Ultrafast Dynamics 107
    6.8.2 Effects of Ultrafast Solvation in Electron-Transfer Reactions 107
    6.8.3 Nonequilibrium Solvation Effects in Chemical Reactions 107
      6.8.3.1 Strong Solvent Forces 109
      6.8.3.2 Weak Solvent Forces 110
  6.9 Solvation Dynamics in Several Related Systems 111
    6.9.1 Solvation in Aqueous Electrolyte Solutions 111
    6.9.2 Dynamics of Electron Solvation 111
    6.9.3 Solvation Dynamics in Supercritical Fluids 112
    6.9.4 Nonpolar Solvation Dynamics 112
  6.10 Computer Simulation Studies: Simple and Complex Systems 113
    6.10.1 Aqueous Micelles 114
    6.10.2 Water Pool in Reverse Micelles 114
    6.10.3 Protein Hydration Layer 114
    6.10.4 DNA Groove Hydration Layer 115
  6.11 Summary 115

7 Activated Barrier-Crossing Dynamics in Liquids 117
  7.1 Introduction 117
  7.2 Microscopic Aspects 119

7.2.1 Stochastic Models: Understanding from Eigenvalue Analysis 119
7.2.2 Validity of a Rate-Law Description: Role of Macroscopic Fluctuations 122
7.2.3 Time-Correlation-Function Approach: Separation of Transient Behavior from Rate Law 124
7.3 Transition-State Theory 126
7.4 Frictional Effects on Barrier-Crossing Rate in Solution: Kramers' Theory 127
7.4.1 Low-Friction Limit 129
7.4.2 Limitations of Kramers' Theory 130
7.4.3 Comparison of Kramers' Theory with Experiments 131
7.4.4 Comparison of Kramers' Theory with Computer Simulations 132
7.5 Memory Effects in Chemical Reactions: Grote–Hynes Generalization of Kramers' Theory 132
7.5.1 Frequency Dependence of Friction: General Aspects 138
7.5.1.1 Frequency-Dependent Friction from Hydrodynamics 138
7.5.1.2 Frequency-Dependent Friction from Mode-Coupling Theory 140
7.5.2 Comparison of Grote–Hynes Theory with Experiments and Computer Simulations 142
7.6 Variational Transition-State Theory 143
7.7 Multidimensional Reaction Surface 144
7.7.1 Multidimensional Kramers' Theory 145
7.8 Transition Path Sampling 146
7.9 Quantum Transition-State Theory 148
7.10 Summary 149
Appendix 150

8 Barrierless Reactions in Solution 155
8.1 Introduction 155
8.2 Standard Model of Barrierless Reactions 158
8.2.1 Exactly Solvable Models for Photochemical Reactions 159
8.2.1.1 Oster-Nishijima Model 160
8.2.1.2 Staircase Model 161
8.2.1.3 Pinhole Sink Model 162
8.2.2 Approximate Solutions of Realistic Models 164
8.2.2.1 Delta Function Sink 164
8.2.2.2 Gaussian Sink 165
8.3 Inertial Effects in Barrierless Reactions: Viscosity Turnover of Rate 166
8.4 Memory Effects in Barrierless Reactions 170
8.5 Unusual Features of Barrierless Chemical Reactions 172
8.5.1 Excitation Wavelength Dependence 172
8.5.2 Negative Activation Energy 172
8.6 Multidimensional Reaction Potential Energy Surface 174
8.7 Analysis of Experimental Results 174

        8.7.1 Photoisomerization and Ground-State Potential Energy Surface 174
        8.7.2 Decay Dynamics of Rhodopsin and Isorhodopsin 175
        8.7.3 Conflicting Crystal Violet Isomerization Mechanism 177
    8.8 Summary 177

9 Dynamical Disorder, Geometric Bottlenecks, and Diffusion-Controlled Bimolecular Reactions 180
    9.1 Introduction 180
    9.2 Passage through Geometric Bottlenecks 181
        9.2.1 Diffusion in a Two-Dimensional Periodic Channel 181
        9.2.2 Diffusion in a Random Lorentz Gas 183
    9.3 Dynamical Disorder 184
    9.4 Diffusion over a Rugged Energy Landscape 186
    9.5 Diffusion-Controlled Bimolecular Reactions 190
    9.6 Summary 193

10 Electron-Transfer Reactions 195
    10.1 Introduction 195
    10.2 Classification of Electron-Transfer Reactions 196
        10.2.1 Classification Based on Ligand Participation 196
        10.2.2 Classification Based on Interactions between Reactant and Product Potential Energy Surfaces 196
    10.3 Marcus Theory 197
        10.3.1 Reaction Coordinate (RC) 198
        10.3.2 Free-Energy Surfaces: Force Constant of Polarization Fluctuation 200
        10.3.3 Derivation of ETR Rate 203
        10.3.4 Experimental Verification of the Marcus Theory 206
    10.4 Dynamical Solvent Effects on ETRs (One-Dimensional Descriptions) 208
    10.5 Role of Vibrational Modes in Weakening Solvent Dependence 210
        10.5.1 Role of Classical Intramolecular Vibrational Modes: Sumi–Marcus Theory 210
        10.5.2 Role of High-Frequency Vibration Modes 213
        10.5.3 Hybrid Model of ETR: Crossover from Solvent to Vibrational Control 215
    10.6 Theoretical Formulation of Multidimensional Electron Transfer 216
    10.7 Effects of Ultrafast Solvation on Electron-Transfer Reactions 220
        10.7.1 Absence of Significant Dynamic Solvent Effects on ETR in Water, Acetonitrile, and Methanol 220
    10.8 Summary 221
        Appendix 222

11 Förster (or, Fluorescence) Resonance Energy Transfer (FRET) 226
    11.1 Introduction 226
    11.2 A Brief Historical Perspective 229
    11.3 Derivation of Förster Expression 230
        11.3.1 Expressions for Emission (or Fluorescence) Spectrum 234

11.3.2 Absorption Spectrum 237
11.3.3 The Final Förster Expression 238
11.4 Applications of Förster Theory to Chemistry, Biology, and Materials Science 239
11.4.1 FRET-Based Glucose Sensor 239
11.4.2 FRET and Macromolecular Dynamics 239
11.4.3 FRET and Single-Molecule Spectroscopy 243
11.4.4 Beyond Organic Dyes as Donor–Acceptor Pairs 247
11.4.5 FRET and Conjugated Polymers 249
11.5 Beyond Förster Formalism 252
11.5.1 Orientation Factor 252
11.5.2 Point-Dipole Approximation 253
11.5.3 Contribution of Optically Dark States 254
11.6 Summary 257

12 Vibrational-Energy Relaxation 259
12.1 Introduction 259
12.2 Isolated Binary Collision (IBC) Model 261
12.3 Landau–Teller Expression: The Classical Limit 263
12.4 Weak-Coupling Model: Time-Correlation-Function Representation of Transition Probability 265
12.5 Vibrational Relaxation at High Frequency: Quantum Effects 268
12.6 Experimental Studies of Vibrational-Energy Relaxation 271
12.7 Computer-Simulation Studies of Vibrational-Energy Relaxation 272
12.7.1 Vibrational-Energy Relaxation of Water 272
12.7.2 Vibrational-Energy Relaxation in Liquid Oxygen and Nitrogen 274
12.8 Quantum Interference Effects on Vibrational-Energy Relaxation in a Three-Level System: Breakdown of the Rate Equation Description 275
12.9 Vibrational Life Time Dynamics in Supercritical Fluids 277
12.10 Summary 279

13 Vibrational-Phase Relaxation 280
13.1 Introduction 280
13.2 Kubo–Oxtoby Theory of Vibrational Line Shapes 282
13.3 Homogeneous vs. Inhomogeneous Linewidths 287
13.4 Relative Role of the Attractive and Repulsive Forces 289
13.5 Vibration–Rotation Coupling 289
13.6 Experimental Results of Vibrational-Phase Relaxation 290
13.6.1 Semiquantitative Aspects of Dephasing Rates in Solution 291
13.6.2 Subquadratic Quantum Number Dependence 291
13.7 Vibrational Dephasing Near the Gas–Liquid Critical Point 292
13.8 Multidimensional IR Spectroscopy 292
13.9 Summary 294

14 Epilogue 296

Index 298

PREFACE

Molecular Relaxation in Liquids grew out of two courses that I taught to chemistry graduate students at the Indian Institute of Science, Bangalore and elsewhere. One of the courses, entitled "Relationship between theory and experiment," was designed to familiarize students with the rich ideas and elegant relations that relate spectroscopic experiments (IR, NMR, Raman, dielectric relaxation, different kinds of fluorescence experiments including resonance energy transfer and time-dependent Stokes shift) to theoretical expressions derived from the time-correlation-function formalism of statistical mechanics. The second course entitled "Chemical dynamics in solutions" covered microscopic aspects of rate laws, activated chemical reactions, barrier less reactions, electron transfer reactions, vibrational relaxation and related topics, again with an emphasis on correlation between theory and experimental results.

The primary aim of this book is to bridge the gap between theory and experiment. This is not an easy task as the theories (and of course equations) to be dealt with are rather involved (like in Navier–Stokes hydrodynamics, mode-coupling theory), and are often foreign to the usual chemistry students. The time-domain experiments employed to study molecular relaxation in liquids are also quite sophisticated, sometimes involving multiple laser pulses measuring nonlinear response, with time scales ranging from a few femtoseconds to hundreds of nanoseconds.

A large number of experimental studies have appeared in the last three decades. We have summarized the main results aimed at giving a general yet quantitative picture and minimized the details. Wherever possible, I have tried to provide a physical picture because it is hard for students to follow the theoretical formalism without a guiding physical picture, particularly for chemistry major students.

While teaching this course, I and the students were acutely aware of the shortage of literature where these topics have been coherently discussed, despite the importance of these areas in modern day physical chemistry/chemical physics.

While I have tried to cover most of the relaxation phenomena we encounter in the study of physical chemistry/chemical physics, preference has been given to topics and subtopics where solvent effects are important and/or solvent plays an important role. As emphasized repeatedly, such phenomena have been studied intensely in the last three to four decades and a certain amount of understanding has evolved so that a book can be attempted.

The writing of the book turned out to be an arduous task. Because this book does not cover topics found in standard text books on statistical mechanics and quantum mechanics, I had to rely quite heavily on research articles and reviews. We hope that this will be of use to junior PhD and advanced Masters students across the globe.

ACKNOWLEDGMENTS

This book would not have been possible without the help of many of the graduate students of our group. These students typed many of my handwritten pages, drew most of the figures, and prepared the references. I am particularly indebted to Mr. Rakesh S. Singh who participated in the writing of Chapters 7 and 8 and diligently read through other chapters several times and gave suggestions for improvement. Ms. Sangeeta Saini helped with the writing of the Förster energy chapter. Mr. Saikat Banerjee drew many of the figures and Mr. Rajib Biswas helped greatly in formatting and referencing. Several other students (Mr. Biman Jana, Ms. Sushmita Roy, Mr. Mantu Santra, Ms. Rikhia Ghosh, and Ms. Sarmistha Sarkar) helped at various stages. I also thank my two sons, Kaushik and Kushal, for obliging when (often) asked to read typed pages and also for constant encouragement.

FOREWORD

GRAHAM R. FLEMING
Vice Chancellor for Research, University of California Berkeley

The past 30 years has seen great progress in the microscopic understanding of dynamical processes in liquids and solutions. Ultrafast spectroscopy, modern statistical mechanics, and numerical computer simulation have been developed with highly productive synergies and feedbacks. Biman Bagchi has been at the center of the interplay of theory, simulation, and experiment during this whole period, making it highly appropriate that he has written a book aimed at placing the theoretical techniques and results in the context of the key experimental results.

The large body of experimental work on liquid dynamics using dynamic light scattering, dielectric relaxation, magnetic resonance, and Kerr relaxation provided the basis for the rapid progress initiated by the advent of reliable sources of ultrafast-light pulses. The two areas that defined the initial and highly productive phase of feedback between theory and experiment in liquid solutions are: activated barrier crossing, epitomized by *trans–cis* isomerization in stilbene, and polar solvation dynamics studied experimentally via the time-resolved Stokes shift. These studies led to ideas such as frequency-dependent friction and inertial solvation becoming part of the language through which chemistry in solution is described, and initiated discussions, which continue today, on the microscopic nature of ultrafast solute–solvent interactions.

From reactions with significant activation barriers, attention turned to barrierless reactions where time-scale separation cannot be assumed; the theoretical models developed by Bagchi, Fleming and Oxtoby, by Sumi and Marcus, and by Agmon and Hopfield, proved remarkably versatile and useful in a wide variety of contexts. Diffusion-controlled reactions, vibrational relaxation, and electron and energy transfer complete the list of elementary chemical processes in solution considered in this book.

From the experimental perspective, ultrafast spectroscopy continues to develop new techniques, while single-molecule spectroscopy enables the removal of the ensemble average and the distinction between "average" and "typical" behavior to be made. For the next generation of theorists and experimentalists alike, *Molecular Relaxation in Liquids* provides a platform on which to complete the agenda of understanding and predicting chemical dynamics in solution with a sophistication rivaling that of gas-phase chemistry.

Molecular Relaxation in Liquids

1

Basic Concepts

1.1 INTRODUCTION

Certain basic concepts and general ideas occur repeatedly in our discussions of dynamics of molecular relaxation processes in condensed phases. Although many of these ideas have their foundation in many-body theories and time-dependent statistical mechanics, there are some that are phenomenological in nature and cannot be precisely derived from first principles considerations. Phenomenology is often based on intuition and experience. It has played an honorable role in understanding the dynamics of liquids, because the complexity of the problems involved often makes a first principles calculation very difficult, even with present-day computer.

A case in point is the study of viscosity dependence of the rate of isomerization of tri-phenyl methane dyes (such as malachite green or crystal violet) in liquid alkanes. If we attempt a fully microscopic description of this reaction starting from a Hamiltonian description (which is of course the ultimate goal of a theory), then the theory involves many coordinates and coupling among them and it quickly becomes too cumbersome to implement in any realistic way. In such cases, a simpler phenomenological description in terms of a diffusion equation with a position-dependent reaction sink (or reaction window) term to model the reaction proves to be more useful and revealing. That is, in complex situations it is often more desirable to use our insight and intuition to first formulate a correct phenomenological description. Subsequently, depending on the success of such a description, a microscopic approach can be initiated.

An additional factor in favor of a phenomenological description is that experimentally measured quantities are often macroscopic, like the viscosity and polarity dependence of the rate of a chemical reaction. Therefore, microscopic calculations would involve theories that would have many layers. In such cases, a phenomenological description based on hydrodynamic, continuum models may provide more insight into the problem.

However, development of a correct phenomenological description is often not easy. There are several fundamental concepts that guide us in developing such descriptions. In this chapter, we shall present some of these concepts so that we can refer to them in subsequent chapters. Many of the detailed theoretical aspects will be discussed along with the problems studied so that the reader can proceed without any break in the continuity.

Perhaps the most important theoretical quantity in understanding liquid state dynamics is a time correlation function (TCF) of a relevant dynamical quantity, such as number density, polarization, dipole moment, or orientation of a molecule. Most of the experimental measurements are expressed and understood in terms of a TCF, as discussed in Chapter 4 where we discuss the relationship between theory and experiments. The TCF formalism was developed largely by Green, Kubo, and Zwanzig [1–5]. In the following we develop the relevant concepts and the motivation behind this beautiful formalism.

Throughout the book, as in this chapter, the emphasis shall be on the fundamental concepts that will be revisited whenever necessary. The detailed derivations of exact relationships have been omitted as they are available in several monographs.

1.2 RESPONSE FUNCTIONS AND FLUCTUATIONS

As the name signifies, response functions provide a quantitative measure of the response of a system to external fluctuations. These functions play important roles in the description of the system. Examples of static response functions are specific heat, compressibility, susceptibility, or dielectric constant. At equilibrium, the thermodynamic response function is given by the second derivatives of the free energy of the system. For example, specific heat is the second derivative of Helmholtz free energy with respect to temperature. Response functions can therefore be considered as the "springs" or the force constants of the system.

It is important to realize that these functions are determined at equilibrium, by the natural fluctuations of the system. For example, the specific heat and isothermal compressibility (κ) are determined by the mean square energy and number fluctuations, as shown by the following expressions:

$$C_V = \frac{\langle (\Delta E)^2 \rangle}{k_B T^2}, \tag{1.1}$$

$$\kappa = \frac{\langle (\Delta N)^2 \rangle}{\rho k_B T \langle N \rangle}, \tag{1.2}$$

where $\langle (\Delta E)^2 \rangle$ is the mean square fluctuation in the total energy E of the system under consideration. Similarly, $\langle (\Delta N)^2 \rangle$ is the mean square number fluctuation, while k_BT is Boltzmann constant (k_B) times the temperature T. If the distribution of energy is Gaussian, then the width of this distribution is proportional to the square root of the specific heat. A similar relation holds true for the width of the distribution of number fluctuation and the compressibility of the system. In fact, a similar fluctuation formula exists for each response function. As mentioned, the response functions can be viewed as the force constants of the system because free energy, to the lowest order, is a quadratic function of these fluctuations. Therefore, these response functions also control the time dependence of the decay of these fluctuations. Under normal stable conditions (away from any phase transition), the respective relaxation functions are well-behaved functions of temperature, pressure, and external electric and magnetic fields.

The response functions can depend on both space and time. The spatial dependence is determined by the correlation between the fluctuations of the relevant quantity at two different positions. In a homogeneous system, this correlation depends only on the separation between the two locations and not on the precise positions of the two locations. Because of this translational invariance it is meaningful to describe the response functions as a function of the wave number (k or q). Many experiments measure the wave-number dependence of response functions directly. The thermodynamic limit of the response functions is recovered in the long-wavelength or small wave-number limit. Similarly, the response functions also depend on time, and when the system is at equilibrium, it makes sense to describe dynamics through its frequency (ω) dependence. Many experiments measure response functions directly as functions of wave number and frequency, as described in Chapter 4.

Dynamic response functions play important roles in describing solvent effects on molecular relaxation in liquids. They provide a quantitative measure of the response of the system to time-dependent perturbations. Examples are viscosity, diffusion, frequency-dependent dielectric constant, and frequency-dependent conductivity.

In fact, viscosity and diffusion coefficient are also expressed as functions of frequency. The dynamic response functions are defined as generalizations of their static counterpart, in terms of integrations over time-correlation functions. For example, viscosity is defined as an integral over stress–stress time-correlation function, diffusion coefficient as integration over velocity–velocity time-correlation function, dielectric function as integration over total dipole moment time-correlation function, and so on.

Note that these time (or, dynamic) correlation functions are also evaluated at equilibrium! Thus, these dynamic functions are determined by the *natural motion* (or natural dynamics) of the system as they describe response to small perturbations. This is the essence of a famous theory, known as linear response theory, formulated first by Kubo in a landmark paper in 1957 [1,2].

The rate of a chemical reaction can also be viewed as a dynamic response function of the reacting system. As an external parameter, such as temperature or pressure, is changed, the rates of both forward and backward reactions change and the equilibrium is restored. Thus, the rate of an elementary chemical reaction also has a time-correlation-function representation, to be discussed in Chapter 7.

Solvation dynamics and vibrational phase relaxation also have respective time-correlation-function representations.

As the dynamic response functions are determined by the relevant time-correlation functions, it is important to understand properties of TCFs.

1.3 TIME-CORRELATION FUNCTIONS

As remarked above, the dynamic response functions can be expressed as integrals over the time-correlation functions of the relevant quantities. The time-correlation function of a dynamic variable is defined in the following way. Let $A(t)$ be the value of a dynamic variable at time t. Then the time-correlation function $C_{AA}(t)$ is defined by the following expression:

$$C_{AA}(t) \equiv \langle A(0)A(t) \rangle = \lim_{T \to \infty} \frac{1}{T} \int_0^T ds\, A(s)A(t+s). \quad (1.3)$$

This expression defines a time average. The average is taken over a time trajectory of the system. Properties of time-correlation functions have been discussed by Hansen and McDonald [3].

Time-correlation functions are immensely useful quantities as they provide microscopic expressions of the transport coefficients. As already emphasized, many time-domain experiments directly measure the time-correlation function of important dynamical variables, such as density, orientation, polarization, or normal coordinate position of a vibration. For example, fluorescence depolarization experiments, where the decay of an initially optically created orientational anisotropy is measured, give information about the time scale of decay of orientational correlations of anisotropic molecules.

The relation between time-correlation functions and the conjugate experimental observables is provided by the linear response theory, discussed next.

1.4 LINEAR RESPONSE THEORY

Linear response theory (LRT) allows us to use the natural dynamics of an unperturbed, equilibrium liquid to describe relaxation of the same under nonequilibrium conditions. Strictly speaking, this theory holds only when the system is slightly perturbed and it remains close to equilibrium. In practice, however, this restriction of closeness to equilibrium does not appear to be too important in most applications. The formal aspect of the theory was developed fully by Ryogo Kubo [2] who used first-order perturbation theory in Liouville equation to obtain a beautiful relation between time-dependent external perturbation on the system (such as an electric field) and the system's response (such as electric polarization).

This theory, in a limited form, is also known as Onsager's linear regression theorem. In order to understand the true meaning of this theory, consider a dipolar liquid subjected to an external electric field in the Z-direction and kept under such a condition for a very long time. Because of the action of the electric field, the

Basic Concepts

liquid system develops a net dipole moment in the Z-direction, denoted by $M_z(t)$, $M_z(t)$ being a constant at times less than zero. Let us now imagine that at time $t = 0$ we turn off the external electric field. At subsequent times, the magnitude of the electric moment will decrease to zero. Let us further assume that this decay is exponential, with a time constant τ_M

$$M_z(t) = M_z(t=0) \exp(-t/\tau_M). \tag{1.4}$$

Note that τ_M is a macroscopic relaxation time.

Let us now consider a small fluctuation in the *average* net dipole moment of the liquid at equilibrium, in the absence of any field when the net electric dipole moment of the system is zero. Although the average moment is zero, there are always fluctuations about the average value. Let us denote this fluctuation by $\delta M_z(t)$. Although the average of these fluctuations over a long time is zero, they are correlated with each other in the short time. This correlation can be expressed in terms of the following normalized time-correlation function:

$$C_M^{eq}(t) = \frac{\langle \delta M_z(0) \delta M_z(t) \rangle}{\langle \delta M_z(0) \delta M_z(0) \rangle}. \tag{1.5}$$

The normalized time-correlation function $C_M^{eq}(t)$ so defined decays from unity to zero as time t is increased from zero to infinity. As mentioned earlier, these time-correlation functions are evaluated at equilibrium, that is, in the absence of the external perturbation that can also produce the effect, that is, a non-zero macroscopic electric polarization.

The central idea of linear response theory is that

$$C_M^{eq}(t) = \frac{M_z(t)}{M_z(t=0)}. \tag{1.6}$$

That is, $C_M^{eq}(t)$ also decays exponentially with the same time constant τ_M of Eq. (1.4)! Of course the decay need not be exponential and Eq. (1.6) is valid even when the decay is nonexponential.

Although experiments often induce rather large perturbation to the system and measure the decay of this perturbation, the linear response theory is found to be valid in a wide range of situations. Linear response theory relates this decay to the decay of the relevant time-correlation function. The reason for the success of the linear response theory is that relaxation subsequent to a macroscopic perturbation can be thought of as relaxation from state to state via small fluctuations.

Thus, the linear response theory provides us with the necessary theoretical tool to understand the experimental results. We can now try to understand non-equilibrium response from equilibrium time-correlation functions which can be obtained from system's dynamics at equilibrium.

1.5 FLUCTUATION-DISSIPATION THEOREM

Linear response theory (LRT) and the fluctuation-dissipation theorem (FDT) are closely related and together they relate response functions to time-correlation functions. Dissipation often accompanies the dynamic response of the system measured in experiments, while fluctuation, as mentioned above, determines this response through a time-correlation function. These two are related.

Fluctuation-dissipation theorem is best understood in terms of the following simple Langevin equation of motion of the velocity $V(t)$ of a tagged particle in solution:

$$\dot{V} = -\zeta V + F(X) + R(t). \qquad (1.7)$$

In this equation, ζ is the friction, $F(X)$ is a systematic force (such as an external field) on the tagged particle, and $R(t)$ is the random force term on the same. The random force is the fluctuation due to the motions of the atoms and molecules of the system and the friction ζ is the dissipation. The (first) fluctuation-dissipation theorem is given by the following relation between the two quantities:

$$\langle R(0)R(t) \rangle = 2k_B T \zeta \delta(t), \qquad (1.8)$$

where $\delta(t)$ is a Dirac delta function. The above equation embodies a remarkable result. It relates the dissipation (that is, the friction ζ) to the fluctuation (that is, the random force $R(t)$). This physically meaningful result can be formally derived from first principles by using the projection operator technique [1].

There is a generalization of Eq. (1.8) where the random force is correlated in time. Then the friction becomes time dependent. This is called the second fluctuation-dissipation theorem, although the reason for such a name is not quite clear. The Langevin equation with time- (or, frequency-) dependent friction is called a generalized Langevin equation. We shall discuss this equation and its application later in Chapters 5 to 7.

Very nice discussions of linear response theory, fluctuation dissipation theorem, and Langevin equations are available in the monographs by Zwanzig [1] and Risken [5]. Therefore, we refrain from discussing them any further here. In subsequent chapters, we shall briefly review the necessary details to make the discussions self-contained.

1.6 DIFFUSION, FRICTION, AND VISCOSITY

The diffusion coefficient of a molecule in the liquid phase is perhaps the most important dynamical quantity in molecular relaxation processes in liquids. Experimental results, starting from photoisomerization reactions of dye molecules to vibrational energy relaxation of a chemical bond immersed in liquids, have all been rationalized in terms of the diffusion coefficient. There are several different types of diffusion coefficients, often causing confusion to a newcomer in the field.

Basic Concepts

The most used and common quantity is the self-diffusion coefficient D_s. The self-diffusion coefficient of a tagged molecule is defined by the following well-known expression, due to Einstein:

$$D_s = \lim_{t \to \infty} \frac{(\Delta \mathbf{r}(t))^2}{6t}, \tag{1.9}$$

where $\Delta \mathbf{r}(t)$ is the displacement during the time interval t. The above definition of the self-diffusion coefficient can be cast into an equally meaningful form involving the velocity time-correlation function:

$$D_s = \frac{1}{3} \int_0^\infty dt\, C_V(t), \tag{1.10}$$

where $C_V(t)$ is the velocity time-correlation function of the tagged molecule, defined in the same way as in Eq. (1.3).

In dense liquid the velocity time-correlation function shows an interesting time dependence which is quite different from that in the gas phase. The decay is exponential-like in the gas phase while there is a pronounced negative region in the liquid phase. The negative in $C_V(t)$ arises from backscattering of the tagged molecule due to collision with its surrounding molecules and is a consequence of the caging of any molecule in a dense liquid by its neighbors. This caging leads to a lowering of the diffusion coefficient in the liquid phase from its gas phase value, which can be understood from the negative value of $C_V(t)$ and Eq. (1.10).

In the long time, a decay of the form $t^{-d/2}$ appears in the velocity time-correlation function. This long-time tail has a hydrodynamic origin and is usually explained in terms the backflow effect. The latter is a fluid current set up by the moving tagged molecule that comes back and acts on it at a later time.

The complex time dependence of the velocity correlation function is reflected in its frequency dependence, which is obtained by Fourier transforming the velocity time-correlation function, given by

$$C_V(\omega) = \int_0^\infty dt\, e^{i\omega t} C_V(t). \tag{1.11}$$

In many theoretical studies of molecular relaxation in liquids, one needs the frequency-dependent friction of the medium on the relevant motion, such as translation or rotation of a molecular moiety, as an input. Fortunately, there is an exact relation between frequency-dependent diffusion and friction that is known as the generalized Einstein relation, given by

$$D_s(\omega) = \frac{k_B T}{m(i\omega + \zeta(\omega))}. \tag{1.12}$$

The zero-frequency limit of this equation is the well-known Stokes–Einstein relation,

$$D_s = \frac{k_B T}{\zeta}. \tag{1.13}$$

However, *friction on a molecule is not the quantity that can be measured experimentally*. The usual measure of frictional force on a molecule is the viscosity η of the liquid. The relation between the friction and the viscosity has proven to be difficult to obtain, as one (the viscosity) is a collective property of the liquid, while the other is a single-particle property. Hydrodynamics gives a simple relation between the two which depends on the boundary condition

$$\zeta = C\eta R, \tag{1.14}$$

where the constant C is equal to 4π for the slip boundary condition, while it is equal to 6π for the stick boundary condition. Eq. (1.14) is strictly valid for translational motion of a large body through a liquid medium. However, one often finds that Eq. (1.14) provides a reliable estimate of friction (and hence of diffusion) of a small molecule, which is comparable in size to (or, larger than) the size of the solvent molecules.

An expression similar to Eq. (1.14) has also been derived for rotational motion. For the rotation of a molecule, there is no friction from the surrounding medium under the slip boundary condition. For the stick boundary condition, the rotational friction (ζ_{rot}) depends strongly on the radius R and the expression is given by

$$\zeta_{rot} = 8\pi \eta R^3. \tag{1.15}$$

Since many of the theoretical descriptions of molecular relaxation in liquids uses diffusion as an ingredient, relations such as Eq. (1.12) and Eq. (1.14) assume additional importance.

1.7 SUMMARY

From a practical point of view, molecular relaxation processes can, potentially, provide us with valuable information on the microscopic processes that control the rate and even the yield of a chemical reaction. We shall discuss examples of such relaxation processes in this book. Since experiments provide us with certain time constants and their dependencies on such properties as temperature, pressure, and viscosity, we need a reliable theory to connect the experimental results with the underlying microscopic processes. Theories developed can be divided into two broad categories: phenomenological and molecular (or, first principles). However, whatever approach one adopts, there are certain basic ideas and concepts whose understanding is essential to progress. In this chapter, we have discussed in simple terms and short space, several such concepts. Fortunately, there are excellent text books that the interested reader can turn to for more details. A list is given below.

In the next two chapters we shall discuss first the phenomenological approach (**Chapter 2**) and the molecular approach (**Chapter 3**) employed to understand relaxation in liquids.

REFERENCES

1. R. Zwanzig, *Non-equilibrium Statistical Mechanics*, Oxford University Press (2001).
2. R. Kubo, *Reports on Progress in Physics*, 29, 255 (1966).
3. J. P. Hansen and I. R. McDonald, *Theory of Simple Liquids*, Academic Press (2006).
4. D. Chandler, *Introduction to Modern Statistical Mechanics*, Oxford University Press (1987).
5. H. Risken, *The Fokker-Planck Equation: Methods of Solution and Applications*, Springer (1989).

2

Phenomenological Description of Relaxation in Liquids

2.1 INTRODUCTION

Many of the systems involved in experimental studies of molecular relaxation processes in liquids are quite complex. The same is true in the real, natural or biological, world. In contrast, theoreticians often model molecules or groups of atoms involved, say, in a chemical reaction, as spheres and ellipsoids. To model dipolar liquids, one models real molecules characterized by a charge distribution by a point dipole located at the center of a sphere. Motivation clearly came from the fact that such simple models can be studied rigorously by using microscopic methods, starting from intermolecular potentials, following the methods prescribed by time-dependent statistical mechanics. The situation has recently improved dramatically because of the development of powerful simulation methods and the availability of fast computers with enormous disk space. Still one is saddled with the accuracy of the intermolecular potentials that are employed and many other computational issues.

Let us take the well-known example of isomerization of stilbene in hexane. In experimental studies, one measures the temperature, pressure, and viscosity dependence of the rate of *cis* to *trans* or *trans* to *cis* isomerization, following the preparation of the initial state by optical excitation. A fully microscopic study for the answers to the desired questions is prohibitively difficult, even using computer simulations. Additionally, a fully microscopic study of such a problem, although often stated to be an ultimate goal, might not really be necessary for each system. This is because the questions explored and answers sought can be of a rather general nature, like the reason for the fractional viscosity dependence of the rate of an isomerization reaction in solution. Therefore, it is more convenient to explore the alternate avenue of using a phenomenological description that can lead to a physically understandable picture of a complex dynamical process.

Phenomenology has a time-honored place in the study of molecular relaxation processes in liquids, starting with the initial studies of Einstein (Brownian motion) and Smoluchowski (precipitation of colloids). Such descriptions are not only elegant but have also been tremendously successful in providing a physical description of the complex processes they model. Additionally and surprisingly, they have been often semiquantitatively successful in explaining observed correlations, for example, the correlation between the observed rate and the viscosity of the medium.

The success of the phenomenological descriptions of molecular relaxation processes is largely because they are based solidly on several fundamental principles, like the fluctuation-dissipation theorem and linear response theory, both discussed in Chapter 1. The relationship between measured quantities and experimental observables is also clear and transparent in such an approach, mainly because the experimental variables like viscosity and dielectric function enter directly through phenomenological expressions.

However, despite all the successes of phenomenological descriptions, there is a limit to which such descriptions can be useful. They are known to fail spectacularly in many cases, like in supercooled liquids, where a microscopic approach (such as mode-coupling theory) becomes essential.

Nevertheless, the vast majority of theoretical descriptions of molecular relaxation processes in liquids employ phenomenological descriptions of one kind or another and we shall use such descriptions extensively in this book. In the following we discuss a few of these equations, starting from their derivation to the presentation of solutions of the equations in a few special cases.

2.2 LANGEVIN EQUATION

In the phenomenological descriptions, one usually models the reactive motion (it may be an angular twist in isomerization or displacement of normal coordinates in vibrational dynamics) as the motion of the solute along the said reactive coordinate in a viscous (or viscoelastic) continuum. The equation of motion is most often a Langevin equation (LE), which was discussed in the preceding chapter but now we will spend more time on this important equation. LE is given by

$$m\dot{V} = -\zeta V + F(X) + R(t), \qquad (2.1)$$

where V is the velocity of the solute, $V = \dot{X}$, where X is the reaction coordinate. ζ is the friction on X, $F(X)$ is called the systematic force, and $R(t)$ is the random force. Friction ζ and random force R are related by the fluctuation-dissipation theorem

$$\langle R(0) \cdot R(t) \rangle = d k_B T \zeta \delta(t), \qquad (2.2)$$

with $\langle R(t)\rangle = 0$. Here d is the dimension of the reaction coordinate, $k_B T$ is Boltzmann constant times temperature and $\delta(t)$ is a Dirac delta function in time. The systematic force term $F(X)$ plays an important role. In a chemical reaction, it is derived from the reaction free-energy surface, with two minima and a maximum, in the simplest possible case. In vibrational relaxation, it describes the harmonic (or anharmonic) potential energy of the normal mode. In the case of motion of an atom on a surface, it describes the surface potential faced by the atom.

The nature of the reaction coordinate determines the nature of the random force term and hence the friction. For example, for translational motion, it is the translational motion that is vastly different from the rotational friction that a molecule experiences during its rotational motion.

The Langevin equation is a stochastic equation. While this can again be solved through computer simulation, it is, in its present form, not amenable to theoretical analysis. The Langevin equation, however, can be used to derive an equation of motion for the probability distribution, $P(X, V, t)$. This equation of motion is known either as Kramer's equation or the Fokker–Planck equation and has proved to be an extremely useful equation of motion and is used in many branches of physics, chemistry, and biology. We shall have several uses for the Fokker–Planck equation, which are detailed below.

In its simplest form, as presented below in Eq. (2.3), the Fokker–Planck equation describes the motion of a single particle whose density is conserved but not its momentum. Both the random force and the friction are the results of the particle's interactions with their nearest neighbors, but the motions of these nearest-neighbor molecules are neglected. Therefore, this is a reduced equation of motion.

There appears to exist in the literature only one microscopic derivation of the Langevin equation from a Hamiltonian description. This is the well-known derivation of Zwanzig for the special case where a solute atom is linearly coupled to a large number of harmonic oscillators [1, 2]. Zwanzig actually derived a *generalized* Langevin equation for the motion of the solute atom where the friction on the solute is time dependent by using projection of an operator technique that allowed him to arrive at a reduced description of the equation of motion only in terms of the solute atom's coordinate (position and momentum). We refer to the original paper of Zwanzig for this derivation [2].

2.3 FOKKER–PLANCK EQUATION

The Fokker–Planck equation in its simplest one-dimensional form is given by [1, 3–7]

$$\frac{\partial P(X, V, t)}{\partial t} = -V\frac{\partial P}{\partial X} - \frac{F}{m}\frac{\partial P}{\partial V} + \zeta\frac{\partial}{\partial V}\left[V + \frac{k_B T}{m}\frac{\partial}{\partial V}\right]P, \qquad (2.3)$$

where F is the systematic force due to the external potential and ζ is the friction. In the description of chemical reactions, the systematic force is usually given by a bistable potential where a barrier separates the two minima corresponding to reactant and product.

Note that if there is no noise or friction, the preceding Fokker–Planck equation reduces to the one-particle Liouville equation for the Hamiltonian,

$$H = \frac{p^2}{2m} + U(X), \qquad (2.4)$$

Here $U(X)$ is the external potential acting on the solute. As is clear from above, this partial differential equation is still quite formidable. This is one of the simplest equations used in the phenomenological description! What makes it complex is the presence of the noise term. However, without the noise term, the distribution function $P(X, V, t)$ conserves energy and never achieves the desired equilibrium Boltzmann distribution in the long time, starting from an initial nonequilibrium distribution. The noise term ensures that the equilibrium Boltzmann distribution is achieved in the long time.

There are several ways to derive the Fokker–Planck (F-P) equation, within a phenomenological description. The direct one starts from the ordinary Langevin equation given above (Eq. (2.1)). Just like the Langevin equation, the F-P equation conserves the single-particle density but not the momentum. For a nice and elegant but at the same time general, derivation we refer to Zwanzig's monograph [1].

2.4 SMOLUCHOWSKI EQUATION

The single-particle Fokker–Planck equation (Eq. (2.3)) is often further simplified if the friction is so large that the momentum of relaxation is much faster than the relaxation of position. In that limit one can assume that momentum relaxation is always at equilibrium. The Fokker–Planck equation then reduces to a simpler equation containing only the position dependent distribution, $P(X, t)$. The resulting equation of motion is called the Smoluchowski equation and is given by

$$\frac{\partial P(X, t)}{\partial t} = D \frac{\partial}{\partial X} \left[\frac{\partial}{\partial X} - \beta F \right] P(X, t), \qquad (2.5)$$

where $F = -\frac{\partial U}{\partial X}$ (U is the potential), $D = \frac{k_B T}{\zeta}$ is the diffusion coefficient along the coordinate X.

Equation (2.5) is still a partial differential equation in two variables. This perhaps provides the simplest description of a relaxation process. Note that Eq. (2.5) becomes a diffusion equation in the absence of the force term. The Smoluchowski equation admits of an analytical solution only when the force term is linear in the coordinate X, which is the case of a harmonic potential. We discuss the special case of harmonic potential below. Let us point out here that even after using several approximations that lead to Eq. (2.5), and despite the fact that the equation provides the simplest description, one can still obtain rich physics and physical insight from this description.

2.5 MASTER EQUATIONS

In many chemical processes, the probability distribution of reactant can change by transitions between *discrete states*. These discrete states might be intermediates along the reactive motion of a chemical reaction or metastable states at various conformations. For such processes, an appropriate theoretical description is provided by a master equation, which is also known as the equation for population distribution in birth and death processes. If $P_i(t)$ is the probability of the system to be in state i at time t, then an equation of motion for rate of change of $P_i(t)$ due to transition between different states is given by

$$\frac{dP_i(t)}{dt} = -\left(\sum_{j \neq i} W_{ij}\right) P_i(t) + \sum_{j \neq i} W_{ji} P_j(t), \qquad (2.6)$$

where W_{ij} is the rate of transition from state i to j. Such an equation is called a master equation. Clearly, we can have a master equation in the continuum space also with x as the variable,

$$\frac{\partial P(X,t)}{\partial t} = -\left(\int dX' W(X \to X')\right) P(X,t) + \int dX' W(X' \to X) P(X',t). \qquad (2.7)$$

The above equation is sometimes called Kolmogorov equation and can be used to derive diffusion equation.

Equation (2.6) has a useful matrix representation. We denote the population $P_i(t)$ by a vector $\mathbf{P}(t)$ and the transition term by a matrix $\mathbf{W}$, then the master equation can be written in a compact form

$$\frac{d\mathbf{P}}{dt} = \mathbf{W} \cdot \mathbf{P}. \qquad (2.8)$$

This equation can be solved to obtain the following formal solution:

$$\mathbf{P}(t) = e^{\mathbf{W}t} \mathbf{P}(t=0). \qquad (2.9)$$

The matrix $\mathbf{W}$ can often be diagonalized to obtain a solution for $\mathbf{P}(t)$. We shall encounter some applications of the solution given by Eq. (2.9) in later chapters.

2.6 THE SPECIAL CASE OF HARMONIC POTENTIAL

In many applications, the potential $U(X)$ is a simple harmonic potential. For convenience, we shall adopt a one-dimensional description and write the potential $U(X)$ as

$$U(X) = \frac{1}{2}\mu\omega^2 X^2, \qquad (2.10)$$

where μ is the effective mass and ω is the harmonic frequency. With the harmonic potential both the Smoluchowski and Fokker–Planck equations have closed-form solutions that have proven to be extremely useful. For the Smoluchowski equation the solution is given by [3]

$$P(X,t|X',t') = \sqrt{\frac{\gamma}{2\pi D\left(1 - e^{-2\gamma(t-t')}\right)}} \exp\left[\frac{-\gamma\left(X - e^{-\gamma(t-t')}X'\right)^2}{2D\left(1 - e^{-2\gamma(t-t')}\right)}\right], \tag{2.11}$$

with initial condition $P(X,t'|X',t') = \delta(X - X')$ and $\gamma = \omega^2/\zeta$. Thus, the time dependence of the probability distribution is non-trivial even in such a simple description of relaxation.

The general solution for the probability density with the initial distribution $P_0(X')$ is given by

$$P(X,t) = \int dX' P(X,t|X',0) P_0(X'). \tag{2.12}$$

Thus, the transition probability serves as a Green's function.

For the Fokker–Planck equation (Eq. (2.3)), the solution is more involved as both position (X) and velocity (V) variables are involved. A complete solution is available in Ref. 4.

2.7 SUMMARY

The phenomenological approach to relaxation phenomena in liquids has evolved over many years, with many celebrated scientists like Einstein, Smoluchowski, Kramers, Kubo, and Zwanzig making valuable contributions to the area. In many cases, the phenomenological equations (Langevin, Fokker–Planck, Smoluchowski) provide the most direct approach to a complex problem. Even with rather bold approximations underlying phenomenology, they provide accurate explanation of the observed results. As emphasized above, even the phenomenological approach to relaxation in liquid can be quite sophisticated, involving non-trivial physics and mathematics. This becomes clear when one attempts to solve a Fokker–Planck equation in two variables. If the external potential is not harmonic, no analytical solution is usually available and numerical solution is often hard. Nevertheless, this is the simplest approach available, and a student of the field may do well to master as many of the available techniques as possible, as these are often the first line of attack when a new problem is studied.

In the next chapter we shall discuss, again in simple terms, the concepts and ideas that drive a molecular approach and also the detailed (and sometimes hard) course that a molecular approach usually takes in addressing relaxation processes in liquids.

REFERENCES

1. R. Zwanzig, *Non-equilibrium Statistical Mechanics*, Oxford University Press (2001).
2. R. Zwanzig, *J. Stat. Phys.* 9, 215 (1973).
3. H. Risken, The *Fokker-Planck Equation: Methods of Solutions and Applications*, Springer Series in Synergetics (1989).
4. S. Chandrasekhar, *Rev. Mod. Phys.* 1, 15 (1943).
5. R. Kubo, *Reports on Progress in Physics*, 29, 255 (1966).
6. N. G. Van Kampen, *Stochastic Processes in Physics and Chemistry*, North-Holland Personal Library (1997).
7. H. Haken, *Synergetics, an Introduction: Nonequilibrium Phase Transitions and Self-Organization in Physics, Chemistry, and Biology*, 3rd edition, Springer-Verlag, 1983.

3

Density and Momentum Relaxation in Liquids

3.1 INTRODUCTION

In **Chapter 2** we discussed phenomenological approaches to relaxation phenomena. In this chapter we shall discuss a few approaches that are based on more microscopic considerations. We shall also discuss how these two approaches can sometimes be combined to obtain a satisfactory description of some of those phenomena which require a semimicroscopic approach.

Molecular relaxation in liquids often requires transport and/or relaxation of mass, momentum, and heat to and from the site of the relaxation. The role of these processes is not obvious in many cases because, for the majority of chemical reactions and molecular relaxation, these processes occur with great efficiency. However, the situation may be different for ultrafast relaxation processes. In such cases local and short-time responses of the liquid are probed. These responses are controlled by the local density, local momentum, and local energy, which are also the basic dynamical variables in liquids, as these quantities are globally conserved (except that number density is locally conserved also). For a complete understanding of the role of solvent on molecular relaxation processes, understanding of dynamics and relaxation of these processes in neat liquid is necessary.

Strictly speaking, molecular relaxation processes relevant to chemical dynamics are mostly determined by the surrounding solvent molecules through *local* density relaxation and heat transfer. By local, we mean distances over the length scale of one or two molecular diameters or so. Liquid dynamics at such small length scales are quite different from those at much longer length scales. The latter is easily accessible by experiments such as light scattering. Let us briefly discuss the three time scales, τ_ρ for density, τ_g for momentum, and τ_H for heat, that are relevant to relaxations in dense liquids. These can be given by three simple expressions

$$\tau_\rho = \sigma^2/6D_s, \tag{3.1}$$

$$\tau_g = \mu/\zeta, \tag{3.2}$$

$$\tau_H = \sigma^2/D_t. \tag{3.3}$$

In the above equations, D_s and D_t are the self- and heat diffusion coefficients, σ is a molecular diameter, μ is an effective mass involved in the momentum relaxation, and ζ is the friction coefficient. For typical values ($\sigma = 3\text{A}^0$, $\mu = 50$ amu, $D_s = 2 \times 10^{-5} \text{cm}^2/\text{s}$, $D_t = 1 \times 10^{-3}$ cm^2/s, $\zeta = 2\pi\eta\sigma = 1.9 \times 10^{-9}$ gm/s (with $\eta = 1$ cP)), these give

$$\tau_\rho = 7.5 \times 10^{-10} \text{ s},$$

$$\tau_g = 4 \times 10^{-14} \text{ s},$$

$$\tau_H = 10^{-13} \text{ s}.$$

Therefore, for most purposes, density relaxation is the slowest mode that controls the rate and needs to be considered. Thus, although friction on a moving molecule is determined largely by the rate of momentum relaxation, density relaxation makes an increasingly important contribution as the density and pressure are increased or the temperature is decreased. This rather perplexing role of density relaxation in determining friction can be understood from *mode-coupling theory*, which is used to understand the emergence of slow relaxation in the supercooled liquid.

Clearly, the preceding numerical estimates are based on macroscopic arguments, which are valid over a rather long length and long time scales. This is the domain of hydrodynamic theory. For molecular processes, these relaxation times can be different and are often considerably longer. We shall first describe the conventional hydrodynamic approach, which will be followed by extensions of the approach to molecular-length scales.

3.2 HYDRODYNAMICS AT LARGE LENGTH SCALES

The dynamics of density, momentum, and heat are studied most simply by using hydrodynamics. The hydrodynamic approach to liquid-state dynamics is based on the assumption that many experimental observables (such as the intensity in a light-scattering experiment) can be rationalized by considering the dynamics of a few slow variables. The natural choice for the slow variables are the densities of the conserved quantities, that is, the number density, $\rho(\mathbf{r}, t)$, the momentum density, $\mathbf{g}(\mathbf{r}, t)$, and the energy density, $e(\mathbf{r}, t)$. The conservation of number, momentum, and energy are expressed locally by the following conservation equations [1, 3]:

$$\dot{\rho}(\mathbf{r}, t) + \frac{1}{m}\nabla \cdot \mathbf{g}(\mathbf{r}, t) = 0, \tag{3.4}$$

$$\dot{\mathbf{g}}(\mathbf{r}, t) + \nabla \cdot \sigma(\mathbf{r}, t) = 0, \tag{3.5}$$

$$\dot{e}(\mathbf{r}, t) + \nabla \cdot \mathbf{J}^e(\mathbf{r}, t) = 0, \tag{3.6}$$

where m is the mass of the particles, σ is the momentum current, or stress tensor, and $\mathbf{J}^e$ is the energy current. These equations apply both to microscopic densities and to locally averaged densities.

The continuity equations are supplemented by the constitutive relations involving the current of number, momentum, and energy. The local velocity field $\mathbf{u}(\mathbf{r}, t)$ is defined via the following relation:

$$\mathbf{g}(\mathbf{r}, t) = m\rho(\mathbf{r}, t)\mathbf{u}(\mathbf{r}, t). \tag{3.7}$$

The stress tensor on a hydrodynamic volume element is obtained from macroscopic considerations, coupled with rotational invariance, and is given by

$$\sigma^{\alpha\beta}(\mathbf{r}, t) = \delta_{\alpha\beta}\mathbf{p}(\mathbf{r}, t) - \eta_s \left(\frac{\partial u_\alpha(\mathbf{r}, t)}{\partial r_\beta} + \frac{\partial u_\beta(\mathbf{r}, t)}{\partial r_\alpha} \right)$$
$$+ \delta_{\alpha\beta}\left(\frac{2}{3}\eta_s - \eta_v\right)\nabla \cdot \mathbf{u}(\mathbf{r}, t), \tag{3.8}$$

where $\mathbf{p}(\mathbf{r}, t)$ is the fluctuating local pressure, η_s is the shear viscosity, and η_v is the bulk viscosity. The last constitutive relation defines the macroscopic energy current as

$$\mathbf{J}^e(\mathbf{r}, t) = h\mathbf{u}(\mathbf{r}, t) - \lambda\nabla T(\mathbf{r}, t), \tag{3.9}$$

where $h = (e + p)$ is the equilibrium enthalpy density, $T(\mathbf{r}, t)$ is the local temperature, p is the average pressure, and λ is the thermal conductivity. The second term on the right-hand side is the diffusive part of the energy flux, which is assumed to be given by the Fourier's law of heat conduction.

The constitutive relations along with the conservation equations give the basic equations of fluid mechanics, which are a set of five nonlinear partial differential equations involving seven variables, ρ, $\mathbf{g}$, e, p, and T. As five equations (Eqs. (3.4)–(3.8)) cannot determine seven quantities, the equations are closed by expressing any two variables of the set (ρ, e, p, T) in terms of the other two remaining variables. This is done by using the assumption of local equilibrium and thermodynamic equations of state.

The density and the temperature can be chosen as independent variables and the continuity equations are linearized assuming that the fluctuations around the equilibrium value are small.

Owing to the isotropic nature of the liquid, the linearized hydrodynamic equations are easily solved when written in the Fourier (wave-number) plane. Thus, the basic equations in fluid mechanics in the wave-number (q) and Laplace frequency (z) plane are written as

$$-iz\,\tilde{\rho}_{\mathbf{q}}(z) + i\mathbf{q}\cdot\tilde{\mathbf{g}}_{\mathbf{q}}(z) = \rho_{\mathbf{k}}, \tag{3.10}$$

$$\left(-iz + aq^2\right)\tilde{T}_{\mathbf{q}}(z) + \frac{iT}{\rho^2 c_V}\left(\frac{\partial P}{\partial T}\right)_{\rho}\mathbf{q}\cdot\tilde{\mathbf{g}}_{\mathbf{q}}(z) = T_{\mathbf{q}}, \tag{3.11}$$

$$\left(-iz + \frac{\eta_s}{\rho m}q^2 + \frac{\frac{1}{3}\eta_s + \eta_v}{\rho m}\mathbf{qq}\right)\tilde{\mathbf{g}}_{\mathbf{q}}(z) + \frac{i\mathbf{q}}{m}\left(\frac{\partial P}{\partial \rho}\right)_T \tilde{\rho}_{\mathbf{q}}(z)$$
$$+ \frac{i\mathbf{q}}{m}\left(\frac{\partial P}{\partial T}\right)_{\rho}\tilde{T}_{\mathbf{q}}(z) = \tilde{\mathbf{g}}_{\mathbf{q}}. \tag{3.12}$$

Here $\mathbf{q}$ is chosen to be along the z-axis. c_V is the specific heat *per particle* at constant volume, and $\rho_{\mathbf{q}}$, $T_{\mathbf{q}}$, and $\mathbf{g}_{\mathbf{q}}$ are the spatial Fourier components at $t = 0$. $a = \lambda/\rho c_V$.

These sets of equations are used to obtain the correlation functions of the hydrodynamic variables. These correlations are measured by Rayleigh–Brillouin spectroscopy (RBS), which we discuss now. RBS provides an amazing source of information on liquid-state dynamics.

3.2.1 Rayleigh–Brillouin Spectrum

As already mentioned, Rayleigh–Brillouin spectroscopy holds a prominent place in the study of liquid-state dynamics as it provides values of several transport properties, like thermal diffusivity and sound attenuation coefficient and also the value of the adiabatic sound velocity.

When Eq. (3.12) is separated into transverse and longitudinal components, the transverse component of the current gets decoupled from all the longitudinal modes, that is, from the fluctuations in the density, the temperature, and the longitudinal current. On the other hand, it is found from Eqs. (3.10)–(3.12) that all the longitudinal modes are coupled together. The coupled longitudinal equations provide the hydrodynamic limiting form of the dynamic structure factor $S(q, \omega)$, which is the spatial and temporal correlation of the density fluctuations in wave-number and Fourier frequency plane:

$$S(q, \omega) = \int dt\, \exp(i\omega t)\langle \rho(\mathbf{q}, t)\rho(-\mathbf{q}, 0)\rangle. \tag{3.13}$$

The spectrum of density fluctuations defined above is measured in the *Rayleigh–Brillouin spectrum*, observed in light-scattering experiments. The derivation of $S(q, \omega)$ constitutes one of the great successes of classical hydrodynamic theory.

The final expression is the well-known Landau–Placzek formula for the dynamic structure factor given by [1, 2]

$$S(q, \omega) = \frac{1}{2\pi} S(q) \left[\left(\frac{\gamma - 1}{\gamma} \right) \frac{2 D_T q^2}{\omega^2 + (D_T q^2)^2} \right.$$
$$\left. + \frac{1}{\gamma} \left(\frac{\Gamma q^2}{(\omega + c_s q)^2 + (\Gamma q^2)^2} + \frac{\Gamma q^2}{(\omega - c_s q)^2 + (\Gamma q^2)^2} \right) \right], \quad (3.14)$$

where $\gamma = c_p/c_V$ and c_s is the adiabatic speed of sound. $D_T = a/\gamma$ is the thermal diffusivity and the acoustic attenuation coefficient Γ is given by

$$\Gamma = \frac{1}{2} \left[(\gamma - 1) D_T + b \right]. \quad (3.15)$$

Here $a = \lambda/\rho c_V$ and the kinematic longitudinal viscosity is

$$b = \left(\frac{4}{3} \eta_s + \eta_v \right) \Big/ \rho m,$$

where η_s and η_v are the shear and bulk viscosities of the liquid, respectively.

Equation (3.14) shows that the R-B spectrum consists of three components. The first term represents an unshifted line called the *Rayleigh line*, which is a Lorentzian with a half-width at half-maximum given by $\Delta\omega_c(q) = D_T q^2$. The next two terms represent a doublet called the *Brillouin doublet*. These two Lorentzian lines are shifted symmetrically from the origin by $\omega = \pm c_s q$, each having a half-width at half-maximum $\Delta\omega_B(q) = \Gamma q^2$. A typical spectrum is shown in Fig. 3.1.

Physically, the *Brillouin* spectrum arises from the inelastic interaction between a photon and the hydrodynamics modes of the fluid. Sometimes, this interaction is referred to as photon–phonon interaction. The doublets can be regarded as the "Stokes" and "anti-Stokes" translational Raman spectrum of the liquid. These lines arise due to the loss of energy of photons to the phonons (the propagating sound modes in the fluid) and thus suffer a frequency shift. The width of the band gives the lifetime $(q^2 \Gamma)^{-1}$ of a classical phonon of wave number q. The Rayleigh band, on the other hand, represents the scattering of the light by the entropy or heat fluctuations, which are purely diffusive or dissipative modes of the fluid. Therefore, the Rayleigh spectrum probes incoherent motions or fluctuations of the liquid while the Brillouin spectrum probes coherent motions.

It should be noted that the above treatment is based on two assumptions. (i) The fluctuations in the conserved variables can be described by the simple linearized hydrodynamic equations and (ii) the widths of the Lorentzians are small compared to the shifts. That is, $\gamma D_T q^2 \ll c_s q$; $\Gamma q^2 \ll c_s q$. This is the case in most of the fluids at small q. For example, in argon at a temperature 235K and mass density 1 gm/cc, $c_s = 6.85 \times 10^4$ cm/sec, $D_T = 1.0 \times 10^{-3}$ cm^2/sec, and $b = 1.6 \times 10^{-3}$ cm^2/sec.

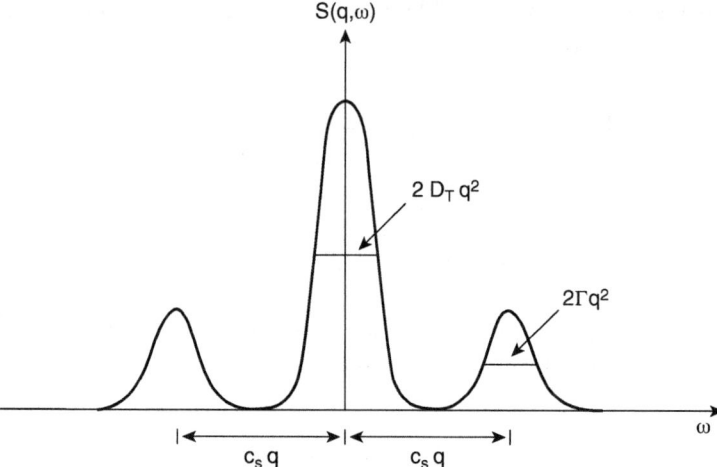

Figure 3.1 A schematic description of the dynamic structure factor (the Rayleigh–Brillouin spectrum), $S(q, \omega)$ at long wavelengths (or, small wave numbers), that is, in the hydrodynamic limit. Here D_T is the thermal diffusivity, c_S is sound velocity. The three peak structure is universal at small wave numbers.

For $\gamma = 1$ and $q = 2.1 \times 10^5 \text{cm}^{-1}$ (which is typical for light scattering), $\dfrac{\gamma D_T q^2}{c_s q} = 1.5 \times 10^{-3}$ and $\dfrac{\Gamma q^2}{c_s q} = 1.2 \times 10^{-3}$. The two values are surprisingly close to each other.

The expression of the transverse current autocorrelation function can also be derived from the linearized hydrodynamic equations. Because it is decoupled from all the longitudinal modes, the derivation is simple and the final expression in the wave-number and Laplace frequency planes can be written as

$$\tilde{C}_t(k, z) = \frac{k_B T}{m(-iz + \nu k^2)}, \qquad (3.16)$$

where $\nu = \eta_s/\rho m$ is the kinematic shear viscosity. Similarly, other correlation functions at small wave numbers are obtained from the linearized hydrodynamic equations.

3.3 HYDRODYNAMIC RELATIONS BETWEEN SELF-DIFFUSION COEFFICIENT AND VISCOSITY

Other than providing expressions for dynamical time-correlation functions, hydrodynamic theory provides explicit expressions for transport properties. In hydrodynamics the diffusion of a tagged particle is defined by the Stokes–Einstein relation, which is given by the following well-known expression between self-diffusion coefficient (D_s) and shear viscosity η_s:

$$D_s = \frac{k_B T}{C\pi \eta_s R}, \tag{3.17}$$

where k_B is the Boltzmann constant. This equation is obtained by combining the Einstein equation that relates the self-diffusion to the friction and the Stokes relation that relates the friction on a spherical molecule of radius R to the shear viscosity of the medium. This Stokes relation is obtained from the Navier–Stokes equation by calculating the frictional force on a sphere with radius R, moving with a constant velocity **u** in a fluid of shear viscosity η_s. The constant C is obtained from the hydrodynamic boundary condition, which is 6 for the "stick" approximation where the fluid velocity everywhere on the surface is considered to be **u** (the same as the velocity of the solute particle) and 4 for the "slip" approximation where the normal component of the fluid velocity is taken to be equal to the normal component of **u**.

The close and useful (to us) relationship between friction and viscosity goes beyond the Stokes relation, and they can be related even in the frequency plane. In an elegant work, the Navier–Stokes hydrodynamics has been generalized by Zwanzig and Bixon [4] to include the viscoelastic response of the medium. This generalization provides an expression for the frequency-dependent friction that depends on the frequency-dependent bulk and shear viscosities and sound velocity.

The frequency-dependent shear and bulk viscosities define the viscoelastic response of the liquid, and are defined by respective relaxation times. We will discuss subsequently the important role played by frequency-dependent viscosity in determining the rate of a chemical reaction.

3.4 SLOW DYNAMICS AT LARGE WAVE NUMBERS: DE GENNES NARROWING

In the previous section we discussed the Rayleigh–Brillouin spectrum, which is observed in light-scattering experiments. The wavelengths involved in these scattering experiments are typically about 5000 Å. It is thus possible to calculate the spectral distribution of scattered light from the macroscopic equations of hydrodynamics. The neutron-scattering experiments, on the other hand, involve wavelengths of the order of the nearest-neighbor separation in the liquid, thus capturing completely different physics, which cannot be explained from the hydrodynamic point of view.

The wave numbers accessible in inelastic neutron-scattering experiments lie typically between 0.1 and 15 Å^{-1}, which is the same range as is usually studied in molecular dynamics simulations. There have been extensive studies of the density fluctuation spectrum by coherent and incoherent inelastic neutron-scattering experiments and also by using molecular dynamics simulations. As discussed above, the dynamic structure factor $S(q, \omega)$ at reduced wave numbers $q\sigma \leq 1$ (where σ is the atomic diameter), has a three peak structure where the two side peaks correspond to propagating sound waves. At shorter wavelengths, the sound waves are strongly damped and the high-frequency structure disappears when $q\sigma \geq 2$, leaving only the Lorentzian-like central peak. The width of the central peak first increases with q, but then shows a *marked decrease at wave numbers close to* q_m, which corresponds to the wave number where the static structure factor has its main peak. At still larger wave numbers the spectrum broadens again, finally reaching its free-particle limit. The sharp decrease in the width near $q = q_m$ is called "de Gennes narrowing" (shown in Fig. 3.2) and cannot be explained from the expression of the dynamic structure factor obtained using linearized hydrodynamic theory (Eq. (3.14)). Many years ago, de Gennes observed that for intermediate wavenumbers (that is, for intermediate momentum transfer) intermolecular correlation between neighboring atoms is important [5]. Often this results in a strong narrowing of the density distribution function. This narrowing can also be explained by using Smoluchowski–Vlasov–type equation, to express the number-density fluctuations.

At wavelengths of the order of nearest-neighbor separation, neither the conservation of the momentum nor the energy is relevant constraint to the dynamics. This is because at these wavelengths there is a rapid exchange of momentum and energy between the nearest neighbors, owing to binary collisions. On the other hand, *the number density must be conserved at all length scales*! Thus, the only slow mode relevant in this intermediate wavenumber regime is the number density.

Now a Smoluchowski–Vlasov–type expression can be easily written down for the number density, which involves a mean-field force term due to the cage effect as

$$\dot{\rho}(\mathbf{r}, t) = D\nabla \cdot [\nabla - \beta \mathbf{F}(\mathbf{r}, t)]\rho(\mathbf{r}, t), \tag{3.18}$$

where $D = k_B T/m\gamma$ is the self-diffusion coefficient of the liquid, γ is the collisional frequency. $\mathbf{F}$ is the mean-field force given by

$$F(\mathbf{r}, t) = -\nabla V_{\textit{eff}}(\mathbf{r}, t). \tag{3.19}$$

Here $V_{\textit{eff}}(\mathbf{r}, t)$ is the effective potential energy that is determined from the mean spherical approximation and is written in terms of the two-particle direct correlation function, $C(\mathbf{r} - \mathbf{r}')$ as [5, 6]

$$\beta V_{\textit{eff}}(\mathbf{r}, t) = -\int d\mathbf{r}' C(\mathbf{r} - \mathbf{r}')\rho(\mathbf{r}', t). \tag{3.20}$$

ρ_0 is the equilibrium solution of Eq. (3.18) and has the following form

$$\rho_0(\mathbf{r}) = c' \exp\left[\int d\mathbf{r}' C(\mathbf{r} - \mathbf{r}')\rho_0(\mathbf{r}')\right]. \tag{3.21}$$

Here c' is a constant determined from the normalization condition. From Eqs. (3.18)–(3.21), the intermediate scattering function $F(q, t)$ can be obtained easily, and it has the following form:

$$F(q, t) = S(q) e^{-\left(\frac{D}{S(q)} q^2 t\right)}, \qquad (3.22)$$

where $S(q)$ is the static structure factor. Note that $S(q)$ and $C(q)$ are related to each other by the simple equation $S(q) = [1 - \rho C(q)]^{-1}$.

Now, the value of self-diffusion coefficient is about three orders of magnitude smaller than the thermal diffusion coefficient. In addition, for wave numbers near q_m, the static structure factor is sharply peaked. These two combine to cause a marked slowing down of $F(q, t)$. This slowing down of $F(q, t)$ in turn leads to a considerable narrowing of the zero-frequency (Rayleigh) peak of the dynamic structure factor, which is the de Gennes narrowing. Underlying all these interpretations is the fact that in a dense liquid the harsh repulsive part of the intermolecular potential plays an important role. Owing to this repulsive potential, a cage is created around each molecule, which inhibits or hinders the movement of the molecule. This makes the self-diffusion the slowest process, thus the transfer of a molecule becomes the bottleneck of any relaxation process.

Sometimes, this sharp peak in $S(q)$ was used to explore the stability of the liquid. This de Gennes narrowing forms the basis for the current mode-coupling theory explanation of the glass transition.

3.5 EXTENDED HYDRODYNAMICS: DYNAMICS AT INTERMEDIATE LENGTH SCALES

In the preceding sections the validity of hydrodynamics at small q and its breakdown at intermediate q have been discussed. Often in the calculation of the transport coefficients, integration of the time-correlation functions over the whole wave number space is required. Thus, to have a unified description over the whole q plane, the extension of the hydrodynamic theory to intermediate wave numbers is essential.

This extension of all the five hydrodynamic modes (heat, sound, and viscous) to intermediate wavelengths was perhaps first done by de Schepper and Cohen [6]. Their theory is based on a model kinetic equation (generalized Enskog theory) for a hard-sphere fluid. They have shown that at low densities all the five modes are increasingly damped with decreasing wavelength until each ceases to exist below a cutoff wavelength. At high densities the extended heat mode behaves differently. Not only does it soften appreciably for wavelengths of the order of the size of the particles and becomes a diffusionlike mode, it also persists until much shorter wavelengths than the other modes. They have also shown that for dense hard-sphere fluids the neutron-scattering structure factor can be qualitatively represented for $0 < q\sigma < 15$ as a superposition of the heat and (two) sound modes. One is thus led to a very important conclusion that the extended hydrodynamic modes can be regarded as a seamless extension of the classical hydrodynamic modes used to explain the $S(q, \omega)$ in the light-scattering experiments.

There exists another prescription to extend the hydrodynamical modes to intermediate wave numbers, which provides similar results for dense fluids. This was done by Kirkpatrick [7], who replaced the transport coefficients appearing in the generalized hydrodynamics by their wave-number and frequency-dependent analogs. He used the standard projection operator technique to derive generalized hydrodynamic equations for the equilibrium time-correlation functions in a hard-sphere fluid. In the short-time approximation the frequency dependence of the memory kernel vanishes. The final result is a closed set of generalized hydrodynamic equations for the time-correlation functions of number density ρ, longitudinal momentum density g_l, temperature T, and transverse momentum density t_i ($i = 1, 2$). $G_{\alpha\beta}(q, z)$, ($\alpha, \beta = \rho, l, T, t_i$) denotes the normalized time-correlation functions in Fourier–Laplace (z) space. The following are the equations of motion:

$$zG_{\rho\beta}(q, z) - \frac{iq}{\sqrt{\beta m S(q)}} G_{l\beta}(q, z) = \delta_{\rho\beta}, \qquad (3.23)$$

$$zG_{l\beta}(q, z) - \frac{iq}{\sqrt{\beta m S(q)}} G_{\rho\beta}(q, z)$$

$$- iq \left[\frac{2}{3\beta m}\right]^{1/2} \left[1 + 2\pi\rho\sigma^3 g(\sigma) \frac{j_1(q, \sigma)}{q\sigma}\right] G_{T\beta}(q, z)$$

$$+ \frac{2}{3t_E} \left[1 - j_0(q, \sigma) + 2j_2(q, \sigma)\right] G_{l\beta}(q, z) = \delta_{l\beta}, \qquad (3.24)$$

$$zG_{T\beta}(q, z) - iq \left[\frac{2}{3\beta m}\right]^{1/2} \left[1 + 2\pi\rho\sigma^3 g(\sigma) \frac{j_1(q, \sigma)}{q\sigma}\right] G_{l\beta}(q, z)$$

$$+ \frac{2}{3t_E} \left[1 - j_0(q, \sigma)\right] G_{T\beta}(q, z) = \delta_{T\beta}, \qquad (3.25)$$

and with ($i = 1, 2$),

$$\left[z + \frac{2}{3t_E}\left[1 - j_0(q, \sigma) - j_2(q, \sigma)\right]\right] G_{t_i\beta}(q, z) = \delta_{T\beta}. \qquad (3.26)$$

Here $j_l(q, \sigma)$ is the spherical Bessel function of order l, $g(\sigma)$ is the radial distribution function at contact, and $t_E = \sqrt{\beta m \pi}/4\pi\rho\sigma^2 g(\sigma)$ is the Enskog mean free time between collisions. The transport coefficients in the above expressions are given only by their Enskog values, that is, only collisional contributions are retained. Because it is only in dense fluid that the Enskog values represent the important contributions to the transport coefficient, the above expressions are reasonable only for dense hard-sphere fluids. The eigenmodes can be obtained by solving the coupled Eqs. (3.23)–(3.26). The extended hydrodynamics also reproduces the softening of the heat mode at intermediate wave numbers.

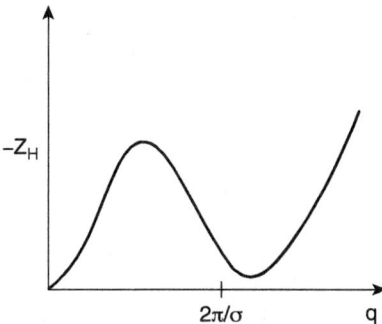

Figure 3.2 The wave-number (q) dependence of the eigenvalue of the heat mode (z_H) obtained from extended hydrodynamics. Note the small value of the frequency (or rate) z_H at intermediate value of q (when $q \approx 2\pi/\sigma$).

In addition, the damping of the propagating sound waves at intermediate wave number was also recovered. This was interpreted as being due to the competition between elasticity and dissipation. There seems to be a trapping of the sound wave on a molecular-length scale. The reappearance of the propagating modes at larger q is similar to deepening of sound waves in porous mediums at larger frequencies, where the effective viscosity or damping becomes smaller.

The complete solution of the heat-mode eigenvalue $z_H(q)$ was found to be well approximated by the form obtained from the revised Enskog kinetic equation, for not too small q. The latter is given by

$$z_H(q) = -\frac{Dq^2}{S(q)} \left[1 - j_0(q\sigma) + 2j_2(q\sigma)\right]^{-1}. \tag{3.27}$$

From the above expression we can say that the softening of the heat mode happens due to the peak in $S(q)$ near q_m and also because the heat mode becomes essentially a density mode that decays via self-diffusion.

As discussed by Kirkpatrick, this slow mode is important in the theories that include mode-coupling effects. Such theories have been used to understand quantitatively the anomalous long-time tails of the stress–stress correlation function and the shear-dependent viscosity, observed in computer simulations [8–10]. As mentioned earlier, a theory of glass transition has also been developed based on the softening of the heat mode.

3.6 MODE-COUPLING THEORY

Mode-coupling theory emerged as a result of sustained efforts on two rather different approaches toward the dynamical theory of gases and liquids. On one side, linear Navier–Stokes hydrodynamics was extended to generalized hydrodynamics that included viscoelasticity and subsequently to molecular hydrodynamics that included wave-number and frequency dependence of transport coefficients. In the above we have described some parts of this effort. A nice description of this approach is available in the classic book by Boon and Yip [3]. On the second front, the well-known Enskog kinetic theory was extended to include many particle dynamical correlations, essentially by including the contributions of the ring collision terms. Major contributions were made by Cohen and Dorfman et al. [8, 9] and by Mazenko [10]. This approach came to be known as renormalized kinetic theory.

In both the approaches the slow down at intermediate wavenumbers, the well-known de Gennes' narrowing, was described accurately. However, what was missing was a self-consistency or boot strapping among the different dynamical quantities.

The necessity of self-consistency can be understood from the following simple example. Both Eqs. (3.22) and (3.27) show that relaxation of the dynamic structure factor is determined, at intermediate to large wavenumbers, by the self-diffusion coefficient D_s. The latter is related to friction by Einstein's relation. But *the friction itself is determined partly by the dynamic structure factor*! Therefore, one must solve both the quantities self-consistently with each other.

In fact, not only diffusion and friction, but most transport properties, including viscosity and ionic conductivity, and time-correlation functions, such as dynamic structure factor, current correlation functions, or orientational time-correlation functions, are all interdependent. Therefore, self-consistent treatment of liquid-state dynamics is often necessary to obtain quantitative results.

3.7 SUMMARY

Molecular relaxation processes in liquids, such as isomerization dynamics and/or vibrational relaxation, are largely determined by local density and momentum relaxation. Nevertheless, we have long used relationships derived from ordinary Navier–Stokes hydrodynamics (such as Stokes–Einstein relation) to correlate experimental findings, such as the dependence of isomerization rate on viscosity.

In this chapter we first described the predictions of Navier–Stokes hydrodynamics. Next, we have described the basic ingredients of a molecular theory that allows one to formulate microscopic expressions for time-correlation functions and also allows calculation of transport coefficients such as friction and diffusion coefficient. As mentioned above, in dense liquids, the value of these transport coefficients is determined by processes at molecular-length scales. Therefore, ordinary Navier–Stokes hydrodynamics is often not useful. This in turn leads to a paradoxical situation because traditionally we use the same Navier–Stokes hydrodynamics to describe dependence of reaction (or, relaxation) rate on viscosity. In recent years this paradox has been solved by using mode-coupling theory, which is an elegant extension of both molecular hydrodynamics and kinetic theory. We refer to Ref. 11 for a recent discussion of the above topics from a physical chemistry/chemical physics perspective.

REFERENCES

1. J. P. Hansen and I. R. McDonald, *Theory of Simple Liquids*, Academic Press (2006).
2. B. Berne and R. Pecora, *Dynamic Light Scattering*, Wiley (1976).
3. J. P. Boon and S. Yip, *Molecular Hydrodynamics*, Dover Publications (1992).
4. R. Zwanzig and M. Bixon, *Phys. Rev. A* 2, 2005 (1970).
5. P. G. de Gennes, *Physica* 25, 825 (1959).
6. I. M. de Schepper and E. G. D. Cohen, *Phys. Rev. A* 22, 287 (1980); *J. Stat. Phys.* 27, 223 (1982).
7. T. R. Kirkpatrick, *Phys. Rev. A* 32, 3130 (1985).
8. E. G. D. Cohen, *Physica A* 194, 229 (1993).
9. J. R. Dorfman, T. R. Kirkpatrick, and J. V. Sengers, *Ann. Rev. Phys. Chem.* 45, 213 (1994).
10. G. Mazenko, *Nonequilibrium Statistical Mechanics*, Wiley-VCH Verlag (2006).
11. B. Bagchi and S. M. Bhattacharyya, *Adv. Chem. Phys.* 116, 67 (2001).

4

Relationship between Theory and Experiment

4.1 INTRODUCTION

The study of molecular relaxation processes in liquids provides a wealth of information regarding the interaction between molecules, their shapes and sizes, the reactivity of the molecules, and their diffusion coefficients, among many other properties. However, probing these properties is not often an easy task. Because we are interested here in the dynamical processes, the main experimental approaches involve exchange of energy and/or momentum between the system and an external perturbation, such as interaction of radiation with matter. This interaction can be in the form of absorption and emission and also in the form of scattering. The exchange of energy and/or momentum between radiation and molecules provides information about the frequency of motion of a molecular property (such as rotation or vibration) and thereby provides information about the time scales of the motion. This still remains the mainstay in our experimental approach.

For a long time, dielectric relaxation, acoustic attenuation, dynamic light scattering, infrared spectroscopy, and a few others had remained the main tools to study molecular relaxation in liquids. These tools cause weak perturbations to the liquid. These perturbations are nonlocal in the sense that they are spread over a large volume, in the scale of an individual molecular size. These tools can detect rather slow responses, often slower than a few nanoseconds (ns). A notable exception is NMR experiments that can probe relaxation times in the picosecond (ps) range, even when the response they measure is in milliseconds (ms). We have discussed the reason for this intriguing behavior below. Nevertheless, the experimental tools available before the 1970s could probe only slow processes. As a result many important processes, like isomerization dynamics, electron transfer

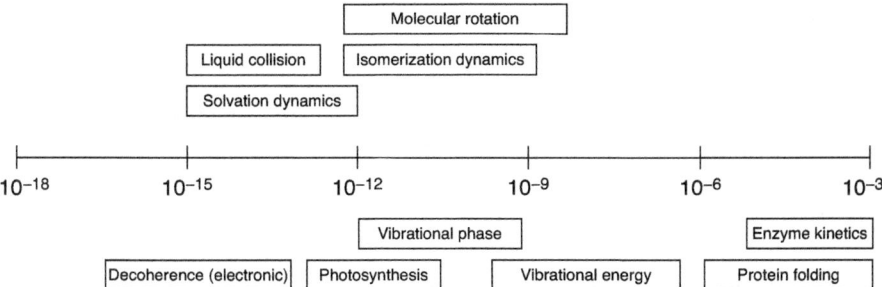

Figure 4.1 Time scales of different molecular relaxation processes in chemical and biological systems.

reactions, and elementary processes involved in photosynthesis could not be studied.

The experimental tools at our disposal have improved tremendously over the last few decades, particularly with the advent of ultrafast laser spectroscopy and various forms of microscopic and scattering techniques that are increasingly becoming available. The approximate time scales of many relaxation processes are shown in Fig. 4.1 that shows that the relaxation processes that we routinely study span time scales over many orders of magnitude. As already mentioned and further discussed below, our ability to study dynamical processes in depth has grown tremendously in the last two decades, with the addition of multidimensional infrared and electronic spectroscopy and single-molecule spectroscopy. However, the basic theoretical approach to understanding these phenomena remains unchanged and was formulated mostly in the second half of the last century, starting mostly in the 1960s when the wonderful work of Kubo and Zwanzig became available.

Note that experimental results are usually obtained in the form of graphs and tables of numbers of intensity of absorbed/emitted/scattered light or neutrons, plotted against frequency, time, or wave number. On the other hand, the information we desire is of a molecular nature, as mentioned above. Therefore, we need theoretical expressions relating these experimental numbers with molecular quantities. *The relationship between the two is not often obvious.* Considerable effort has gone into bridging the gap between the theory and the experiment.

Establishment of the required relationship between theory and experiment involves three distinct steps (also shown in Fig. 4.2).

> I. The first step involves use of a quantum (or classical in the appropriate limit) theory to derive an expression between observables, like the scattering cross-section (as in neutron scattering), or the absorption or emission intensity and the appropriate molecular transition probability. Such an expression usually involves use of the Fermi Golden Rule (FGR) for the transition probability.

II. The second step involves the translation of the right-hand side of the derived expression, involving the FGR to an appropriate time-correlation function (TCF). For example, the neutron-scattering cross-section (which is directly measured in experiments) from a liquid is given as a time and space integral over density–density time- and space-dependent correlation functions.

III. The third and final step involves evaluation of the time-correlation function in terms of molecular properties. This step allows us to extract the information we want from experimental measurements.

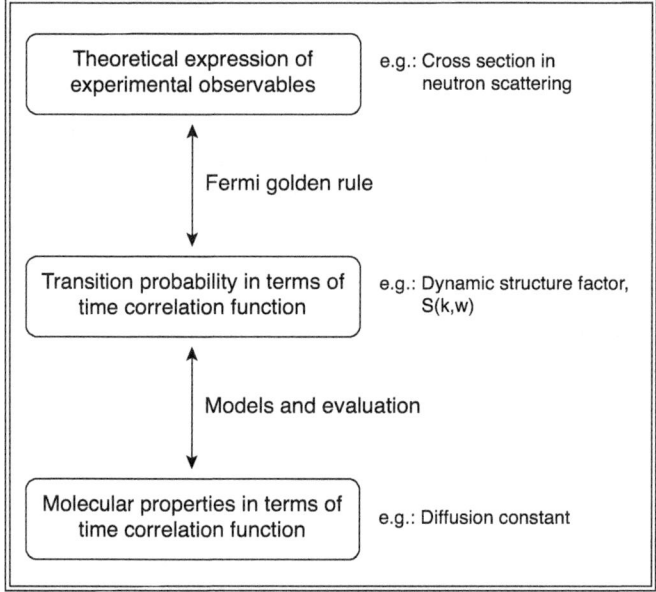

Figure 4.2 A pictorial representation of different steps involved in establishing the relationship between theory and experiment.

We next discuss implementation of the above scheme in deriving relationships between theory and experiments for several important cases.

4.2 DYNAMIC LIGHT SCATTERING: PROBE OF DENSITY FLUCTUATION AT LONG LENGTH SCALES

Light scattering occurs because of local fluctuations in the dielectric constant of the medium. These fluctuations in turn arise because the small molecules in solutions undergo Brownian motion and so the distance between them in the solution is constantly varying with time. As a result, the number of molecules in a small volume element fluctuates in time and also with location. Such fluctuation in density gives rise to fluctuations in dielectric constant, which can be measured by carrying out scattering of light or neutron, by choosing an appropriate experimental method.

Let us consider a medium with fluctuating local dielectric constant:

$$\varepsilon(\mathbf{r}, t) = \varepsilon_0 \mathbf{I} + \delta\varepsilon(\mathbf{r}, t), \quad (4.1)$$

where ε_0 is the average static dielectric constant of the liquid and $\delta\varepsilon(\mathbf{r}, t)$ is the dielectric constant fluctuation tensor at a position $\mathbf{r}$ and time t and $\mathbf{I}$ is the second rank unit tensor.

An incident electric field can be given as

$$E_i(\mathbf{r}, t) = \mathbf{n}_i E_0 \exp\left[i\left(\mathbf{k}_i \cdot \mathbf{r} - \omega_i t\right)\right], \quad (4.2)$$

where $\mathbf{n}_i$ is a unit vector in the direction of the incident electric field, E_0 is the electric field amplitude, $\mathbf{k}_i$ is the wave vector, and ω_i is the angular frequency of the incident electric field. The corresponding scattered electric field of wave vector $\mathbf{k}_f$ and angular frequency ω_f, at a large distance R from the scattering volume, owing to the interaction of the electric field and local dielectric constant $\varepsilon(\mathbf{r}, t)$, is given by [1]

$$E_s(R, t) = \frac{E_0}{4\pi R \varepsilon_0} e^{ik_f R} \int_V d^3 r \exp i(\mathbf{q} \cdot \mathbf{r} - \omega_i t)$$

$$\times \left[\mathbf{n}_f \cdot \left[\mathbf{k}_f \times \left(\mathbf{k}_f \times (\delta\varepsilon(\mathbf{r}, t) \cdot \mathbf{n}_i)\right)\right]\right], \quad (4.3)$$

where $\mathbf{q} = \mathbf{k}_i - \mathbf{k}_f$. The integration in this expression is over the scattering volume.

In light-scattering experiments one measures the spectral density of the electric field autocorrelation function of the scattered light wave, $I_E(\omega)$, given as

$$I_E(R, \omega) = \frac{1}{2\pi} \int_{-\infty}^{\infty} d\tau \langle E_s(R, t) E_s(R, t + \tau)\rangle e^{i\omega\tau}. \quad (4.4)$$

The spectral density of the light scattered by the medium, obtained by using Eqs. (4.4)(4.5) is [1]

$$I_{if}(\mathbf{q}, \omega_f, R) = \left[\frac{I_0 k_f^4}{16\pi^2 R^2 \varepsilon_0^2}\right] \frac{1}{2\pi} \int_{-\infty}^{\infty} dt \langle \delta\varepsilon_{if}(\mathbf{q}, 0) \delta\varepsilon_{if}(\mathbf{q}, t)\rangle \exp\left(i(\omega_f - \omega_i)t\right), \quad (4.5)$$

where $I_0 \equiv |E_0|^2$. $\delta\varepsilon_{if}(\mathbf{q}, t) = \mathbf{n}_f \cdot \delta\varepsilon(\mathbf{q}, t) \cdot \mathbf{n}_i$ is the component of the dielectric constant fluctuation tensor along initial and final electric field directions. As is clear from Eq. (4.5), the light-scattering event that produces the wave vector change **q** and frequency shift ω is entirely due to dielectric constant fluctuations.

A light-scattering experiment measures the long-wavelength ($k \to 0$ limit) response of a liquid and, therefore, it measures the *collective* density relaxation, that is, the scattering cross-section derives contributions from a large number of molecules. The spectrum obtained by dynamic light-scattering experiments at long wavelengths and low frequencies is known as the Rayleigh–Brillouin spectrum. It provides a wealth of information about many properties of the liquid.

Like the spatial counterpart, orientational density fluctuations also lead to light scattering. Orientational fluctuations are best studied by depolarized (VH) Rayleigh scattering, because pure transverse fluctuations are observed in this geometry and the spectrum is less complicated. In general, the evaluation of the correlation function is complicated. Because light scattering involves only small wave-vector processes, the translational contribution to orientational relaxation can be neglected and the error that is involved in assuming statistical independence of molecular rotation and translation may not be significant. If we further assume that interactions between different rotating molecules are negligible, then the correlation function that is measured by the light-scattering experiments corresponds to only single-particle motion. The relevant correlation functions for several different models have been given in Berne and Pecora [1].

The assumption of "no interaction" between the rotating molecules is not correct in a dense dipolar liquid where the molecules may possess a dipole moment in addition to being polarizable. In such a situation, dynamic light-scattering experiments measure the collective property. The polarizability can be expanded in terms of spherical harmonics and, for a molecule of ellipsoidal or cylindrical symmetry, the only terms that appear are Y_{2m}. Therefore, the dynamical quantity is $Y_{2m}(\mathbf{k}, t)$ and the correlation functions, $C_{2m}(\mathbf{k}, t)$, are defined as

$$C_{2m}(\mathbf{k}, t) = \langle Y_{2m}(-\mathbf{k}, 0) Y_{2m}(\mathbf{k}, t) \rangle. \quad (4.6)$$

In the long-wavelength limit, the quantities of interest are $C_{2m}(\mathbf{k} = 0, t)$, which can be significantly different from the single-particle correlation function defined as

$$S_{2m}(t) = \langle Y_{2m}(\mathbf{\Omega}(0)) Y_{2m}(\mathbf{\Omega}(t)) \rangle, \quad (4.7)$$

where $\mathbf{\Omega}(t)$ is the orientation of the dipole at time t.

An excellent introduction to the theory of dynamic light scattering and the relevant correlation functions is given by Berne and Pecora [1].

4.3 MAGNETIC RESONANCE EXPERIMENTS: PROBE OF SINGLE-PARTICLE DYNAMICS

Nuclear magnetic relaxation due to magnetic dipole–dipole interactions in a liquid can provide direct information both about the orientational relaxation time of a rotating molecule and the translational diffusion coefficient, as discussed below. In nuclear magnetic resonance experiments, usually *single-particle* relaxation is measured. The spin-lattice relaxation time and the spin–spin relaxation time are essentially the time integrals over various components of $S_{2m}(t)$.

There is an intriguing point about NMR relaxation that deserves special attention. Although NMR experiments measure time constants in the second or millisecond range, they can provide accurate information about rotation time constants of molecules that are in the picosecond range. We elucidate the reason below.

In NMR, the dipolar correlation function for the case of two interacting nuclear spins separated by distance r_{12} is defined as [2]

$$G(t) = \frac{3\pi}{5} \left(\frac{\mu_0}{4\pi}\right)^2 \frac{\hbar^2 \gamma_1^2 \gamma_2^2}{r_{12}^6} S_{2m}(t), \qquad (4.8)$$

where $S_{2m}(t)$ is the single-particle correlation function defined in Eq. (4.7). γ_1 and γ_2 are the gyromagnetic ratios of the nuclei and r_{12} is the distance between them. For a simple Debye model of rotational diffusion, Eq. (4.8) reduces to the following simple equation [2]:

$$G(t) = \frac{3}{20} \left(\frac{\mu_0}{4\pi}\right)^2 \frac{\hbar^2 \gamma_1^2 \gamma_2^2}{r_{12}^6} \exp(-t/\tau_c), \qquad (4.9)$$

with the rotational correlation time τ_c:

$$\tau_c = a^2/6D_R, \qquad (4.10)$$

where a is the radius of the molecule and D_R is the rotational diffusion constant that is related to the viscosity, η, by the Stokes–Einstein relation as

$$D_R = \frac{k_B T}{8\pi \eta a^3}. \qquad (4.11)$$

For water at room temperature τ_c is around 3 ps. This is a very short correlation time and in such cases the relaxation time is given by

$$\frac{1}{T_1} = \frac{3}{2} \left(\frac{\mu_0}{4\pi}\right)^2 \frac{\hbar^2 \gamma_1^2 \gamma_2^2}{r_{12}^6} \tau_c. \qquad (4.12)$$

For water this relaxation time (relaxation from the intramolecular interaction in the water) is 4.78 s. The above equation shows that *even though the time observed in NMR experiments could be in the second or millisecond range, the microscopic correlation time τ_c can be in the picosecond range.* This is simply because the prefactors in Eq. (4.12) combine to give a value in the range of 10^{15} s^{-2} or so.

The translational diffusion can also have an influence on the relaxation process and the measurement of the relaxation time can be used to study the translation diffusion process. If the molecules are spherical in shape and translational molecular motions are fast, then the relaxation rate is [3]

$$\frac{1}{T_1} = \left(\frac{\mu_0}{4\pi}\right)^2 \frac{8\hbar^2 S(S+1)}{45 D_T r_{IS}} N_S \gamma_I^2 \gamma_S^2, \qquad (4.13)$$

where N_S is the concentration of S spins, r_{IS} the closest internuclear distance, and D_T is the translational diffusion coefficient.

4.4 KERR RELAXATION

In Kerr relaxation, one studies optically the decay of transient birefringence created initially by an intense polarized light pulse sent through the medium. The polarized light pulse forces a partial alignment of the anisotropic molecules because of the torque generated by the light field on the anisotropic molecule. The induced orientational anisotropy leads to a difference in the refractive index in the parallel and perpendicular light polarization directions. The relaxation of this anisotropy is now probed by a second light pulse that is polarized at 45° to the pump pulse. The optical Kerr effect essentially measures the nonlinear polarization $(\mathbf{P}^{NL})$ created by the electric field of the laser pulse. This nonlinear polarization depends on the polarizability–polarizability time correlation function defined by

$$C_{ijkl}(t) = \langle \alpha_{ij}(0) \alpha_{kl}(t) \rangle. \qquad (4.14)$$

Note that it is the same correlation that is measured in depolarized light scattering. *The difference is that the Kerr relaxation studies can be carried out directly in the time domain with ultrashort laser pulses, so that the short-time dynamics of the liquid can be studied* [6, 7]. It is again important to note that Eq. (4.14) is a *collective* orientational correlation function because $\alpha_{ij}(t)$ is a sum over all the polarizabilities of the system. Usually $C_{ijkl}(t)$ contains an ultrashort part with time constant of the order of 500 fs or so and a slow time constant on the order of a few tens of picoseconds. The ultrashort part contains information from intermolecular vibrations due to interaction-induced effects that can make a significant contribution toward the magnitude of the polarizability tensor. Therefore, interpretation of the short-time measurements becomes a little tricky.

4.5 DIELECTRIC RELAXATION

Conventional dielectric relaxation experiments measure the relaxation of a macroscopic electric polarization of the liquid and therefore give us information about the *collective* orientational motion of the dipolar molecules. In contrast to the light-scattering experiments, dielectric relaxation measures the first rank spherical harmonic, $Y_{1m}(\mathbf{k}, t)$, with the value of the $\mathbf{k}$ very small ($\mathbf{k} \sim 0$). If we define the dipolar correlation function $\phi(\mathbf{k}, t)$ by

$$\phi(\mathbf{k}, t) = \frac{\sum_{m=-1}^{1} C_{1m}(\mathbf{k}, t)}{\sum_{m=-1}^{1} C_{1m}(\mathbf{k}, 0)}, \quad (4.15)$$

then the $k \to 0$ limit of $\phi(\mathbf{k}, t)$ is related to the frequency-dependent dielectric function $\varepsilon(\omega)$ by the following relation [8]:

$$L\left[-\frac{d\phi(t)}{dt}\right] = \frac{[\varepsilon(\omega) - 1][2\varepsilon(\omega) + 1]\varepsilon_0}{[\varepsilon_0 - 1][2\varepsilon_0 + 1]\varepsilon(\omega)}, \quad (4.16)$$

where ε_0 is the static dielectric constant of the liquid and $L[f(t)]$ denotes the Laplace transformation of the function $f(t)$. As is shown by Eq. (4.16), all the molecules participate in the $k = 0$ limit. Thus, dielectric relaxation measures the collective relaxation. In view of this, it is rather surprising that the simple noninteracting model introduced by Debye many years ago, which gives the following expression of $\varepsilon(\omega)$

$$\varepsilon(\omega) = \varepsilon_\infty + \frac{\varepsilon_0 - \varepsilon_\infty}{1 + i\omega\tau_D}, \quad (4.17)$$

where ε_∞ is the infinite-frequency dielectric constant and τ_D is the Debye relaxation time, is rather successful in describing the polar response of many strongly interacting dipolar liquids. Note that because all the molecules are involved equally in $\varepsilon(\omega)$, information on the collective relaxation of only a few molecules is not available from $\varepsilon(\omega)$.

4.6 FLUORESCENCE DEPOLARIZATION

Fluorescence depolarization is an important technique to study the rotational relaxation of molecules in liquids. This experimental technique is based on a very simple observation — when a fluorescent molecule is excited with a plane-polarized light, the electric field of the optical pulse creates an *orientational anisotropy* in the angular distribution of the excited molecule. This anisotropy decays due to orientational Brownian motion of the participating molecules. Orientational diffusive motion of the probe molecule can thus be monitored by observing both the parallel $(I_\parallel)$ and the perpendicular $(I_\perp)$ components (with respect to the plane of the electric vectors of the exciting light) of the intensity of the emitted light.

In experiments one measures the time-dependent anisotropy, $r(t, \gamma)$, defined in terms of the parallel $(I_{\parallel})$ and perpendicular $(I_{\perp})$ components of the intensity of emitted light as

$$r(t, \gamma) = \frac{I_{\parallel}(t, \gamma) - I_{\perp}(t, \gamma)}{I_{\parallel}(t, \gamma) + 2I_{\perp}(t, \gamma)}, \quad (4.18)$$

where γ is transition dipole associated with the probing event.

Now, our next aim is to relate this experimentally observed quantity to the relevant time-correlation function that describes the orientational motion of the probe molecules and one proceeds as follows. We define a probability density, $f^{(i)}(\Omega, t)$, of finding a molecule in the ith electronic state with its orientation between Ω and $\Omega + d\Omega$ (with respect to the laboratory fixed axis) at time t. The parallel and the perpendicular concentrations of molecules giving rise to electric dipole emission from i are given by [5]

$$N^{(i)}_{\perp}(t, \gamma) = \int d\Omega P_{\parallel}(\Omega, \gamma) f^{(i)}(\Omega, t), \quad (4.19)$$

where $P_{\parallel}(\Omega, \gamma)$ and $P_{\perp}(\Omega, \gamma)$ are proportional to the probabilities of a molecule with orientation Ω and transition dipole moment γ, interacting with a light field polarized in the parallel and the perpendicular directions. Parallel $(I_{\parallel})$ and perpendicular $(I_{\perp})$ components of the intensity of emitted light will be proportional to the parallel and perpendicular concentrations defined above.

If one considers the case where only one electronic state is initially prepared, then one can define the anisotropy associated with the transition dipole γ as

$$r(t, \gamma) = \frac{I_{\parallel}(t, \gamma) - I_{\perp}(t, \gamma)}{I_{\parallel}(t, \gamma) + 2I_{\perp}(t, \gamma)} = \frac{N_{\parallel}(t, \gamma) - N_{\perp}(t, \gamma)}{N_{\parallel}(t, \gamma) + 2N_{\perp}(t, \gamma)}. \quad (4.20)$$

Therefore, in this case, anisotropy is given by the following expression [5]:

$$r(t, \gamma) = \frac{2}{5} \langle P_2(\boldsymbol{\mu}(0) \cdot \boldsymbol{\gamma}(t)) \rangle, \quad (4.21)$$

where P_2 is the second-order Legendre polynomial of the correlation of the transition moment vector at time zero and at time t. Thus $r(t, \gamma)$ attains the value 0.4 at $t = 0$. In experiments, one often gets slightly less than 0.4 which could be attributed either to limited time resolution or to the presence of decay to the other electronic states.

4.7 SOLVATION DYNAMICS (TIME-DEPENDENT FLUORESCENCE STOKES SHIFT)

Solvation dynamics will be discussed in more detail in Chapter 6; here we just mention a few pertinent facts. For many organic molecules, the dipole moment (both the magnitude and the direction) is significantly different in the excited state from its value in the ground state. Because of their photophysics (such as long fluorescence lifetime), some of these molecules can be used as a good probe of polar solvent dynamics. This is conventionally done by exciting the molecule and subsequently monitoring the time-dependent fluorescence Stokes shift of the emission spectrum, which usually shifts to longer wavelengths because of the increased dipolar stabilization of the excited state. If the shapes of the emission spectra do not change significantly with time, then the experimental data can be analyzed in terms of a solvation time-correlation function $C_S(t)$ defined by [9, 10]

$$C_S(t) = \frac{\bar{\nu}(t) - \bar{\nu}(\infty)}{\bar{\nu}(0) - \bar{\nu}(\infty)}, \qquad (4.22)$$

where $\bar{\nu}(t)$ is the average frequency of the spectrum at time t, defined as the first moment of the spectrum. Thus, Eq. (4.22) gives an average estimate of the time dependence of the progress of solvation.

In order to understand the role of collective orientational relaxation in solvation dynamics, a fairly accurate expression for the energy of interaction of the bare electric field, $\mathbf{E}_0(\mathbf{r})$, of a polar solute molecule with the polarization, $\mathbf{P}(\mathbf{r}, t)$, of the dipolar molecule is given by

$$E_{solv}(t) = -\frac{1}{2} \int d\mathbf{r} \mathbf{E}_0(\mathbf{r}) \cdot \mathbf{P}(\mathbf{r}, t). \qquad (4.23)$$

Equation (4.23) can be written in the Fourier space (with $\mathbf{k}$ as the Fourier variable conjugate to $\mathbf{r}$) as

$$E_{solv}(t) = -\frac{(2\pi)^{-3}}{2} \int d\mathbf{k} \mathbf{E}_0(\mathbf{k}) \cdot \mathbf{P}(\mathbf{k}, t). \qquad (4.24)$$

The time dependence of $\mathbf{P}(\mathbf{k}, t)$ is controlled by the collective orientational relaxation [11]. Under favorable conditions (small dipolar solute in a strongly dipolar liquid of larger solvent molecules), the integration in Eq. (4.24) may be dominated by the intermediate wave-vector processes, so that $E_{solv}(t)$ contains nontrivial information on collective orientational dynamics, but also mixed with translational motion of the neighboring solvent molecules. Thus, as in Kerr relaxation, understanding of the experimental results again becomes a bit tricky. One needs a good theory and also the help of computer simulations.

4.8 NEUTRON SCATTERING: COHERENT AND INCOHERENT

Thermal neutrons have wavelengths of the order of 1–2 Å, which is comparable to the distance between neighboring particles or molecules in solution. This is to be contrasted with light-scattering experiments where wavelength of the scattered light is between 4000 and 7000 Å. Therefore, thermal neutrons are used to study small-length-scale phenomena, both structure and dynamics. As in dynamic light scattering, neutron scattering is also coupled to density fluctuations in liquids. However, this is one of the few techniques where it is possible to gain information about wave-vector and frequency-dependent fluctuations in liquids at a molecular-length scale.

Neutron-scattering experiments can be classified as (i) elastic and (ii) inelastic. Elastic scattering occurs when a neutron interacts with a nucleus but does not leave it in an excited state, meaning the emitted neutron has the same energy as the incident neutron and no exchange of energy between neutron and the nucleus of the molecule has taken place. Scattering processes that involve an energetic excitation by the neutron are termed inelastic. Inelastic neutron scattering can be used to study dynamics of liquids.

As in any scattering experiment, in neutron-scattering experiments also one measures the scattering cross-section of the incident beam. In case of inelastic neutron scattering, the scattering cross-section per molecule is related to the *dynamic structure factor* as [4]

$$\frac{d^2\sigma}{d\Omega d\omega} = b^2 \left(\frac{k_1}{k_0}\right) S(\mathbf{k}, \omega), \qquad (4.25)$$

where b is the scattering length of the nucleus. k_0 and k_1 are the magnitude of the wave vectors corresponding to the incident and the scattered neutrons, respectively. $S(\mathbf{k}, \omega)$ is the dynamic structure factor, which is related to the density autocorrelation function as

$$S(\mathbf{k}, \omega) = \frac{1}{2\pi N} \int_{-\infty}^{\infty} \exp(i\omega t) \langle \rho_\mathbf{k}(t) \rho_{-\mathbf{k}} \rangle \, dt, \qquad (4.26)$$

where N is the total number of molecules in the system. One can also write the inelastic cross-section as the sum of *coherent* and *incoherent* parts:

$$\frac{d^2\sigma}{d\Omega d\omega} = \left(\frac{d^2\sigma}{d\Omega d\omega}\right)^{coh} + \left(\frac{d^2\sigma}{d\Omega d\omega}\right)^{incoh}, \qquad (4.27)$$

with

$$\left(\frac{d^2\sigma}{d\Omega d\omega}\right)^{coh} = \left(\frac{k_1}{k_0}\right) \langle b^2 \rangle S(\mathbf{k}, \omega), \qquad (4.28)$$

$$\left(\frac{d^2\sigma}{d\Omega d\omega}\right)^{incoh} = \left(\frac{k_1}{k_0}\right) \left[\langle b^2 \rangle - \langle b \rangle^2\right] S_s(\mathbf{k}, \omega), \qquad (4.29)$$

where $S(\mathbf{k}, \omega)$ is the self-part of the dynamic structure factor. Thus, it is possible to separate interparticle and self-particle space-time correlation functions.

Study of dynamic structure factor of a liquid by scattering of neutrons offers a very powerful experimental technique to study dynamics of liquids.

4.9 RAMAN LINE-SHAPE MEASUREMENTS

The traditional method for studying the phase relaxation of a given vibration (a chemical bond) has been infrared and isotropic Raman line shape analyses. The reason that these two techniques are useful in the study of dephasing is that they couple directly to molecular vibration via molecular polarizability. The infrared line shape is given by the Fourier transform of the time-correlation function of the total dipole moment operator of the system:

$$I_{IR}(\omega) = \frac{1}{2\pi} \int_{-\infty}^{+\infty} dt\, e^{i\omega t} \langle \mu(t)\mu(0) \rangle. \quad (4.30)$$

The connection with vibrational linewidth can be seen by the following steps. First, the dipole moment of the system is approximated as the sum of dipole moments of the individual molecules. Second, the dipole moment vector of an individual molecule, $\langle \mu_i(t) \rangle$, is expanded in the normal coordinate Q and only the first-order terms in Q are retained. Third, cross-correlations between vibrations in different molecules are neglected. These steps can be expressed by the following equations:

$$\begin{aligned}
I(\omega) &\propto \int_{-\infty}^{\infty} dt\, e^{i\omega t} \left\langle \left(\mu_i(Q(t)) + \frac{\partial \mu_i(t)}{\partial Q_i} Q_i(t) \right) \cdot \left(\mu_j(Q(0)) + \frac{\partial \mu_j(0)}{\partial Q_j} Q_j(0) \right) \right\rangle, \\
&\propto \int_{-\infty}^{\infty} dt\, e^{i\omega t} \sum_{j=1}^{N} \left\langle \frac{\partial \mu_i(t)}{\partial Q_i} \cdot \frac{\partial \mu_j(0)}{\partial Q_j} Q_i(t) Q_j(0) \right\rangle, \\
&\propto \int_{-\infty}^{\infty} dt\, e^{i\omega t} \sum_{j=1}^{N} \langle \mathbf{u}_i(t) \cdot \mathbf{u}_j(0) Q_i(t) Q_j(0) \rangle, \\
&\approx \int_{-\infty}^{\infty} dt\, e^{i\omega t} \langle \mathbf{u}_i(t) \cdot \mathbf{u}_i(0) Q_i(t) Q_i(0) \rangle.
\end{aligned} \quad (4.31)$$

Thus, the infrared line shape includes contributions from both rotational and vibrational motions. Note that in going from the first to the second step, use has been made of the fact that line shape is measured around the transition frequency ω_0.

The most straightforward way to obtain information about vibrational phase relaxation is via the isotropic Raman line shape. This line shape is given by the isotropic polarizability time-correlation function. The isotropic polarizability, α, is a scalar quantity, defined by the following operation on the polarizability tensor $\underline{\alpha}$:

$$\alpha = \frac{1}{3} \text{Tr}\, \underline{\alpha}, \quad (4.32)$$

where the operator **Tr** indicates sum over the diagonal elements. One uses steps similar to those outlined to obtain a simple expression for this line shape [12]:

$$I_{iso}(\omega) \propto \int_{-\infty}^{\infty} dt e^{i\omega t} \langle \alpha(t)\alpha(0) \rangle, \qquad (4.33)$$

$$\propto \left(\frac{\partial \alpha}{\partial Q}\right)^2 \int_{-\infty}^{\infty} dt e^{i\omega t} \sum_{j=1}^{N} \langle Q_i(t)Q_j(0) \rangle. \qquad (4.34)$$

Neglecting the phase correlation between neighboring molecules as a first approximation, the line shape becomes

$$I_{iso}(\omega) \propto \left(\frac{\partial \alpha}{\partial Q}\right)^2 \int_{-\infty}^{\infty} dt e^{i\omega t} \langle Q_i(0)Q_i(t) \rangle, \qquad (4.35)$$

and depends only on the separate dephasing of each molecule. In the case where the decay of time-correlation function $\langle Q_i(0)Q_i(t) \rangle$ is exponential with time constant τ_v,

$$\langle Q_i(t)Q_i(0) \rangle = \langle Q_i^2 \rangle \exp\left(-\frac{t}{\tau_v}\right), \qquad (4.36)$$

the isotropic Raman spectrum is a Lorentzian with *full width at half maxima* (FWHM) $2/\tau_v$.

Anisotropic Raman linewidths also contain information about vibrational phase relaxation (a detailed discussion is given in Chapter 13). In this case the relevant time-correlation function involves the correlation of the anisotropic part of the polarizability tensor, defined by

$$\underline{\beta} = \underline{\alpha} - \alpha \underline{I}. \qquad (4.37)$$

Anisotropic Raman line shape obviously contains contributions from molecular rotations. The steps relevant for its reduction to a tractable form are similar to the one used for IR and isotropic line shape and are given below [12]:

$$I_{aniso}(\omega) \propto \int_{-\infty}^{\infty} dt e^{i\omega t} \langle \mathrm{Tr}\underline{\beta}(t) \cdot \underline{\beta}(0) \rangle,$$

$$\propto \int_{-\infty}^{\infty} \left\langle \mathrm{Tr} \sum_{j=1}^{N} \frac{\partial \underline{\beta}_i(t)}{\partial Q_i} \cdot \frac{\partial \underline{\beta}_j(0)}{\partial Q_j} Q_i(t)Q_j(0) \right\rangle, \qquad (4.38)$$

$$\propto \int_{-\infty}^{\infty} \langle P_2(\mathbf{u}_i(t)\mathbf{u}_i(0))Q_i(t)Q_i(0) \rangle,$$

where P_2 is the second-order Legendre polynomial.

For slow dephasing, such as in N_2, this is a valid scheme. The dephasing times here are typically above 100 ps.

4.10 COHERENT ANTI-STOKES RAMAN SCATTERING (CARS)

In contrast to the frequency-domain techniques discussed above, time-domain techniques have been used widely after the short laser pulses have become available since the late 1970s. One such method is the so-called coherent anti-Stokes Raman scattering (CARS). This technique can be particularly useful in separating the homogeneous line width from the inhomogeneous ones. The principle of this experiment is fairly simple and consists of two pulses. The first pulse creates a coherent superposition of vibrationally excited molecules, while the second pulse measures the amplitude of coherence remaining after a time delay t_D, through coherent anti-Stokes Raman scattering. A detailed discussion of this technique is given in Ref. [5].

4.11 ECHO TECHNIQUES

Let us discuss some of the difficulties associated with the conventional techniques to measure the ultrafast vibrational dynamics. (a) The experimental observations based on Raman isotropic line shape analysis and coherent anti-Stokes Raman scattering measurements are not able to separate the pure dephasing (that is, elastic) contribution from the resonant transfer and population relaxation contributions. (b) Additionally, these techniques cannot separate between homogeneous and inhomogeneous contributions to line broadening. We remind the reader that inhomogeneous broadening arises from different vibrational energy spacing for different molecules located in different locations of the liquid. Often vibrational phase relaxation is faster than the relaxation of the environment, and as a result line shapes from individual molecules are placed at different frequencies and can be characterized by different linewidths. In many liquids at low temperature, this inhomogeneous broadening can be larger than the homogeneous broadening. (c) With such techniques, only fundamentals and lower overtones of the vibrational modes can be studied. However, with recent developments in ultrafast higher-order nonlinear optical spectroscopic techniques such as resonance Raman and photon echo experiments, higher-order quantum levels are now accessible. In these experiments a laser pulse excites a coherent superposition of the vibrational modes. A second probe pulse sent after a delay time τ then measures the decay of relaxation processes that occur on the time scale of τ. These techniques are used *to distinguish inhomogeneous from homogeneous line broadening*. There are several echo techniques developed to study the ultrafast dynamics in liquid. Below we discuss perhaps the most popular technique, the three-pulse photon echo technique (3PEPS).

Echo techniques have also been used to study electronic solvation dynamics. While the technique of time-dependent fluorescence Stokes shift (TDFSS) has been used widely in experimental studies, it suffers from the following limitations. (a) The short-time limit is often not accessible because the excitation pulse has a duration comparable to the inverse of the initial solvation rate. (b) It is not possible to separate the vibrational relaxation of the solute from the solvation energy progression because at short times the two processes can occur on the same time scale.

To overcome these difficulties, three-pulse photon echo technique has been developed to measure ultrafast solvation dynamics [13]. The key idea of photon echo spectroscopy is based on the fact that, subsequent to optical excitation in an inhomogeneous ensemble of absorbers, the transition frequencies in different absorbers become uncorrelated in time. A second pulse is then used to initiate rephasing. If the inhomogeneous width greatly exceeds the homogeneous width, an echo will be produced at a fixed time and the echo will narrow in time.

Echo signals are most often discussed in terms of the frequency-fluctuation time-correlation function, $M(t)$, defined as (for more detailed discussion see **Section 2** of Chapter 6)

$$M(t) \equiv \frac{\langle \Delta\omega(0) \Delta\omega(t) \rangle}{\langle \Delta\omega^2 \rangle},$$

$$= \frac{1}{\langle \Delta\omega^2 \rangle} \int_0^\infty d\omega \omega^2 \rho(\omega) \coth\left(\frac{\hbar\beta\omega}{2}\right) \cos \omega t, \quad (4.39)$$

where $\langle \Delta\omega^2 \rangle$, the mean-square fluctuation amplitude of the vibrational mode with frequency ω, is commonly referred to as the coupling strength:

$$\langle \Delta\omega^2 \rangle = \int_0^\infty d\omega \omega^2 \rho(\omega) \coth\left(\frac{\hbar\beta\omega}{2}\right), \quad (4.40)$$

where $\hbar$ is Planck's constant divided by 2π and $\beta = 1/(k_B T)$. In Eq. (4.39), $\cos \omega t$ is the short time decay of an undamped harmonic oscillator.

The second expression in Eq. (4.39) has been derived under harmonic approximation for the time dependence of the modes. $\rho(\omega)$ gives the density of vibrational modes. Now, the fluorescence Stokes shift function $C_S(t)$ defined by Eq. (4.22) can be written in terms of the spectral density as

$$C_S(t) = \frac{\hbar}{\lambda} \int_0^\infty d\omega \omega \rho(\omega) \cos(\omega t). \quad (4.41)$$

Note that Eqs. (4.39) and (4.41) are equivalent in the high-temperature limit. Thus, in the high-temperature limit photon echo experiments provide information about frequency-fluctuation time-correlation function, $M(t)$, which is essentially the same as the equilibrium solvation time-correlation function $C_S(t)$ obtained from the fluorescence Stokes shift experiments.

The same technique has also been employed to study the vibrational dephasing of the O–H stretch mode of HOD dissolved in D_2O and the distinction between inhomogeneous and homogeneous broadening has also been made [14].

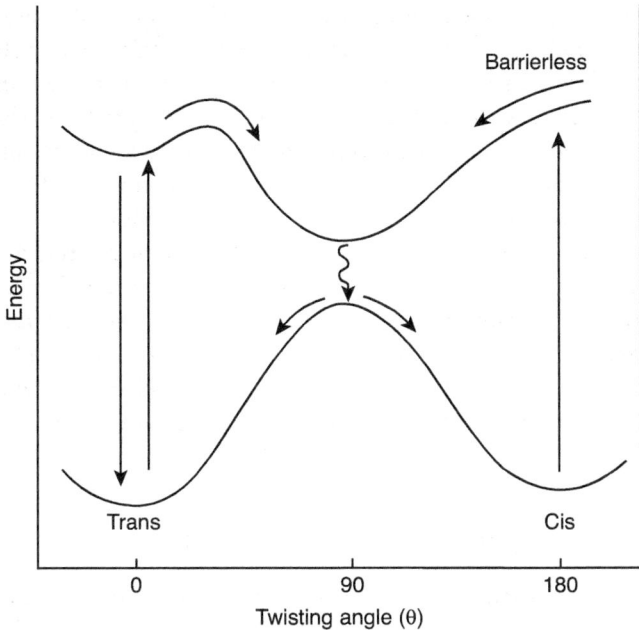

Figure 4.3 A schematic one-dimensional potential energy surface for *cis–trans* isomerization of stilbene.

4.12 ULTRAFAST CHEMICAL REACTIONS

In the 1950s and early 1960s, the standard methods of finding rates of fast chemical reactions were flash photolysis [15] and the temperature jump method. These methods could provide rates with time constants in milliseconds and even microseconds. With the development of laser spectroscopy in the late 1960s, optical techniques became more popular. One can now study chemical reactions with time constants in the few picoseconds or even femtosecond range. Thus, processes like electron-transfer reactions in the excited state can now be studied. These reactions are often strongly coupled to dynamical properties of the liquid such as dielectric relaxation or solvation dynamics. Another useful example is provided by the *cis–trans* isomerization in the stilbene molecule, as shown below in Fig. 4.3. Such reactions provide useful information about the coupling of the reactive motion with solvent properties like viscosity or viscoelasticity. Note the time scales of the reactions depicted in the figure.

4.13 FLUORESCENCE QUENCHING

Fluorescence quenching is a process in which the fluorescence intensity of a molecule decreases. A variety of processes can result in fluorescence quenching, such as excited-state reactions, energy transfer, complex formation, and collisional quenching. The decay kinetics of fluorescence provides important information about the rate of excited-state reactions.

In the fluorescence-quenching process first an absorption of a photon takes place that excites a molecule (fluorophore) to its lowest singlet excited state, for which three internal decay pathways exist: fluorescence, in which the molecule returns to the ground state with the emission of radiation; internal conversion, in which the energy of the molecule is converted into vibrational energy; and intersystem crossing, in which the singlet state is converted to the triplet state. If certain other molecules (quenchers) are present along with the fluorophore, external decay pathway(s) may also become available in addition to the internal decay pathways. Such external pathways facilitate the transfer of energy to a molecule with a similar energy gap or the transfer of an electron to or from another molecule. All of these relaxation processes competitively contribute to the decay of the fluorophore excited state. An example is given in Fig. 4.3. The efficiency of the external decay pathways determines the fluorescence yield and hence the rate of the excited state reactions.

4.14 TWO-DIMENSIONAL INFRARED (2D-IR) SPECTROSCOPY

We have already discussed three experimental probes of vibrational relaxation, namely, Raman linewidth, IR absorption spectroscopy, and CARS. However, these well-known and conventional experiments provide only average information about vibrational relaxation and cannot separate homogeneous from inhomogeneous contributions, and also cannot unearth information about coupling between different vibrational modes due to anharmonicity. As discussed in detail later, inhomogeneous broadening occurs from long-lived diagonal disorder that forces both vibrational phase and energy relaxations to occur at different rates at different regions of the system, thus masking the results intended for. The effects of vibrational anharmonicity are felt on several fronts. First, the energy gap between different vibrational levels change as we probe higher vibrational levels. Second, it allows coupling between different modes. Such a coupling also shifts the vibrational energy levels, in addition to allowing energy exchange between different modes. An important example of the latter is provided by vibrational energy relaxation of the O–H stretch, which follows a pathway through other low-frequency modes, like bending and librations.

Two-dimensional infrared (2D-IR) spectroscopy aims at removing these lacunae [17–20]. Such experiments have become possible with the availability of a femtosecond infrared laser pulse that allows excitation of ground-state vibrations without the involving excited electronic states. By 2D-IR spectroscopy one can have access to more detailed information about vibrational dynamics of molecules in the condensed phases.

In 2D-IR spectrum, we plot experimentally observed absorption intensity as a function of two frequency coordinates, in contrast to conventional linear experiments where intensity is plotted against one frequency. In these two-frequency plots, the intensity of the emitted light is plotted as contours and different intensities are often denoted by different color codes. These intensities are obtained by using three pulses with time delays between them and then Fourier transforming with respect to time delay.

Along the diagonal, one finds the same features as in linear, conventional spectroscopes. But the off-diagonal terms reveal the presence of correlations between different modes due to anharmonicity and the presence of inhomogeneity.

Theoretical 2D-IR spectroscopy is still not fully developed. The first two steps of Fig. 4.2 in the relationship are complete. We do have an expression for the intensity in terms of the nonlinear polarization. We also have an expression for this nonlinear polarization in terms of the third-order susceptibility, $\chi^{(3)}$, which is a four-time-correlation function. However, the evaluation of these time-correlation functions is highly nontrivial. Thus, extraction of accurate molecular and solvent properties from such measurements is an issue that remains to be solved.

4.15 SINGLE-MOLECULE SPECTROSCOPY

All the experiments discussed above involve averaging over many molecules, in principle, over all the molecules present in a solution. This averaging process can sometimes obscure or even remove interesting features that are present due to a distribution of configurations. Take, for example, the case of an active enzyme catalyzing a reaction. If we follow the reaction, say, for example, by some kind of titration in the solution phase where many enzymes participate in the catalysis, we would obtain an average rate. This average rate is often found to follow the Michaelis–Menten substrate concentration dependence of rate. However, if one studies the time trajectory of the catalytic reaction of one single enzyme under a steady flow of substrates into the enzyme, and obtains a rate by averaging over a constant time block, one finds that the rate has a rather wide distribution [21–23]. Such a distribution clearly contains far more information than just the average rate. In addition, one can find the correlation between rates in different time slabs, which provides valuable information about the state of the enzyme during its catalytic cycle. Such detailed information has motivated further theoretical development [24].

The ability to perform experiments at a single molecular level has been acquired only in the last two decades [25, 26]. This amazing development combined confocal microscopy with laser-spectroscopic techniques to obtain dynamical information on biopolymers and also on slow-moving molecules on solid surfaces. The result is always a distribution over rate constants or times, which, as already mentioned, can then be used to obtain further information about the reaction mechanism and also the conformational fluctuations [21, 26]. This experimental method has already given rise to many new results, but it is still in the process of development and much more is expected from this method.

4.16 SUMMARY

At the simplest level of description, one wants an analytical expression relating an experimental observable, such as the width of a line shape (such as NMR or IR) to a time constant describing natural dynamics of the liquid (like Eqs. (4.9) and (4.10) presented above). Another example is the relationship given by the continuum model between solvation time-correlation function and the dielectric relaxation time. These two are good, successful examples of the relationship between theory and experiment. However, for more complex situations, such as in sophisticated experiments (examples are 2D-IR or electronic spectroscopies), the relationships are not simple and often need careful analysis, of both theory and experiment.

In this chapter, we have attempted to provide a general perspective on the relationship between theory and experiment. Our goal was to familiarize the reader with the steps involved, with examples. In the following chapters we shall go into details of such relationships.

REFERENCES

1. B. Berne and R. Pecora, *Dynamic Light Scattering*, Wiley (1976).
2. B. Cowan, *Nuclear Magnetic Resonance and Relaxation*, Cambridge University Press (1997).
3. V. I. Bakhmutov, *Practical NMR Relaxation for Chemists*, Wiley (2004).
4. J. P. Hansen and I. R. McDonald, *Theory of Simple Liquids*, Academic Press (2006).
5. G. R. Fleming, *Chemical Applications of Ultrafast Spectroscopy*, Oxford University Press (1986).
6. Kalpouzos, D. McMorrow, W. T. Lotshaw, and G. A. Kenney-Wallace, *Chem. Phys. Lett.* 150, 138 (1988).
7. D. McMorrow, W. T. Lotshaw, and G. A. Kenney-Wallace, *J. Quantum Elec.* 24, 443 (1988).
8. T. W. Nee and R. Zwanzig, *J. Chem. Phys.* 52, 6353 (1970).
9. B. Bagchi, D. W. Oxtoby, and G. R. Fleming, *Chem. Phys.* 86, 257 (1984).
10. G. van der Zwan and J. T. Hynes, *J. Phys. Chem.* 89, 4181 (1985).
11. A. Chandra and B. Bagchi, *Chem. Phys. Lett.* 151, 47 (1988).
12. D. W. Oxtoby, *Adv. Chem. Phys.* 40, 1 (1979).
13. G. R. Fleming, T. Joo, and M. Cho, *Adv. Chem. Phys.* 101, 141 (1997).
14. W. P. de Boeij, M. S. Pshenichnikov and D. A. Wiersma, *Annu. Rev. Phys. Chem.* 49, 99 (1998).
15. R. G. W. Norrish and G. Porter, *Nature* 164, 658 (1949).
16. G. R. Fleming and P. G. Wolynes, *Phys. Today* 43, 36 (1990).
17. M. Zanni and R. M. Hochstrasser, *Current Opinion in Structural Biology*, 11, 516 (2001).
18. R. M. Hochstrasser, *Proc. Nat. Ac. Sc.* 104, 14190 (2007).
19. S. Mukamel, *Principles of Nonlinear Optics and Spectroscopy*, Oxford University Press (1995).
20. J. Zheng, K. Kwak, and M. D. Fayer, *Acc. Chem. Res.* 40, 75 (2007).
21. H. P. Lu, L. Xun, and X. S. Xie, *Science* 282, 1877 (1998).
22. P. C. Blainey, G. Luo, S. C. Kou, W. F. Mangel, G. L. Verdine, B. Bagchi, and X. S. Xie, *Nature Struct. Mol. Bio.* 16, 1224–29 (2009).
23. W. Min, X. S. Xie, and B. Bagchi, *J. Chem. Phys.* 131, 065104 (2009).
24. G. Hummer and Attila Szabo, *Acc. Chem. Res.* 38, 504 (2005).
25. (a) W. E. Moerner and L. Kador, *Phys. Rev. Lett.* 62, 2535 (1989). (b) M. Orrit and J. Bernard, *Phys. Rev. Lett.* 65, 2716 (1990).
26. X. S. Xie and J. K. Trautman, *Ann. Rev. Phys. Chem.* 49, 441 (1998).

5

Orientational and Dielectric Relaxation

5.1 INTRODUCTION

In a liquid, as a result of intermolecular interactions (such as collisions), molecules are continuously rotating and translating. Thus, if we could tag a molecule and study its detailed motion, we would find it executing a random Brownian motion not only in the three-dimensional positional space, but also a similar motion in the three-dimensional orientational space. For simplicity, let us first consider the motion of a tagged prolate-shaped molecule in a solvent of spherical molecules, as shown in Fig. 5.1.

We shall refer to this simple model as an "ellipsoid in a sea of spheres (EISS)." In this model, both the position and the orientation of the tagged molecule undergo continual change due to the interactions with the surrounding solvent molecules. For a molecule of ellipsoidal symmetry, we need two angles (θ and φ) to denote the orientation in the space-fixed frame. In this chapter we shall denote the two angles by an orientation vector Ω. In liquids, both the position vector r and the orientation vector Ω execute small-amplitude Brownian motion in the respective configuration space. Quantitative knowledge about the rate of orientation of anisotropic molecules in liquid is essential to understand many relaxation processes and chemical reactions. As discussed in Chapter 4, a large number of experimental techniques have been developed and used to understand the details of molecular-orientational process in liquids. As expected, there are features that are common to both translational and rotational motions. There are also many features that are quite distinct. In general, orientational time-correlation functions (which are measured in most experiments discussed in Chapter 4) decay at a time scale faster than the corresponding time-correlation functions of the linear motion

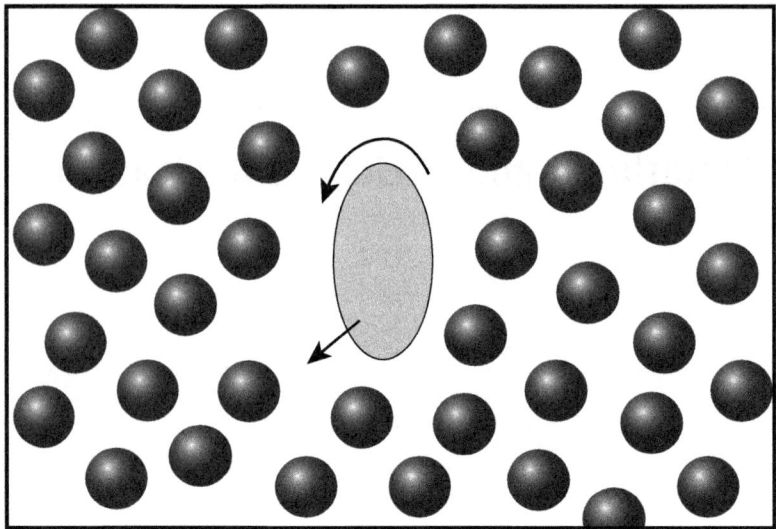

Figure 5.1 Motion of a prolate-shaped solute molecule in a solvent of spherical molecules. We refer to this model as "ellipsoid in a sea of spheres."

(such as the dynamic structure factors). For example, the time constant of decay of the orientational time-correlation function of a tagged water molecule is 2.6 ps during which a water molecule diffuses only 20% of its molecular diameter, which is quite meager. However, we shall describe subsequently how translational motions can deeply affect the rates of orientational relaxation in dense dipolar liquids, in one of the most interesting aspects of orientational relaxation.

An important aspect of rotational relaxation is the dependence of rotational diffusion on the shape of the molecule. This dependence is much stronger than that in the case of translational diffusion. The reason is easy to understand. A perfect sphere can rotate freely inside a liquid without causing any disturbance/displacement to the liquid. The situation changes if the liquid surrounding the sphere sticks to it. Then the friction becomes significant. Usually small molecules do not have liquids sticking to it unless there is a certain specific interaction (like hydrogen bonding).

The difference between the slip (when the surrounding liquid does not stick to the surface of the rotating sphere) and the stick boundary conditions is enormous for rotation but rather small for translation. Translational motion always causes displacement of the liquid. In the language of hydrodynamics, a moving solute creates a flow field which causes energy dissipation, and hence causes frictional resistance. But rotational motion is far more sensitive to the shape of the solute and also to the nature of interactions of the solute with the surrounding solvent/liquid.

One can thus gather a considerable amount of insight by studying orientational motion of molecules in computer simulations by monitoring both the spatial and the orientational trajectories of the tagged molecule in simple model systems, such as our model of "isolated ellipsoid in the sea of spheres (EISS)" discussed above.

At long times, the motions are diffusive and the rate of displacement in both the spaces can be each described by a diffusion equation, with respective diffusion coefficients,

$$\frac{\partial \rho_S(\mathbf{r}, t)}{\partial t} = D_T \nabla^2 \rho_s(\mathbf{r}, t), \qquad (5.1)$$

$$\frac{\partial \rho_S(\mathbf{\Omega}, t)}{\partial t} = D_R \nabla^2 \rho_s(\mathbf{\Omega}, t), \qquad (5.2)$$

where $\rho_s(\mathbf{r}, t)$ and $\rho_s(\mathbf{\Omega}, t)$ are position- and orientation-dependent **single-**particle densities, respectively, defined as

$$\rho_s(\mathbf{r}, t) = \langle \delta(\mathbf{r} - \mathbf{r}_s(t)) \rangle, \qquad (5.3)$$

$$\rho_s(\mathbf{\Omega}, t) = \langle \delta(\mathbf{\Omega} - \mathbf{\Omega}_s(t)) \rangle. \qquad (5.4)$$

D_T and D_R are the translational and rotational diffusion coefficients, respectively. The analogy between translational and rotational motion can be extended further. We can define not only position- ($\mathbf{r}$) dependent isotropic collective number density, $\rho(\mathbf{r}, t)$, but also position- ($\mathbf{r}$) and orientation- ($\mathbf{\Omega}$) dependent **collective** orientational density, $\rho(\mathbf{r}, \mathbf{\Omega}, t)$. These collective quantities are different from tagged (single-) particle densities, as they count not just the tagged but all the molecules in the system present in a volume element around $\mathbf{r}$ and $\mathbf{\Omega}$:

$$\rho(\mathbf{r}, \mathbf{\Omega}, t) = \left\langle \sum_i \delta(\mathbf{r} - \mathbf{r}_i(t)) \delta(\mathbf{\Omega} - \mathbf{\Omega}_i(t)) \right\rangle. \qquad (5.5)$$

We would like to emphasize, at the very outset, the difference between the single-particle and collective quantities because (i) different experiments measure different quantities; (ii) they can have vastly different dynamics; and (iii) this fact is sometimes overlooked. However, the difference might not be great when the molecules are nonpolar and not near any orientational phase transition, like the isotropic–nematic transition in mesogens that form liquid crystals.

We have discussed the dynamics of isotropic density fluctuations in Chapter 3. Here we shall discuss the orientational relaxation—both single molecule and collective.

As we already discussed in Chapter 4, orientational relaxation plays a key role in many relaxation processes, such as polarization and dielectric relaxation, solvation dynamics, and quadrupolar relaxation. It profoundly influences the dynamics of the many important chemical reactions, like the electron and proton transfer reactions in a polar liquid. The orientational relaxation of an anisotropic molecule in liquid clearly depends on the density and the temperature of the liquid and also on the nature of the anisotropy of the interaction potential. It may also be coupled to the translational modes of the liquid and the internal modes of the molecule. Orientational relaxation in a liquid can show rich and diverse dynamical behavior.

As also discussed in Chapter 4, orientational motions are studied directly and indirectly by many different experimental techniques. There are several issues that arise in the study of orientational relaxation. For example, the relationship between orientational relaxation of a single molecule and the collective relaxation, where orientation of many molecules is probed, remains a frontier problem in the area of orientational relaxation. Many experiments, such as NMR relaxation, incoherent neutron scattering and Raman line shapes, are sensitive to the single-particle orientational motion. The other limit of collective orientational motion involving all the molecules of the system is probed by experiments like dielectric relaxation, depolarized light scattering, and coherent neutron scattering. It has not yet been possible to study successfully and directly collective relaxation involving a limited number of molecules (say, of the order of 10–100). However, indirect information about this intermediate limit can be obtained from such linear spectroscopic techniques as time-dependent fluorescence Stokes shift (TDFSS) measurement of newly created ion or dipole or such nonlinear optical techniques as Kerr relaxation. In these measurements, some moments of the inverse distances of the solvent molecules from the probe over time- and space-dependent solvent polarization are studied. If the orientational relaxation of the nearest-neighbor molecule is significantly different from those that are far off in the bulk, then these indirect methods can provide reliable information on collective orientational relaxation in the intermediate regime.

The collective orientational relaxation in the long-wavelength limit involving all the molecules of the system was studied initially by borrowing concepts from the continuum description of electrostatics. In these theories, the dipolar molecule is replaced by a cavity of some simple shape and with a point dipole at its center and the liquid is replaced by a frequency-dependent dielectric continuum. The interaction between the rotating dipoles and the bulk of the liquid is included through a time-dependent reaction field, which arises from the electric polarization of the solvent by the rotating dipole. The reaction field is obtained by a quasistationary boundary-value calculation. This field gives rise to a dielectric friction, which retards the rotation of the dipole and gives rise to a nonexponential decay of the dipolar correlation function. The classic example of such a theory is the work of Nee and Zwanzig [1].

The advantage of the continuum-model theories [1, 2] is that they provide simple expressions for the orientational correlation functions that can be tested against experiments. This is especially true for a complex liquid where a microscopic theory is bound to be complicated. The main drawback of the continuum-model theories is that they ignore the intermolecular correlations that are present in a dense dipolar liquid. If these correlations are important in a relaxation process, then the continuum model is obviously inadequate.

Molecular theory of orientational correlation, either space or time dependent, or both, is rather complicated. This is because orientational correlation between two dipolar or nonspherical molecules is nontrivial primarily because the orientational part of the interaction potential is nontrivial. Imagine describing position and orientation dependence of the interaction potential between two hard ellipsoids and you will appreciate the difficulty. This is a bit unfortunate because the orientational correlation functions, especially their time dependence, are so easily accessible in experiments. On the positive side, there have been several interesting theoretical studies that throw much light on the role of orientational correlation in the dynamics of liquids.

5.2 EQUILIBRIUM AND TIME-DEPENDENT ORIENTATIONAL CORRELATION FUNCTIONS

In linear optical spectroscopy, the quantity that is usually measured is a two-point time-correlation function of a dynamical quantity. For the clarity of the following discussion, let us consider a dipolar liquid, such as acetonitrile or water, although much of the following discussion holds true for nondipolar anisotropic molecules. For a dipolar liquid, a dynamical quantity that is of interest is the collective dipole moment of a dipolar liquid. A dense dipolar liquid is a strongly correlated system, and this is clearly reflected in the value of the equilibrium correlation functions at short to intermediate distances. The equilibrium correlation functions that are of interest here are the radial distribution functions $g(\mathbf{r}, \Omega, \mathbf{r}', \Omega')$, the pair correlation function $h(\mathbf{r}, \Omega, \mathbf{r}', \Omega')$, and the two-particle direct correlation function $c(\mathbf{r}, \Omega, \mathbf{r}', \Omega')$, where r denotes the position and Ω is the orientation of a molecule, both measured in a laboratory fixed frame. Here we shall assume that the molecules are rigid so that the orientation of the molecule is given by a unit vector with orientation. The radial distribution function is defined as [3,5]

$$g\left(\mathbf{r}, \Omega, \mathbf{r}', \Omega'\right) = \frac{\Omega_0^2}{\rho_0^2} \sum_{i \neq j} \langle \delta\left(\mathbf{r} - \mathbf{r}_i\right) \delta\left(\mathbf{r}' - \mathbf{r}_j\right) \delta\left(\Omega - \Omega_i\right) \delta\left(\Omega' - \Omega_j\right)\rangle,$$

(5.6)

where $\langle \ldots \rangle$ denotes averaging over an equilibrium ensemble and ρ_0 is the number density of the liquid. Ω_0 is the measure of an angular space. Thus, $\Omega_0 = 4\pi$ for axially symmetric molecules, and $8\pi^2$ for nonsymmetric molecules. The pair correlation function is related to the radial distribution function by the usual relation:

$$h\left(\mathbf{r}, \Omega, \mathbf{r}', \Omega'\right) = g\left(\mathbf{r}, \Omega, \mathbf{r}', \Omega'\right) - 1.$$

(5.7)

The two distribution functions $h\left(\mathbf{r}, \Omega, \mathbf{r}', \Omega'\right)$ and $c\left(\mathbf{r}, \Omega, \mathbf{r}', \Omega'\right)$ are related to each other by the molecular (generalized) Ornstein–Zernike relation [3,4]

$$h\left(\mathbf{r}, \Omega, \mathbf{r}', \Omega'\right) = c\left(\mathbf{r}, \Omega, \mathbf{r}', \Omega'\right) \\ + \frac{\rho_0}{4\pi} \int d\mathbf{r}'' d\Omega'' c\left(\mathbf{r}, \Omega, \mathbf{r}', \Omega'\right) h\left(\mathbf{r}'', \mathbf{r}', \Omega'', \Omega'\right). \quad (5.8)$$

It is usual to expand the two-particle distribution functions in terms of Wigner matrices [3] as

$$g(\mathbf{r}-\mathbf{r}', \Omega, \Omega') = \sum_{l_1 l_2 l} \sum_{m_1 m_2 m} \sum_{n_1 n_2} \left[g(l_1 l_2 l; n_1 n_2; \mathbf{r}-\mathbf{r}') C_g(l_1 l_2 l; m_1 m_2 m) \right.$$
$$\left. \times D_{l_1 m_1}^{n_1}(\Omega) D_{l_2 m_2}^{n_2}(\Omega') Y_{lm}(\Omega'') \right], \qquad (5.9)$$

where Ω'' gives the orientation of the vector $\mathbf{r}-\mathbf{r}'$ in the laboratory fixed frame, $C_g(l_1 l_2 l; m_1 m_2 m)$ is the usual Clebsch–Gordan coefficient, and $D_{lm}^n(\Omega(t))$ are the Wigner rotation matrices. The **single-particle** orientational time-correlation functions are defined by

$$S_{l_1 l_2 m_1 m_2}^{n_1 n_2}(t) = \left\langle D_{l_1 m_1}^{n_1}(\Omega(0)) D_{l_2 m_2}^{n_2}(\Omega(t)) \right\rangle. \qquad (5.10)$$

Although only the angular coordinate of the particle in question appears in the above equation, it depends implicitly on the dynamics of the other variables of the system through the time dependence of $\Omega(t)$.

The collective orientational correlation function depends on the relative positions of all the particles of the system. For an isotropic liquid, because of translational invariance, it is convenient to work in the Fourier space, and we define the collective dynamical variable

$$D_{lm}^n(\mathbf{k}, t) = \sum_i e^{i\mathbf{k} \cdot \mathbf{r}_i} D_{lm}^n(\Omega_i(t)), \qquad (5.11)$$

where the sum is over all the molecules of the system. Here $\mathbf{k}$ is the wave vector conjugate to the position vector, $\mathbf{r}$. The collective orientational time-correlation function is now defined as

$$C_{l_1 l_2 m_1 m_2}^{n_1 n_2}(\mathbf{k}, t) = \left\langle D_{l_1 m_1}^{n_1}(-\mathbf{k}) D_{l_2 m_2}^{n_2}(\mathbf{k}, t) \right\rangle. \qquad (5.12)$$

We shall be concerned mostly with homogeneous systems, in the absence of any external field.

In this case we have considerable simplifications,

$$S_{l_1 l_2 m_1 m_2}^{n_1 n_2}(t) = S_{l_1 m_1}^{n_1}(t) \delta_{l_1 l_2} \delta_{m_1 m_2} \delta_{n_1 n_2}, \qquad (5.13)$$

and

$$C_{l_1 l_2 m_1 m_2}^{n_1 n_2}(\mathbf{k}, t) = C_{l_1 m_1}^{n_1}(\mathbf{k}, t) \delta_{l_1 l_2} \delta_{m_1 m_2} \delta_{n_1 n_2}. \qquad (5.14)$$

The correlation functions become rather simple for a spherical diffuser, because the Wigner matrices are now replaced by spherical harmonics,

$$S_{lm}(t) = \langle Y_{lm}(\Omega(0)) Y_{lm}(\Omega(t)) \rangle, \qquad (5.15)$$

$$C_{lm}(\mathbf{k}, t) = \langle Y_{lm}(-\mathbf{k}) Y_{lm}(\mathbf{k}, t) \rangle. \qquad (5.16)$$

Experimental techniques that measure the functions $S_{lm}(t)$ and $C_{lm}(\mathbf{k}, t)$ are discussed in Chapter 4.

5.3 RELATIONSHIP WITH EXPERIMENTAL OBSERVABLES

As we mentioned in the Introduction, both the **single-particle** correlation function $S_{lm}(t)$ and the **collective** correlation function $C_{lm}(\mathbf{k}, t)$ are routinely measured in different experiments. In Chapter 4, the relevant expressions that relate the experimental observables to the orientational correlation functions have been discussed. For simplicity, the correlation functions are written in terms of spherical harmonics, so they are applicable to molecules of ellipsoidal and spherical shapes. These can be generalized easily to the case of anisotropic diffusers.

5.4 MOLECULAR HYDRODYNAMIC DESCRIPTION OF ORIENTATIONAL MOTION

The main dynamical quantities involved in orientational relaxation are the number density, $\rho(\mathbf{r}, \Omega, t)$, and the spatial and angular momenta densities $g_T(\mathbf{r}, \Omega, t)$ and $g_\Omega(\mathbf{r}, \Omega, t)$, respectively. In a dense liquid, the relaxation of energy fluctuation may not be rate-determining and will not be considered here. Because we are interested in the collective orientational relaxation involving not only large but also few (for example, nearest-neighbor) molecules, the traditional Navier–Stokes hydrodynamics description [4], valid only in the long-wavelength limit, is not adequate. However, this time-honored description can be extended meaningfully to treat dynamics at a molecular-length scale. Such an extension has been carried out recently by several groups and is described here. The theory described below is perhaps the simplest and is an extension of molecular hydrodynamic theory initiated by de Gennes that has been discussed in Chapter 3.

The main idea in this extended hydrodynamic description is as follows. At molecular-length scales, the momentum relaxation depends on the force field experienced by the molecules. This force field comes from intermolecular interaction. Thus, the gradient of the pressure term in Navier–Stokes equation should be modified to include the effects of intermolecular interactions, which depend on the molecular arrangements. The force field acting on a molecule at position $\mathbf{r}$ with orientation Ω can be obtained from the density functional theory, which gives a general expression for the free-energy functional of an inhomogeneous system. As already mentioned, the theory is an extension of de Gennes' molecular hydrodynamic theory to treat orientation.

5.4.1 The Equations of Motion

The extended hydrodynamic equations for the number density and the momenta densities consist of the continuity equation for the conservation of number density and then a generalized Langevin equation type description that is akin to the Navier–Stokes equation

$$\frac{\partial}{\partial t}\rho(\mathbf{X},t) + \frac{1}{m}\boldsymbol{\nabla}_T \cdot \mathbf{g}_T(\mathbf{X},t) + \frac{1}{I}\boldsymbol{\nabla}_\Omega \cdot \mathbf{g}_\Omega(\mathbf{X},t) = 0, \quad (5.17)$$

$$\frac{\partial}{\partial t}g_i(\mathbf{X},t) = -\rho(\mathbf{X},t)\boldsymbol{\nabla}_i \frac{\delta F'[\rho(t)]}{\delta\rho(\mathbf{X},t)}$$

$$-\sum_j dt_1 d\mathbf{X}_1 \Gamma_{ij}(\mathbf{X},\mathbf{X}_1,t-t_1;\rho(t))\frac{\delta F}{\delta g_j(\mathbf{X},t)} + f_i(\mathbf{X},t), \quad (5.18)$$

where $\mathbf{X} = (\mathbf{r},\Omega)$, i stands for T or Ω, m and I are the mass and the moment of inertia, respectively, $\boldsymbol{\nabla}_T$ and $\boldsymbol{\nabla}_\Omega$ are the usual spatial and the angular gradient operators, $F'[\rho(t)]$ is the interaction part of the free energy for an inhomogeneous fluid. F and F' are given by the following expressions:

$$F'[\rho(t)] = k_B T \left\{ \int d\mathbf{X} \rho(\mathbf{X},t)[\ln\rho(\mathbf{X},t) - 1] \right.$$

$$-\frac{1}{2}\int d\mathbf{X}d\mathbf{X}_1 c_2(\mathbf{X},\mathbf{X}_1)\delta\rho(\mathbf{X},t)\delta\rho(\mathbf{X}_1,t)$$

$$\left. -\frac{1}{6}\int d\mathbf{X}d\mathbf{X}_1 d\mathbf{X}_2 c_3(\mathbf{X},\mathbf{X}_1,\mathbf{X}_2)\delta\rho(\mathbf{X},t)\delta\rho(\mathbf{X}_1,t)\delta\rho(\mathbf{X}_2,t) + \ldots \right\}, \quad (5.19)$$

and

$$F[\rho(t)] = F'[\rho(t)] + \int d\mathbf{X} \left[\frac{g_T^2}{2m\rho} + \frac{g_\Omega^2}{2I\rho}\right], \quad (5.20)$$

where $\delta\rho(\mathbf{X},t)$ is the density fluctuation and $c_n(\mathbf{X},\mathbf{X}_1,\ldots)$ is the n-particle direct correlation function, which is the nth expansion coefficient in the density expansion of free energy. Equation (5.19) is exact. In Eq. (5.18), $f_i(\mathbf{X},t)$'s are the random force and random torque, which are assumed to be Gaussian and related to the dissipative kernels Γ_{ij} by the second fluctuation–dissipation theorem:

$$\langle f_T(\mathbf{X},0)f_T(\mathbf{X}_1,t)\rangle = k_B T \Gamma_{TT}(\mathbf{X},\mathbf{X}_1,t;\rho(t)), \quad (5.21)$$

$$\langle f_\Omega(\mathbf{X},0)f_\Omega(\mathbf{X}_1,t)\rangle = k_B T \Gamma_{\Omega\Omega}(\mathbf{X},\mathbf{X}_1,t;\rho(t)), \quad (5.22)$$

$$\langle f_T(\mathbf{X},0)f_\Omega(\mathbf{X}_1,t)\rangle = k_B T \Gamma_{T\Omega}(\mathbf{X},\mathbf{X}_1,t;\rho(t)), \text{ with, } \Gamma_{T\Omega} = \Gamma_{\Omega T}. \quad (5.23)$$

Equations (5.17)–(5.23) allow a general description of the coupled translational and orientational dynamics in dense liquid, provided reasonable forms of the dissipative kernels are available. Clearly, similar equations can also be derived from the Mori–Zwanzig projection operator technique with $\{\rho, g_T, g_\Omega\}$ as the subset of slow variables [5]. However, we regard the hydrodynamic approach more straightforward and intuitive. This approach is also closely related to the well-known time-dependent Ginzburg–Landau (TDGL) approach to collective dynamics.

5.4.2 Limiting Situations

The complexity of molecular hydrodynamic equations do not allow a general solution of the problem. However, different limits of these equations have been studied and they reveal the rich dynamical behavior of orientational relaxation. The following limiting situations have been discussed:

(a) The Markovian limit where the random forces are delta-correlated in time. Also, the random forces are assumed to be totally local and the dissipative kernels are diagonal. That is,

$$\Gamma_{ij}(\mathbf{X}, \mathbf{X}_1, t - t_1; \rho(t)) = \Gamma_{ij}(\bar{\rho}) \delta(\mathbf{X} - \mathbf{X}_1) \delta(t - t_1) \delta_{ij}, \qquad (5.24)$$

where $\bar{\rho}$ is the average density, equal to $\rho_0/4\pi$ for a spherical diffuser, and ρ_0 is the number density of the liquid. Both the overdamped and the inertial limits of this approximation have been studied.

(b) The non-Markovian form for the angular dissipative kernel, but the Markovian form for the spatial kernel. The reason for studying this situation is that the long-range nature of dipolar interactions makes the angular kernel frequency dependent at a lower frequency where the viscoelastic effects are yet to be important; therefore, that the frequency dependence of the spatial dissipative kernel can be neglected at such frequencies.

Such an extended hydrodynamic approach as discussed above can be valid at small length scales. The basic idea is that at the molecular length scale (that is, at small separations), the relaxation of density is very slow and it is the most important dynamical quantity. The reason is that at this length only the number density is a conserved quantity and both momentum and energy relaxations decay much faster than the density relaxation. Thus, a generalized diffusion equation description of the dynamics is expected to be fairly reliable.

5.5 MARKOVIAN THEORY OF COLLECTIVE ORIENTATIONAL RELAXATION: BERNE TREATMENT

A self-consistent Markovian theory of the collective orientational relaxation in a dipolar liquid was first presented by Berne [6]. This treatment was based on an equation for rotational and translation diffusion with an interaction term that arises from the torques and forces on a molecule resulting from the electric field generated by other charges in the system. The equation of motion is given by [6]

$$\frac{\partial}{\partial t}\rho(\mathbf{r}, \mathbf{\Omega}, t) = D_R \nabla_\Omega^2 \rho(\mathbf{r}, \mathbf{\Omega}, t) + D_T \nabla_T^2 \rho(\mathbf{r}, \mathbf{\Omega}, t)$$
$$- \beta D_R \nabla_\Omega \cdot [N(\mathbf{r}, \mathbf{\Omega}, t)\rho(\mathbf{r}, \mathbf{\Omega}, t)]$$
$$- \beta D_T \nabla_T \cdot [F(\mathbf{r}, \mathbf{\Omega}, t)\rho(\mathbf{r}, \mathbf{\Omega}, t)], \qquad (5.25)$$

where D_T and D_R are the translational and rotational diffusion coefficients, respectively. ∇_T and ∇_Ω are the spatial and rotational gradient operators. The force and torque are expressed in terms of the molecular charge distribution $Z(s)$ and the electric field E by the following equations [6]:

$$F(\mathbf{r}, \mathbf{\Omega}, t) = \int ds Z(s) E(\mathbf{r} + s\mathbf{\Omega}, t), \qquad (5.26)$$

$$N(\mathbf{r}, \mathbf{\Omega}, t) = \int ds Z(s) s\mathbf{\Omega} \cdot E(\mathbf{r} + s\mathbf{\Omega}, t). \qquad (5.27)$$

From Eqs. (5.25)–(5.27), it can be shown that in the limit of zero wave vector ($k \to 0$) and for a liquid of nonpolarizable molecules the correlation function $\phi(t)$ (defined by Eq. (4.15) in Chapter 4) is bi-exponential and is given by [6]

$$\phi(t) = \mu^2 \left[\frac{2}{3} e^{-2D_R t} + \frac{1}{3} e^{-2D_R(1+3Y)t} \right], \qquad (5.28)$$

with

$$Y = (4\pi/9)\beta\mu^2\rho_0. \qquad (5.29)$$

This result indicates that the dipole–dipole interaction can give rise to a marked difference between the decay of the longitudinal and the transverse fluctuations. If the dipole–dipole interaction is neglected, as was originally done by Debye, then $\phi(t)$ is single exponential and is given by

$$\phi(t) = \exp[-2D_R t]. \qquad (5.30)$$

5.5.1 Generalized Smoluchowski Equation Description

Building on the effort of Berne, a microscopic theory of orientational relaxation was developed later [7]. This theory is based on a generalized Smoluchowski equation (GSE) to describe the dynamics of the position- and orientation-dependent density of the dipolar liquid molecules. The derivation of the kinetic equation is based on the fact that in a dense polar liquid the relaxations of both the spatial momentum and the angular momentum are very fast and only the number conservation is important on the time scale of interest. Therefore, the starting equation is the continuity equation given by Eq. (5.17). One now neglects the non-Markovian effects and the random fluctuations in the momentum relaxation. That means equation of motion of g_T is now given by [7]

$$\frac{\partial}{\partial t} g_T(\mathbf{r}, \mathbf{\Omega}, t) = -\rho(\mathbf{r}, \mathbf{\Omega}, t) \nabla_T \frac{\delta F'[\rho(t)]}{\delta \rho(\mathbf{r}, \mathbf{\Omega}, t)}$$

$$- \int d\mathbf{r}' d\mathbf{\Omega}' \Gamma_T(\mathbf{r}, \mathbf{r}', \mathbf{\Omega}, \mathbf{\Omega}'; \rho(t)) \frac{\delta F[\rho(t)]}{\delta g_T}, \quad (5.31)$$

and a similar equation for g_Ω. $F'[\rho(t)]$ is the free energy of an inhomogeneous equilibrium fluid minus its kinetic energy contribution. The functional derivative on the right-hand side of Eq. (5.31) is given by [7]

$$\frac{\delta F'[\rho(t)]}{\delta \rho(\mathbf{r}, \mathbf{\Omega}, t)} = k_B T \left\{ \ln \rho(\mathbf{r}, \mathbf{\Omega}, t) - \int d\mathbf{r}' d\mathbf{\Omega}' c(\mathbf{r}, \mathbf{\Omega}, \mathbf{r}', \mathbf{\Omega}') \delta \rho(\mathbf{r}', \mathbf{\Omega}', t) \right.$$

$$- \frac{1}{2} \int d\mathbf{r}' d\mathbf{\Omega}' d\mathbf{r}'' d\mathbf{\Omega}'' c_3(\mathbf{r}, \mathbf{r}', \mathbf{r}'', \mathbf{\Omega}, \mathbf{\Omega}', \mathbf{\Omega}'')$$

$$\left. \times \delta \rho(\mathbf{r}', \mathbf{\Omega}', t) \delta \rho(\mathbf{r}'', \mathbf{\Omega}'', t) + \dots \right\}, \quad (5.32)$$

where $c(\mathbf{r}, \mathbf{r}', \mathbf{\Omega}, \mathbf{\Omega}')$ and $c_3(\mathbf{r}, \mathbf{r}', \mathbf{r}'', \mathbf{\Omega}, \mathbf{\Omega}', \mathbf{\Omega}'')$ are the two-particle and the three-particle direct correlation functions of the liquid, respectively. To proceed further, one needs the bare nonlinear dissipative kernel Γ_T and Γ_Ω. It is assumed that [7]

$$\Gamma_T^{ij} = \delta_{ij} \delta(\mathbf{r} - \mathbf{r}') \delta(\mathbf{\Omega} - \mathbf{\Omega}') k_B T / D_T, \quad (5.33)$$

$$\Gamma_\Omega^{ij} = \delta_{ij} \delta(\mathbf{r} - \mathbf{r}') \delta(\mathbf{\Omega} - \mathbf{\Omega}') k_B T / D_R, \quad (5.34)$$

where i, j denote the Cartesian components. Next one neglects the momenta relaxation, in the spirit of time-dependent Ginzburg–Landau theories, and uses the resultant expressions for g_T and g_Ω in the continuity equation to obtain the following nonlinear Smoluchowski equation:

$$\frac{\partial}{\partial t} \delta \rho(\mathbf{r}, \mathbf{\Omega}, t) = D_R \nabla_\Omega^2 \delta \rho(\mathbf{r}, \mathbf{\Omega}, t) + D_T \nabla_T^2 \delta \rho(\mathbf{r}, \mathbf{\Omega}, t)$$

$$- D_R \nabla_\Omega . \rho(\mathbf{r}, \mathbf{\Omega}, t) \nabla_\Omega \int d\mathbf{r}' d\mathbf{\Omega}' c(\mathbf{r}, \mathbf{r}', \mathbf{\Omega}, \mathbf{\Omega}') \delta \rho(\mathbf{r}', \mathbf{\Omega}', t)$$

$$- D_T \nabla_T . \rho(\mathbf{r}, \mathbf{\Omega}, t) \nabla_T \int d\mathbf{r}' d\mathbf{\Omega}' c(\mathbf{r}, \mathbf{r}', \mathbf{\Omega}, \mathbf{\Omega}') \delta \rho(\mathbf{r}', \mathbf{\Omega}', t), \quad (5.35)$$

where for pure liquid, $\delta \rho(\mathbf{r}, \mathbf{\Omega}, t) = \rho(\mathbf{r}, \mathbf{\Omega}, t) - \rho_0/4\pi$. In writing Eq. (5.35), the terms containing the triplet and higher-order direct correlation functions are neglected. It should be made clear here that Eq. (5.35) describes decay of fluctuations of a homogeneous state. Thus, we are studying orientational relaxation in a homogeneous liquid. It is of interest to compare Eq. (5.35) with Eq. (5.25). By comparing Eq. (5.25) and Eq. (5.35) one can easily obtain molecular expressions (in terms of the direct correlation function) for the force and torque terms. The important point here is that Eq. (5.35) is microscopic as it contains the microstructural information of the liquid. Note that previous investigations have shown that a Smoluchowski equation with a mean-field force term due to intermolecular interactions can describe large wave-number processes rather accurately in a dense liquid. In the long-wavelength limit, the translational contribution to the density relaxation becomes insignificant compared to the rotational relaxation. The limit is correctly described by Eq. (5.35).

The nonlinear Smoluchowski Eq. (5.35) is next linearized in density fluctuations δp to obtain

$$\frac{\partial}{\partial t} \delta \rho \left(\mathbf{r}, \mathbf{\Omega}, t \right) = D_R \nabla_\Omega^2 \delta \rho \left(\mathbf{r}, \mathbf{\Omega}, t \right) + D_T \nabla_T^2 \delta \rho \left(\mathbf{r}, \mathbf{\Omega}, t \right)$$
$$- D_R \left(\rho_0 / 4\pi \right) \nabla_\Omega^2 \int d\mathbf{r}' d\mathbf{\Omega}' c \left(\mathbf{r} - \mathbf{r}', \mathbf{\Omega}, \mathbf{\Omega}' \right) \delta \rho \left(\mathbf{r}', \mathbf{\Omega}', t \right)$$
$$- D_T \left(\rho_0 / 4\pi \right) \nabla_T^2 \int d\mathbf{r}' d\mathbf{\Omega}' c \left(\mathbf{r} - \mathbf{r}', \mathbf{\Omega}, \mathbf{\Omega}' \right) \delta \rho \left(\mathbf{r}', \mathbf{\Omega}', t \right). \quad (5.36)$$

Because it is convenient to work in wave-vector space, Eq. (5.36) is Fourier-transformed in space (with $\mathbf{k}$ as the Fourier variable) to obtain

$$\frac{\partial}{\partial t} \delta \rho \left(\mathbf{k}, \mathbf{\Omega}, t \right) = D_R \nabla_\Omega^2 \delta \rho \left(\mathbf{k}, \mathbf{\Omega}, t \right) - D_T k^2 \delta \rho \left(\mathbf{k}, \mathbf{\Omega}, t \right)$$
$$- D_R \left(\rho_0 / 4\pi \right) \nabla_\Omega^2 \int d\mathbf{\Omega}' c \left(\mathbf{k}, \mathbf{\Omega}, \mathbf{\Omega}' \right) \delta \rho \left(\mathbf{k}, \mathbf{\Omega}', t \right)$$
$$+ D_T \left(\rho_0 / 4\pi \right) k^2 \int d\mathbf{\Omega}' c \left(\mathbf{k}, \mathbf{\Omega}, \mathbf{\Omega}' \right) \delta \rho \left(\mathbf{k}, \mathbf{\Omega}', t \right). \quad (5.37)$$

5.5.2 Solution by Spherical Harmonic Expansion

Now, one expands $\delta \rho \left(\mathbf{k}, \mathbf{\Omega}, t \right)$ in terms of spherical harmonics:

$$\delta \rho \left(\mathbf{k}, \mathbf{\Omega}, t \right) = \sum_{lm} a_{lm} \left(\mathbf{k}, t \right) Y_{lm} \left(\mathbf{\Omega} \right), \quad (5.38)$$

where $a_{lm} \left(\mathbf{k}, t \right) = \int d\mathbf{\Omega} Y_{lm}^* \left(\mathbf{\Omega} \right) \delta \rho \left(\mathbf{k}, \mathbf{\Omega}, t \right)$. It is straightforward to show that correlation function $C_{lm} \left(\mathbf{k}, t \right)$ is given by

$$C_{lm} \left(\mathbf{k}, t \right) = \left\langle a_{lm} \left(\mathbf{k}, t \right) a_{lm} \left(-\mathbf{k}, 0 \right) \right\rangle, \quad (5.39)$$

where $\langle\ldots\rangle$ means average over an equilibrium ensemble of homogeneous states of the liquid. The direct correlation function $c(\mathbf{k}, \mathbf{\Omega}, \mathbf{\Omega}')$ can also be expanded in terms of the spherical harmonics:

$$c(\mathbf{k}, \mathbf{\Omega}, \mathbf{\Omega}') = \sum_{l_1 l_2 m} c(l_1 l_2 m; \mathbf{k}) Y_{l_1 m}(\mathbf{\Omega}) Y_{l_2 m}(\mathbf{\Omega}'), \qquad (5.40)$$

where $\mathbf{k}$ is taken parallel to the z axis. If one substitutes Eq. (5.40) and Eq. (5.38) into Eq. (5.37) and takes the scalar product of the resulting equation with $Y_{lm}(\mathbf{\Omega})$, one obtains

$$\frac{\partial}{\partial t} a_{lm}(\mathbf{k}, t) = -\left[D_R l(l+1) + D_T k^2\right] a_{lm}(\mathbf{k}, t)$$
$$+ \frac{\rho_0}{4\pi}(-1)^m \left[D_R l(l+1) + D_T k^2\right] \sum_{l_2} c(ll_2 m; \mathbf{k}) a_{l_2 m}(\mathbf{k}, t). \qquad (5.41)$$

It is clear that each $a_{lm}(\mathbf{k}, t)$ is coupled to several other coefficients. The equation of motion for the correlation function relevant for dielectric relaxation is obtained by setting $l = 1$ in Eq. (5.41).

A considerable simplification in Eq. (5.41) results if one assumes that the two-particle direct correlation function of the dipolar liquid is given by the linearized equilibrium theory. Examples of such linear theories are the mean-spherical-approximation model (MSA) or the linearized hypernetted-chain model (LHNC) model [3,4]. These models predict that the only nonvanishing $c(l_1 l_2 m; k)$'s are $c(000; k)$, $c(110; k)$, and $c(111; k)$. Within this approximation, Eq. (5.41) reduces to a simpler equation of the form

$$\frac{\partial}{\partial t} a_{lm}(\mathbf{k}, t) = -\left[D_R l(l+1) + D_T k^2\right] a_{lm}(\mathbf{k}, t)$$
$$+ \frac{\rho_0}{4\pi}(-1)^m \left[D_R l(l+1) + D_T k^2\right] \sum_{l_2} c(llm; k) a_{lm}(\mathbf{k})[\delta_{l1} + \delta_{l0}]. \qquad (5.42)$$

Therefore, the orientational correlation function, $C_{lm}(\mathbf{k}, t)$ is given in this Markovian limit by [7]

$$C_{lm}(\mathbf{k}, t) = \left\langle |a_{lm}(\mathbf{k})|^2 \right\rangle \exp\left[-\{D_R l(l+1) + D_T k^2\}\right.$$
$$\left. \times \{1 - (\rho_0/4\pi)(-1)^m c(llm; k)(\delta_{l0} + \delta_{l1})\}\right]. \qquad (5.43)$$

Equation (5.43) is an important result of this chapter. It shows that the only correlation functions that are affected by the dipolar forces are $C_{lm}(\mathbf{k}, t)$; the higher-order correlation functions are totally unaffected. This conclusion, of course, depends on the assumption that $c(l_1 l_2 m; k)$ can be given by linearized theory. This assumption breaks down for strongly dipolar liquid.

5.5.3 Relaxation of Longitudinal and Transverse Components

Because of linearization in density fluctuation, Eq. (5.43) is valid when the dipolar interactions are weak. The dipolar correlation function $\phi(\mathbf{k}, t)$ can be calculated from Eq. (5.43) and is given by

$$\phi(\mathbf{k}, t) = A_1 \exp\left[-\left(2D_R + D_T k^2\right)\left(1 - (\rho_0/4\pi) c(110; k)\right) t\right] + A_2 \exp\left[-\left(2D_R + D_T k^2\right)\left(1 + (\rho_0/4\pi) c(111; k)\right) t\right], \quad (5.44)$$

where

$$A_1 = \left\langle |a_{10}(k)|^2 \right\rangle \Big/ \left(\left\langle |a_{10}(k)|^2 \right\rangle + 2\left\langle |a_{11}(k)|^2 \right\rangle\right), \quad (5.45)$$

$$A_2 = 2\left\langle |a_{11}(k)|^2 \right\rangle \Big/ \left(\left\langle |a_{10}(k)|^2 \right\rangle + 2\left\langle |a_{11}(k)|^2 \right\rangle\right). \quad (5.46)$$

Equation (5.43) predicts that the only multiple moment that contributes to dipolar orientational relaxation is the dipole moment. Note that there are two relaxation times for the relaxation of the dipolar correlation function. The longitudinal dipolar correlations are given by $C_{10}(\mathbf{k}, t)$ and the transverse correlations by $C_{11}(\mathbf{k}, t)$. The two relaxation times are

$$\tau_{10}^{-1}(k) = 2D_R \left\{\left(1 + p'(k\sigma)^2\right)\left(1 - (\rho_0/4\pi) c(110; k)\right)\right\}, \quad (5.47)$$

and

$$\tau_{11}^{-1}(k) = 2D_R \left\{\left(1 + p'(k\sigma)^2\right)\left(1 + (\rho_0/4\pi) c(111; k)\right)\right\}, \quad (5.48)$$

where σ is the molecular diameter and $p' = D_T/2D_R\sigma^2$. τ_{10} is the longitudinal relaxation time and τ_{11} is the transverse relaxation time. *Thus, p' is a measure of the relative importance of the translational modes in the orientational relaxation.*

Note that we refer to the (110) component as longitudinal and the (111) component as transverse, and this can cause confusion. These are not to be confused as longitudinal and transverse fields of a propagating electromagnetic field. The (10) and (11) components refer to spherical harmonic coefficients in the intermolecular frame (with k along the Z-axis) and the (ij) components come from the angles that the two dipoles in the two molecules make in the intermolecular frame.

5.5.4 Molecular Theory of Dielectric Relaxation

In the simple Debye expression, dielectric function, $\varepsilon(\omega)$, is given by

$$\varepsilon(\omega) = \varepsilon_\infty + \frac{\varepsilon_0 - \varepsilon_\infty}{1 + i\omega\tau_D}, \quad (5.49)$$

where τ_D is the well-known Debye relaxation time. As a function of wave number, the dielectric function of the dipolar liquid becomes a tensorial quantity and is described by two diagonal components. These two components are sometimes referred to as longitudinal, $\varepsilon_L(\mathbf{k}, \omega)$, and transverse, $\varepsilon_T(\mathbf{k}, \omega)$. It is to be noted that this nomenclature does not mean the transverse or longitudinal components in the electrodynamics sense. Rather, they refer to the components of dielectric tensor along and perpendicular to the vector $\mathbf{k}$ in the intermolecular frame [3].

By assuming the definition, one can derive the following expressions for $\varepsilon_L(\mathbf{k}, \omega)$ and $\varepsilon_T(\mathbf{k}, \omega)$:

$$\varepsilon_L(\mathbf{k}, \omega) = [1 - 4\pi\alpha_L(\mathbf{k}, \omega)]^{-1}, \tag{5.50}$$

$$\varepsilon_T(\mathbf{k}, \omega) = 1 + 4\pi\alpha_T(\mathbf{k}, \omega), \tag{5.51}$$

where

$$\alpha_L(\mathbf{k}, \omega) = \frac{\beta}{V}\left[C_{ML}(\mathbf{k}, 0) - i\omega\tilde{C}_{ML}(\mathbf{k}, \omega)\right], \tag{5.52}$$

$$\alpha_T(\mathbf{k}, \omega) = \frac{\beta}{V}\left[C_{MT}(\mathbf{k}, 0) - i\omega\tilde{C}_{MT}(\mathbf{k}, \omega)\right], \tag{5.53}$$

with

$$\tilde{C}_{ML}(\mathbf{k}, \omega) = \hat{k} \cdot \tilde{C}_M(\mathbf{k}, \omega) \cdot \hat{k}, \tag{5.54}$$

$$\tilde{C}_{MT}(\mathbf{k}, \omega) = \hat{u} \cdot \tilde{C}_M(\mathbf{k}, \omega) \cdot \hat{u}, \tag{5.55}$$

where $\hat{k}$ is a unit vector parallel to $\mathbf{k}$ and $\hat{u}$ is a unit vector orthogonal to $\mathbf{k}$ and $\tilde{C}_M(\mathbf{k}, \omega)$ is the Laplace transform of $C_M(\mathbf{k}, t)$. Molecular hydrodynamic theory provides the following final expressions

$$\varepsilon_L(\mathbf{k}, \omega) = 1 + \frac{\varepsilon_L(\mathbf{k}) - 1}{1 + i\omega\tau_L(\mathbf{k})\varepsilon_L(\mathbf{k})}, \tag{5.56}$$

$$\varepsilon_T(\mathbf{k}, \omega) = 1 + \frac{\varepsilon_T(\mathbf{k}) - 1}{1 + i\omega\tau_T(\mathbf{k})}. \tag{5.57}$$

The above expressions are valid in the Markovian limit. In the long-wavelength ($\mathbf{k} \to 0$) limit, they reduce to the form

$$\varepsilon_L(\mathbf{k} = 0, \omega) = \varepsilon_T(\mathbf{k} = 0, \omega) = 1 + \frac{\varepsilon_0 - 1}{1 + i\omega\tau_D}. \tag{5.58}$$

5.5.5 Hidden Role of Translational Motion in Orientational Relaxation

Equations (5.47) and (5.48) give the decay of the transverse and the longitudinal fluctuations, respectively. This result is similar to that obtained by Berne, but in his calculation τ_{11} was simply $(2D_R)^{-1}$; there was no wavevector dependence of the rate constants. It has been found that both τ_{10} and τ_{11} are modified by polar interactions, the former is affected more strongly than the latter. For the complete determination of $\phi(\mathbf{k}, t)$ one needs the expressions for the equilibrium correlation functions $\langle |a_{10}|^2 \rangle$ and $\langle |a_{11}|^2 \rangle$. These are determined by the microscopic structure of the liquid and are given by

$$\langle |a_{10}(k)|^2 \rangle = (N/4\pi)\{1 + (\rho_0/4\pi)h(110;k)\}, \qquad (5.59)$$

$$\langle |a_{11}(k)|^2 \rangle = (N/4\pi)\{1 - (\rho_0/4\pi)h(111;k)\}, \qquad (5.60)$$

where N is the total number of molecules in the system and $h(110;k)$ and $h(111;k)$ are the different components in the spherical harmonic expansion of the total pair-correlation function. $h(llm;k)$'s are related to $c(llm;k)$'s by the following Ornstein–Zernike relation:

$$h(llm;k) = c(llm;k) + \frac{\rho_0}{4\pi}(-1)^m \sum_{l_1} c(ll_1 m;k) h(l_1 lm;k). \qquad (5.61)$$

Berne assumed that $\langle |a_{10}|^2 \rangle = \langle |a_{11}|^2 \rangle$. Numerical calculations show that this equality is not exact. As was emphasized previously, the theory discussed is valid when a linearized equilibrium theory is used for the two-particle direct correlation function.

In Fig. 5.2 the wave-vector dependence of the relaxation times, $\tau_{10}(k)$ and $\tau_{11}(k)$, is shown for several different values of parameter p'. For comparison, the case $p' = 0$ (when translational contribution is totally absent) is also included. It is seen clearly that the translational diffusion plays an important role in the intermediate values of the wave vector. The relaxation times decrease with increase in the translational contribution.

The dielectric function can be calculated from the $k = 0$ part of the polarization correlation function. The real and imaginary parts of the dielectric function $(\varepsilon(\omega))$ are shown in Fig. 5.3 [9]. In the absence of any translation contribution $(p' = 0)$ the plot is significantly non-Debye. As the translation diffusion contribution is increased, the Cole–Cole plot approaches Debye behavior. Thus, the translational modes, always present in a liquid but neglected in many earlier models, can weaken the non-Debye behavior predicted in the absence of translational contribution.

5.5.6 Orientational de Gennes Narrowing at Intermediate Wave Numbers

It is interesting to note that when translation motions are neglected, then $C_{10}(k, t)$ has a pronounced slowing down at the intermediate value of the wave vector k. This slowing down of relaxation at intermediate wave vectors is similar to the de Gennes' narrowing of the dynamic structure factor of a dense liquid at intermediate wave vectors. In the latter case, the slowing down occurs near the wave vector where the static structure factor is sharply peaked and is because of the

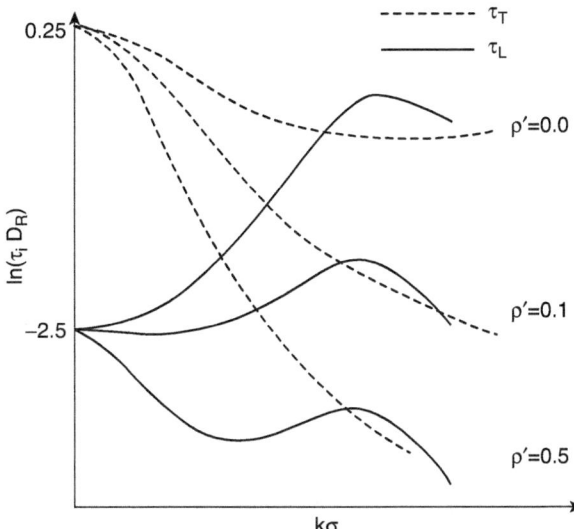

Figure 5.2 The dependence of longitudinal (solid lines) and transverse (dashed lines) relaxation times on wave vector k and translational diffusion, given by the parameter p' defined in the text. As the translational parameter p' is increased, the relaxation becomes considerably faster at intermediate wave vectors. The values of the static dielectric constant ε_0 and reduced density $\rho_0 \sigma^3$ are, respectively, 18 and 0.8. Here ρ_0 is the solvent number density and σ is solvent molecular diameter. Reprinted with permission from *J. Chem. Phys.* **91** 1829 (1989). Copyright (1989) American Institute of Physics.

short-range order that is present in a dense liquid. In the case of orientational relaxation, the slowing down of relaxation also occurs near the first peak of static structure factor and this may be attributed to the short-range *orientational correlations* present in a dense polar liquid. The slowing down of relaxation becomes progressively weaker when the translation modes are important in the density relaxation.

In their classic study on dielectric relaxation, Titulaer and Deutch [10] pointed out that linearized Smoluchowski equation treatment of orientational relaxation may not be reliable if the torque–torque correlation function decays on the same time scale as the orientational correlation function $\phi(k, t)$. This calls for two additional studies. First, a non-Markovian theory may be needed in some cases. Second, it is thus important to obtain an estimate of the contributions of the nonlinear terms.

5.5.7 Reduction to the Continuum Limit

It is an interesting exercise to find out how the above microscopic theory reduces to the continuum-limit result, as it must in the long-wavelength, or $k = 0$, limit.

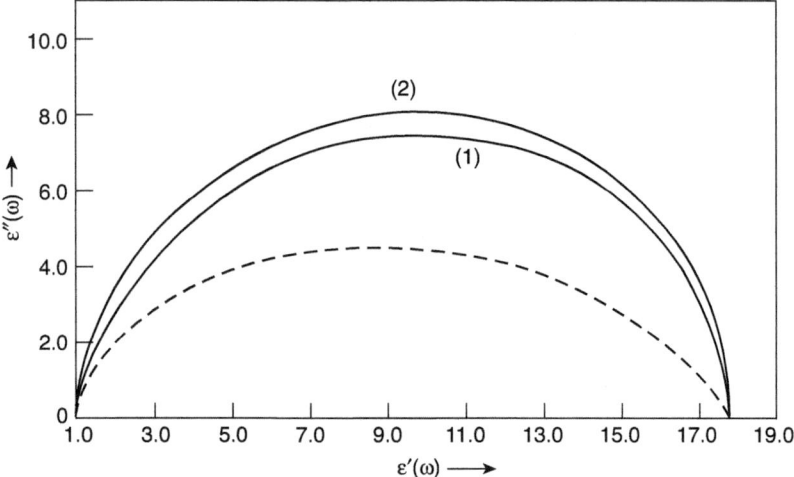

Figure 5.3 Cole–Cole plot of the frequency-dependent dielectric function is shown for different values of the dimensionless translational parameter p' (defined in the text). The X-axis is the real part and the Y-axis is the imaginary part of the dielectric function. The dashed curve is for $p' = 0$, i.e., no translation contribution. The curves (1) and (2) are for $p' = 0.05$ and 0.25, respectively. Reprinted with permission from B. Bagchi and A. Chandra, *Phys. Rev. Lett.* **64**, 455–458 (1990). Copyright (1990) American Physical Society.

Such a reduction also requires the assumption that the memory effects are neglected. One also needs to use the appropriate limiting forms of the two-particle pair- and direct-correlation functions. When such a limiting procedure is carried out, one indeed finds that the microscopic longitudinal relaxation time reduces exactly to the continuum-model expression of the longitudinal relaxation time, and the transverse relaxation time reduces to the Debye relaxation time.

One can also derive a microscopic expression of the dielectric friction by evaluating the torque–torque time-correlation function, following the definition of Nee and Zwanzig [1], except that we use the microscopic expression of the torque in terms of the angular gradient of the free energy. The expression of this friction has the form of a typical mode-coupling theory expression of friction on a tagged molecule. The most interesting aspect of this expression is the strong effect of the translational motion in reducing the magnitude of the rotational dielectric friction. That is, a significant translational contribution can make the relaxation single exponential. As a result, the dielectric relaxation becomes Debye-like. In addition, the microscopic expression of friction also reduces exactly to the continuum expression of Nee and Zwanzig [1].

5.6 MEMORY EFFECTS IN ORIENTATIONAL RELAXATION

Because of the presence of long-range orientational correlations in a dipolar liquid, each molecule experiences a dielectric friction (ζ_{DF}), in addition to the usual Stokes friction (ζ_s) due to viscosity, which arises from short-range interactions.

Because of the long-range nature of the dipolar interaction, the frictional response of the solvent often lags behind the solute's rotation. This has been discussed elegantly by Nee and Zwanzig [1]. This effect makes the *dielectric friction time or frequency dependent*. This, in turn can lead to a nonexponential decay of the correlation function $C_{lm}(k, t)$ and a strong non-Debye form of $\varepsilon(\omega)$.

As discussed above, the Markovian treatment of the orientational relaxation gives rise to a simple exponential decay of the orientational correlation function and a Debye form for the dielectric function $\varepsilon(\omega)$. A dense, highly polar, dipolar liquid is a system where molecules interact strongly and the Markovian theory for the polarization relaxation may not be reliable.

The non-Markovian theory is based on the molecular hydrodynamic equations with the memory terms. In the treatment that we discuss here, the dissipative kernels $\Gamma_T(\mathbf{X}, \mathbf{X}'; t - t'; \rho[t])$ and $\Gamma_\Omega(\mathbf{X}, \mathbf{X}'; t - t'; \rho[t])$ are approximated by their values at equilibrium number density, ρ_0, that is, the nonlinearities in Γ's are neglected. The non-Markovian effect is included only in the angular kernel, but neglected in the spatial part. The rationale for this approximation is that one is mostly interested in the frequency range where the viscoelastic effects are still not important in the liquids usually considered. In the overdamped limit, it is straightforward to solve the non-Markovian hydrodynamic equations in the frequency domain, and the final expression of the wave-vector- and the frequency-dependent longitudinal and transverse orientational-correlation functions, $C_{10}(k, \omega)$ and $C_{11}(k, \omega)$, are given by

$$C_{10}(\mathbf{k}, \omega) = C_{10}(\mathbf{k}, t = 0) \left[i\omega + \left(2k_B T/\zeta(\omega) + D_T k^2\right) \right.$$
$$\left. \times \left(1 - (\rho_0/4\pi) c(110; k)\right) \right]^{-1}, \quad (5.62)$$

$$C_{11}(\mathbf{k}, \omega) = C_{11}(\mathbf{k}, t = 0) \left[i\omega + \left(2k_B T/\zeta(\omega) + D_T k^2\right) \right.$$
$$\left. \times \left(1 + (\rho_0/4\pi) c(111; k)\right) \right]^{-1}. \quad (5.63)$$

The equilibrium correlation functions, $C_{10}(\mathbf{k}, \omega)$ and $C_{11}(\mathbf{k}, \omega)$, are given by Eq. (5.59) and Eq. (5.60). $\zeta(\omega)$ is the total rotational friction (Stokes + dielectric). If $\zeta_{DF}(\omega)$ is replaced by $\zeta_{DF}(\omega = 0)$, we recover the Markovian limit and a Debye form for $\varepsilon(\omega)$. An expression for $\zeta_{DF}(\omega)$ can be obtained from the torque–torque correlation functions. Its calculation simplifies if we consider the limit where the **single-particle** orientation is slower than the torque–torque correlation function. In the limit $\zeta_{DF}(\omega)$ is given by

$$\zeta_{DF}(\omega) = (32\pi^4 \beta)^{-1} \int dk \sum_{m=-1}^{1} \langle |a_{1m}(k)|^2 \rangle c^2(11m; k) \tau_{1m}(k, \omega)$$
$$\times \left[1 + i\omega \tau_{1m}(k, \omega)\right]^{-1}, \quad (5.64)$$

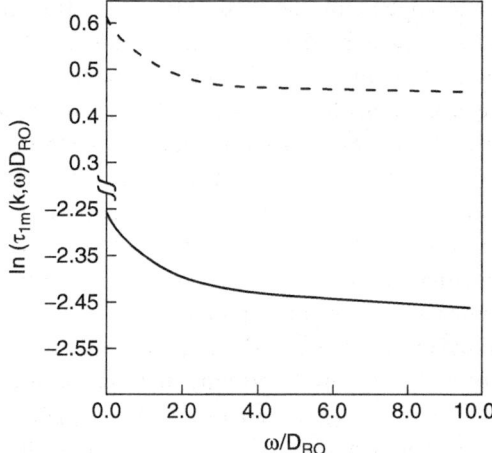

Figure 5.4 Frequency dependence of real parts of longitudinal τ_{10} ($k=0, \omega$) (shown by bold line) and transverse τ_{11} ($k=0, \omega$) (shown by dotted line) parts of the complex relaxation times.

where $a_{1m}(k)$ and $c(11m; k)$ have been defined previously, the transverse (τ_{11}) and longitudinal (τ_{10}) relaxation times are given by

$$\tau_{1m}^{-1}(k,\omega) = 2D_{RO}\left[\frac{k_B T}{\zeta(\omega) D_{RO}} + p'(k\sigma)^2\right]\left[1 - \frac{\rho_0}{(\omega) 4\pi}(-1)^m c(11m; k)\right], \quad (5.65)$$

where $p' = D_T/(2D_{RO}\sigma^2)$, D_T is the translational diffusion coefficient $D_{RO} = k_B T/\zeta_S$, and σ is the molecular diameter. p' is a relative measure of the importance of translational diffusion in the polarization relaxation of the dipolar liquid. Equations (5.64) and (5.65) are to be solved self-consistently for $\zeta_{DF}(\omega)$ and $\tau_{1m}(k, \omega)$. Numerical results of the frequency-dependent relaxation times are shown in Fig. 5.4.

In Fig. 5.4 the real part of the longitudinal and transverse relaxation times are plotted against frequency at the long-wavelength limit. It is seen that both the relaxation times decrease with increase in frequency. Thus, the short-time dynamics of the orientational relaxation may be quite different from that at long times.

5.7 RELATIONSHIP BETWEEN MACROSCOPIC AND MICROSCOPIC ORIENTATIONAL RELAXATIONS

In the case of liquid water, the single-particle relaxation time is 2.6 ps while the dielectric relaxation time is 8.4 ps. Both measure the rotational motion of water. However, dielectric relaxation probes a collective response. This relationship between single particle and collective constitute an important aspect of relaxation phenomena in general, and orientational relaxation in particular where the difference is often large, as in the case of water.

As discussed in Chapter 4, different experimental techniques measure different aspects of orientational relaxation. Most of the experiments in neat liquid measure the collective orientational response. For example, dielectric relaxation measures total moment–moment time correlation function. Therefore, it is necessary to extract the single-particle rotational relaxation time from the collective time-correlation function. The relationship between the single-particle orientational relaxation time, τ_m, and the many-body orientational relaxation time, τ_M, of a dense dipolar liquid has been a subject of much recent discussion. Another related question is the relationship between τ_m and the dielectric relaxation time τ_D of a pure dipolar liquid. The dielectric relaxation time contains many-body effects. The precise relations among τ_m, τ_M, and τ_D are still not clearly understood.

The first macro–micro relation was perhaps proposed by Debye who used his continuum theory of dielectric relaxation to propose the following relation:

$$\tau_m = \frac{n^2 + 2}{\varepsilon_0 + 2} \tau_D, \qquad (5.66)$$

where ε_0 is the static dielectric constant and n is the refractive index. The Debye form is inadequate for strongly polar solvents, in particular for water. Glarum [11] modified Eq. (5.66) for strongly polar liquids and the modified expression is given by

$$\tau_m = \frac{2\varepsilon_0 + \varepsilon_\infty}{3\varepsilon_0} \tau_D, \qquad (5.67)$$

where ε_∞ is the infinite frequency dielectric constant. Equation (5.67) is based on the Onsager's model of static dielectric constant, and it is referred as the Onsager–Glarum expression. Equation (5.67) predicts that for large ε_0, $\tau_m = 0.67\tau_D$. In fact, Onsager–Glarum expression is a limiting form of a more general relation derived by Powles [12] several years earlier. This form is given by

$$\tau_m = \frac{2\varepsilon_0 + \varepsilon_\infty}{3\varepsilon_0} \frac{\tau_D}{g}, \qquad (5.68)$$

where g is the well-known Kirkwood's g factor, which is a measure of short-range correlations in the dense dipolar liquids; these correlations are neglected in an Onsager-type theory. An analytic expression for g can be given in terms of integration over the anisotropic part of the radial distribution function of the dipolar liquid. In the literature, one finds a value of g in the range of 2.5–3. Using $g = 3$ gives value of $\tau_m = 0.23\,\tau_D$. Because τ_D is 8.3 ps, we have $\tau_m = 1.9$ ps, which is in the correct range. An important point about Eq. (5.68) is that it predicts that intermolecular orientational correlations can significantly slow down dielectric relaxation and that single-particle orientational correlation can be much different from the dielectric relaxation time, as has been assumed tacitly since the initial work of Debye.

Madden and Kivelson made several interesting observations in the understanding of macro–micro relations [13]. Let us define macroscopic and microscopic correlation functions $C_M(t)$ and $C_m(t)$ by

$$C_m(t) = \langle \mu_i(0) \cdot \mu_i(t) \rangle, \tag{5.69}$$

$$C_M(t) = \left\langle \sum_i \mu_i(0) \cdot \sum_j \mu_j(t) \right\rangle, \tag{5.70}$$

where μ_i is the dipole moment of the ith molecule and the sum is over all the N molecules of the system and the average is over an equilibrium ensemble. Madden and Kivelson have shown that τ_D and τ_m are related by the following equation:

$$\tau_m = \frac{\beta \mu^2 \rho_0}{3\varepsilon_0(\varepsilon_0 - 1)} g' \tau_D, \tag{5.71}$$

where $\beta = (k_B T)^{-1}$, k_B is the Boltzmann's constant, and T is the temperature. μ is the magnitude of the dipole moment of the liquid molecules and ρ_0 is the equilibrium density of the liquid. g' is the dynamical coupling parameter. Recently, it has been shown that the macro–micro theorem of Madden and Kivelson holds in a Markovian theory based on a generalized Smoluchowski equation [7, 14]. This theory shows that neither Eq. (5.66) nor Eq. (5.67) provides a correct description at large ε_0, and it also shows that, although that dipolar correlation function is bi-exponential, the frequency-dependent dielectric function has a simple Debye form with τ_D equal to the transverse polarization relaxation time of the dipolar liquid.

The dependence of the dielectric relaxation time τ_D, on the static dielectric constant ε_0 of the dipolar liquid is shown in Fig. 5.5.

The microscopic theory indeed gives, for liquid water, $\tau_D(\equiv \tau_M) \approx 4\tau_m(\equiv \tau_R)$, in good agreement with experimental results discussed above.

5.8 THE SPECIAL CASE OF ORIENTATIONAL RELAXATION OF WATER

Orientational relaxation of water has been studied by both NMR and dielectric relaxation measurements. NMR provides an estimate of 2.5 ps at 298 K for the single-particle orientational correlation function. Dielectric relaxation experiments, which measure the orientation of dipole moment vector, give a very similar value (2.0–2.5 ps) of the time constant, when corrected by the micro–macro relation. The latter procedure, however, leaves room for uncertainty. Moreover, these experiments give only the long-time decay constant. Dielectric-dispersion experiments clearly show the presence of faster time constants but they are of rather

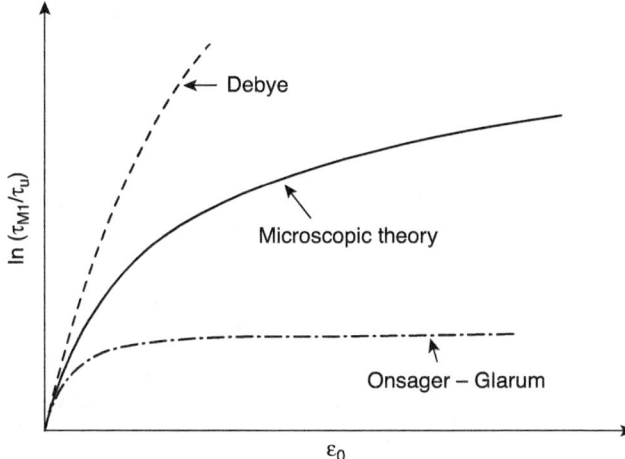

Figure 5.5 The dependence of dielectric relaxation time, τ_D, on the static dielectric constant ε_0 of the dipolar liquid, as predicted by three different theoretical models. Reprinted with permission from *J. Phys. Chem.* 94, 3159 (1990). Copyright (1990) American Chemical Society.

smaller amplitude. As the NMR measurements give only the average time constant, any information about the faster time constants is naturally missed in NMR.

Surprisingly, orientational motion of water molecules is found to defy the simple picture of Brownian motion. In the latter picture, displacements are envisaged to occur in small steps. Recent computer simulation studies have shown that water molecules not only undergo nearly continuous small amplitude motions, but the main mode of rotational change is a jump motion where the angle changes by nearly 60°. Rotation occurs by a concerted mechanism where a water molecule from the second-neighbor shell forms a bifurcated hydrogen bond (that is, for a transient 5-coordinated species) with the central water molecule whose rotation is in question. In the subsequent steps one of the water molecules from the first shell leaves, which is followed by a large amplitude motion of the central water molecule. This mechanism of water rotation has been verified in computer simulations. This was first proposed by Laage and Hynes [15]. If one calculates the time constant of rotational correlation from jump diffusion, one indeed finds a time constant that is in close agreement with experiments.

Recently, the short-time orientational relaxation of water was studied by ultrafast infrared pump-probe spectroscopy of the hydroxyl stretching mode (OD of dilute HOD in water) [6]. The anisotropic decay displays a sharp drop at very short times caused by inertial orientational motion, followed by a much slower decay that fully randomizes the orientation. The time dependence of the anisotropic decay for various frequencies at 25°C is shown in Fig. 5.6.

An interesting aspect of Fig. 5.6 is that the amplitude of the inertial component (extent of inertial angular displacement) depends strongly on the stretching frequency of the OD oscillator, more so at higher frequencies, although the

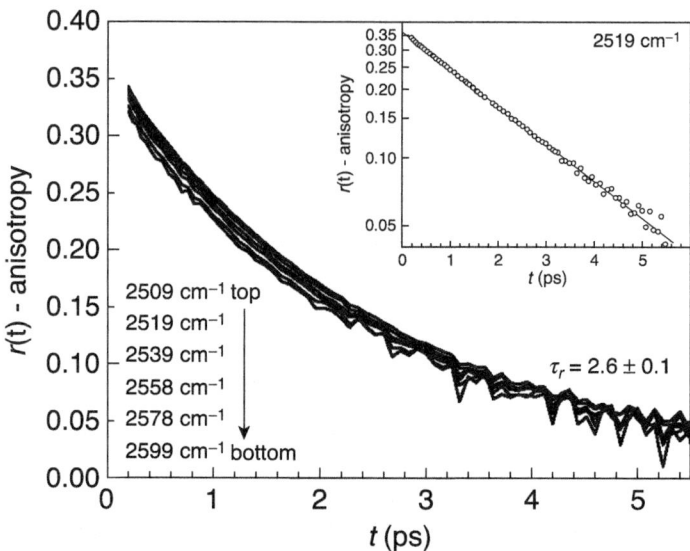

Figure 5.6 Plot of long-time anisotropy decay curves probed at various frequencies of the O-D stretch of water HOD at 25°C. (*Inset*) A decay curve on a semilog plot with single exponential fit. Reprinted with permission from *Proc. Natl. Acad. Sci. U.S.A.*, **105**, 5295 (2008). Copyright (2008) National Academy of Sciences, U.S.A.

slow component is completely frequency independent. The inertial component also becomes frequency independent at lower temperatures. At high temperatures, there is a correlation between the amplitude of the inertial decay and the strength of the O-D - - O hydrogen bond, but at low temperatures the correlation disappears, showing that a single hydrogen bond is no longer a significant determinant of the inertial angular motion. It is suggested that the loss of correlation at lower temperature is caused by the increased importance of collective effects of the extended hydrogen bond network. Temperature dependence of the experimentally measured orientational correlation function at $t = 100$ fs as a function of O-D stretching frequency is shown in Fig. 5.7.

The short time inertial motions of water molecules are important indicators of the nature of the local hydrogen-bonding structure and of the underlying potential energy landscape.

5.9 LATTICE MODELS OF ORIENTATIONAL RELAXATION

In order to avoid some of the complexity that arises due to the coupling between the simultaneous presence of both orientational and translational motions of the molecules, and also to model some realistic situations, a lattice model of orientational relaxation was introduced by Zwanzig in 1963 [17]. In this model, point dipoles were kept fixed at the lattice sites of a simple cubic lattice. The point dipoles were allowed to execute Brownian rotational motion in the force field of all other dipoles. The interaction potential between the dipole

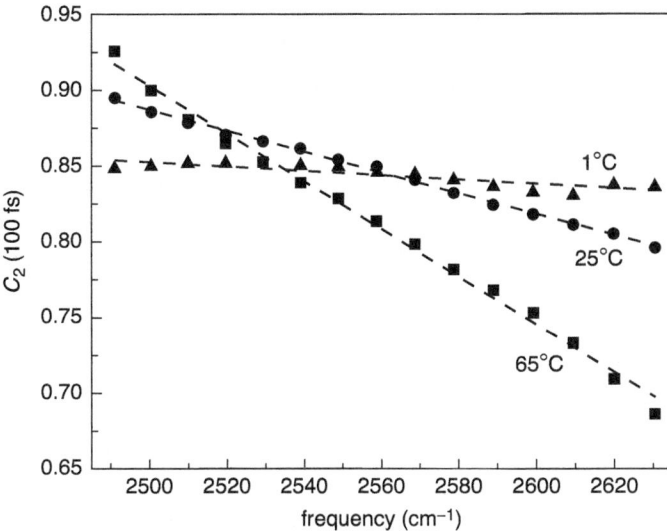

Figure 5.7 Plot of temperature dependence of the experimentally measured value of the orientational correlation function at $t = 100$ fs as a function of OD stretching frequency. Note the strong temperature dependence, especially the flattening of the curve at 1°C. Reprinted with permission from *Proc. Natl. Acad. Sci. U.S.A.* **105**, 5295 (2008). Copyright (2008) National Academy of Sciences, U.S.A.

was the well-known dipole–dipole $1/r^3$ potential, involving orientation of both the dipoles. The model undergoes a phase transition to an oriented phase at larger value of the dielectric constant. This limits the study of the model to low polarity.

Because of the relative simplicity of the lattice model, it was possible to study it by perturbative means with $\mu^2/a^3 k_B T$ as the smallness parameter. Zwanzig found that dipolar interactions gave rise to nonexponential orientational correlation function and non-Debye dielectric function. This was one of the first explicit demonstrations of such effects starting from a Hamiltonian description.

Later Zhou and Bagchi carried out a detailed numerical simulation of this model where all the approximation of the dielectric continuum model was tested [18]. The main limitation of the continuum model was found to lie in the expression of the dielectric friction employed to relate the friction to the frequency dependent dielectric function.

5.10 NONASSOCIATED LIQUIDS

Orientational and dielectric relaxation of many nonassociated liquids like acetonitrile, dimethyl sulfoxide (DMSO), chloroform, and several others have been studied extensively.

Owing to the absence of the extensive hydrogen bond network present in water and many associated liquids, molecules in a nonassociated liquid can rotate quite fast. When the molecules are not-too-aspherical, dielectric friction can retard the motion significantly. However, dielectric friction is usually smaller than the Stokes friction due to viscosity; for example, in the case of acetonitrile, the average rotational time constant of the first spherical harmonic is less than 500 fs, with a significant inertial component. Dimethyl sulfoxide is another important and interesting dipolar liquid which has a larger dipole moment and that explains why rotation here is somewhat slower than that in acetonitrile.

5.11 SUMMARY

It is a bit paradoxical that while experimental study of orientational relaxation of solute molecules in liquids can be carried out with relative ease by a host of techniques that are now indeed at a rather advanced stage, theoretical study is fraught with difficulties, and is still at a primitive stage. The reasons, as already mentioned, are the complexity and the long-range nature of dipolar interaction energy (even under point dipole approximation) which is responsible for the non-availability of static two-point correlation functions. Also, theory is rather hard even when the correlation functions are available because dynamic correlations at different ranks of spherical harmonics are coupled. In such a situation, linearized molecular hydrodynamic theories and continuum model descriptions provide answers that are semiquantitatively reliable.

In the present chapter we described a linearized molecular hydrodynamic theory that reduces to the continuum-model-based theory. This reduction brings out many interesting aspects of both the theories. While simple theories work moderately well for nonpolar and nonassociated liquids, orientation of water remains a difficult problem. We also described a novel mechanism of jump orientation motion of water which has been proposed recently.

REFERENCES

1. T. Nee and R. Zwanzig, *J. Chem. Phys.* **52**, 6353 (1970).
2. E. Fatuzzo and P. R. Mason, *Proc. Phys. Soc. (London)*, **90**, 741 (1967).
3. C. G. Grey and K. E. Gubbins, *Theory of Molecular Fluids Fundamentals*, Claredon, 1984, Vol. 1.
4. J. P. Hansen and I. R. McDonald, *Theory of Simple Liquids*, Academic Press (1976).
5. B. Berne and R. Pecora, *Dynamic Light Scattering*, Wiley (1976).
6. B. Berne, *J. Chem. Phys.* **62**, 1154 (1975).
7. B. Bagchi and A. Chandra, *Adv. Chem. Phys.* **LXXX**, 1 (1991).
8. A. Chandra and B. Bagchi, *J. Chem. Phys.* **91**, 1829 (1989).
9. B. Bagchi and A. Chandra, *Phys. Rev. Lett.* **64**, 455 (1990).
10. U. M. Titulaer and J. Deutch, *J. Chem. Phys.* **60**, 1502 (1974).
11. S. H. Glarum, in *Dielectric Properties and Molecular Behavior*, N. E. Hill et. al., eds., Van Nostrand (1969).
12. J. G. Powles, *J. Chem. Phys.* **21**, 633 (1953).
13. P. Madden and D. Kivelson, *Adv. Chem. Phys.* **LVI**, 467 (1984).

14. A. Chandra and B. Bagchi, *J. Phys. Chem.* **94**, 3159 (1990).
15. D. Laage and J. T. Hynes, *Science* **311**, 832–835 (2006).
16. D. E. Moilanen, E. E. Fenn, Yu-Shan Lin, J. L. Skinner, B. Bagchi, and M. D. Fayer, *PNAS*, **105**, 5295 (2008).
17. R. Zwanzig, *J. Chem. Phys.* **38**, 2766 (1963).
18. H.-X. Zhou and B. Bagchi, *J. Chem. Phys.* **97** (5), 3610 (1992).

6

Solvation Dynamics in Dipolar Liquid

6.1 INTRODUCTION

In the last chapter we discussed single-particle and collective orientational relaxation of molecules in liquids, with emphasis on dipolar liquids, as they constitute common solvents for reactions in chemistry. We discuss below solvation dynamics, which is intimately connected with orientational relaxation of dipolar solvent molecules.

Solvation dynamics directly probes the time-dependent polarization response of a dipolar liquid to a changing charge distribution in suitably chosen solute probes. The area of solvation dynamics has enjoyed enormous attention in the last two decades. The reason for such continued attention arises from the profound influence that solvation can have on various chemical reactions, especially on charge-transfer processes [1]. While attention in the earlier period was focused primarily on equilibrium solvent effects (such as the effect of solvent polarity on the reaction potential energy surface), more recently the focus has changed to understand the role of solvation dynamics on the changing charge distribution that often takes place during a reaction. Unlike dielectric relaxation, which also gives information about polarization relaxation of a dipolar solvent, solvation dynamics is believed to provide information which is more *local* and therefore, more relevant to chemical reactions.

Such local response can be achieved by studying the dynamics of solvation of a newly created ion or of an instantaneously changed dipole *inside* a polar liquid. Alternatively, one can study solvation energy time-correlation function of a molecule by electronic and/or vibrational spectroscopy. In all these cases, it is the electronic and/or the vibration energy of either a solute or a specific chemical bond of a solute that is studied. In some cases, highly sophisticated spectroscopic techniques have been developed to answer some of the probing questions. Most commonly, however, solvation dynamics is studied by time dependence of fluorescence Stokes shift (TDFSS) of an excited dye molecule.

The area of solvation and solvation dynamics underwent a renaissance in the late 1980s when the ultrashort laser pulses became available that made it possible to study solvation dynamics directly with a time resolution that was hitherto impossible. Successful studies were carried out on important solvents, like water, acetonitrile, and normal alcohols [2]. Many new and unexpected results were obtained. Subsequently, buoyed by the success of solvation measurements, a vigorous study of the same in complex systems was undertaken. However, the later studies turned out to be more difficult to understand and interpret.

Because solvation energy involves long-range interactions, the initial theoretical development employed continuum models, popular in electrostatics, except here the theories needed to be extended to treat dynamics. These theories predict simple, closed-form expressions for the time dependence of the solvation energy in terms of macroscopic properties of the solvent, such as the dielectric constant and the Debye dielectric relaxation time. The continuum models provided a set of initial predictions that could be tested against experiments. Detailed experiments showed that more complex models are required to explain the results. However, continuum models are still the starting point for understanding the experimental results.

Theoretical activity in this field was also motivated by an interesting comment of Onsager at the 1976 Banff conference on solvated electrons. Onsager suggested that *the polarization structure around an electron would form from outside in*. This is the famous *inverse snow ball effect*. The inverse snow ball picture of Onsager suggests that the relaxation of solvation energy is intrinsically nonexponential because several length scale dependent time scales are involved. Microscopic theories of solvation have been developed recently to check this conjecture and to explain new experimental results. These theories constitute a major improvement over the continuum models.

In this chapter, we will articulate, in simple terms, the motivation behind the study of solvation dynamics and also the reason for its importance in chemistry, particularly physical and inorganic chemistry. We will summarize the main experimental results of the last two decades. We will discuss the theoretical developments and also compare their success and failure in explaining the experimental results. We will also describe studies of complex systems, and the difficulties in unambiguously interpreting the experimental results. Another objective of this chapter is to articulate the understanding of relaxation that has emerged from the study of this problem.

6.2 PHYSICAL CONCEPTS AND MEASUREMENT

In experiments, the time-dependent progress of solvation of a newly created charge distribution in a solute probe is usually followed by measuring the time-dependent fluorescence Stokes shift (TDFSS) of the emission spectrum of the solute molecule [3]. *TDFSS measures the instantaneous vibronic energy of the solute.* A dilute solution of probe dye molecules (such as coumarin, Nile red, or prodan) is employed in solvation experiments. These molecules act as efficient probes for the following reasons. First, they undergo a large change in the dipole moment upon optical excitation, or may even photoionize. This gives rise to large polar stabilization energy of the molecule in the excited state that is absent in the ground state. However, solvent dipoles are required to undergo rotational (and, to a smaller extent) translational motion to achieve the maximum stabilization energy. Second, these dye molecules exhibit fluorescence in the excited state with a long lifetime (on the order of nanoseconds). This allows detection of Stokes shift with time of the fluorescence toward lower frequency as the solute probe molecule gets stabilized by dipolar solvent molecules. More recently, higher-order nonlinear optical measurements such as a three-pulse photon echo peak shift (3PEPS) measurement have been carried out to study the solvation dynamics. The understanding of 3PEPS is a bit more complicated, as it involves both phase and energy relaxation.

Physically speaking, the process of solvation of a solute probe is rather simple and may be described as follows. Consider that a solute probe in its ground state is in equilibrium with the surrounding solvent molecules and the equilibrium charge distribution of the former is instantaneously altered by a radiation field. Ideally, when the solute–solvent system undergoes an optical Franck–Condon transition upon excitation, the equilibrium charge distribution of the solute is instantaneously altered. The solvent molecules, however, still remain in their previous spatial and orientational configuration. This is a highly nonequilibrium situation for the system. The energy of the Franck–Condon state is higher than the minimum of the potential energy in the excited state, in equilibrium with the solvent (Fig. 6.1). Subsequent to the excitation, the solvent molecules rearrange and reorient themselves to stabilize the new charge distribution in the excited state. The resultant energy is the solvation energy of the solute. The time dependence of the rearrangement of the solvent environment (solvation) is reflected in the continuous red shift of the emission spectrum. The temporal characteristics of solvation are then followed by monitoring the spectral response function.

The solvation time correlation function (STCF) can be defined as [1, 4, 5]

$$S(t) = \frac{\nu(t) - \nu(\infty)}{\nu(0) - \nu(\infty)}, \qquad (6.1)$$

where $\nu(t)$ denotes the time-dependent emission frequency of the solute chromophore. This function is properly normalized and decays from unity at $t = 0$ to zero at $t = \infty$. In Fig. 6.2, the typical time dependence of the solvation time correlation function is shown. $S(t)$ is accessible experimentally with relative ease.

The STCF may also be written as

$$S(t) = \frac{E_{sol}(t) - E_{sol}(\infty)}{E_{sol}(0) - E_{sol}(\infty)}, \qquad (6.2)$$

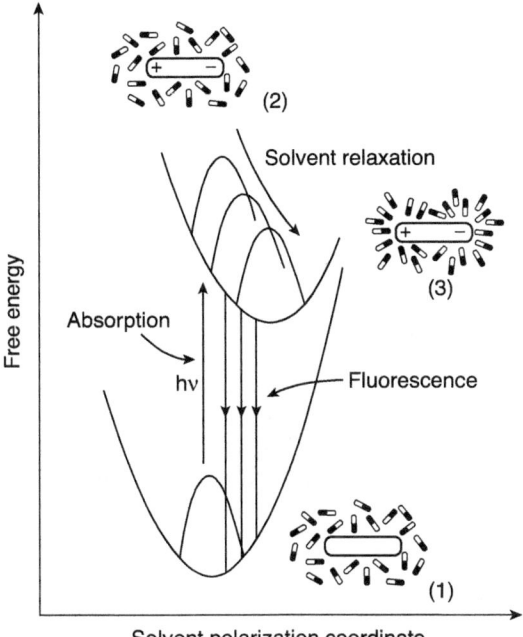

Figure 6.1 A schematic illustration of the physical processes involved in the experimental study of solvation dynamics. The initial state (marked by 1) is a nonpolar ground state. Laser excitation prepares it instantaneously in a charge-transfer (CT) state, which, however, is a highly nonequilibrium state because, at time $t = 0$, the solvent molecules are still orientationally at equilibrium with the ground state of the solute molecule. This is denoted as state (2). Subsequent solvation of the CT state brings the system to the final equilibrium (denoted by 3) and the evolution is detected by the time-dependent fluorescence Stokes shift.

where $E_{sol}(t)$ is the solvation energy of the solute at time t. The solvent response function $S(t)$ is accessible experimentally with relative ease. When the solvent response is linear to the external perturbation, $S(t)$ is equivalent to the *equilibrium* energy–energy time-correlation function, $C_S(t)$, according to fluctuation–dissipation theorem. This is defined as [6]

$$C_S(t) = \frac{\langle \delta E_{sol}(0) \delta E_{sol}(t) \rangle}{\langle (\delta E_{sol}(0))^2 \rangle}, \tag{6.3}$$

where $\delta E = E(t) - \langle E \rangle_{eqm}$ is the fluctuation in electronic energy of the solute from the equilibrium average. A note of caution is warranted here. The equilibrium average depends on the electronic state of the system. This point is discussed below in more detail. This assumption substantially simplifies the description of

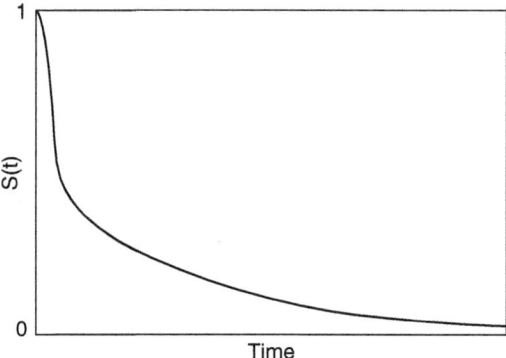

Figure 6.2 A schematic illustration of the time dependence of the solvation time-correlation function defined by Eqs. (6.1) and (6.2), and often observed in real world experiments.

dynamics of solvation because one now needs to consider only the equilibrium time-correlation function.

As mentioned, there is a possible caveat in equating Eq. (6.3) with Eq. (6.2). Because the equilibrium time-correlation function in Eq. (6.3) depends on the state of the system, it needs to be calculated with respect to either the initial or the final state of the solute–solvent system. However, the system is not stationary during solvation—it evolves from one polar state to another. Therefore, while equating Eq. (6.3) with Eq. (6.2) it is assumed that the state of the solvent is minimally perturbed so that the equilibrium and dynamic correlation functions of the solvent are not different in the initial and final states. As the perturbation is shared by many solvent molecules and could be small for each, except perhaps those closest to the solute, the state dependence of the equilibrium average $\langle E \rangle_{eqm}$ may be neglected. Computer simulations seem to find that this assumption is generally true.

The time-dependent solvation energy, $E_{sol}(t)$, may be decomposed (in a somewhat ad hoc but physically meaningful manner) into several components, such as (i) the time-dependent electrostatic interaction energy between the polar solute and the dipolar solvent molecules, $E_P(t)$, (ii) the solute–solvent nonpolar interaction energy $E_{NP}(t)$, (iii) the relaxation of intramolecular vibrational modes of the solute, denoted by $E_{intra}(t)$, and (iv) the change in the interaction among the solvent molecules in the excited state $E_{solv}(t)$.

The observed spectral dynamics can be attributed to $E_P(t)$ only when the contributions of the other three terms are negligible. This is justified under the following conditions. First, the intramolecular vibrational energy relaxations occur on a much faster time scale. Second, the nonpolar solvent response is usually not probed because there is little change upon excitation in shape and size of the solute probe. Third, the interactions among the solvent molecules remain the same in both the states.

The process of solvation of a newly created charge surrounding dipolar molecules involves primarily (a) rotational motion, (b) translational displacement, and (c) librational motion of the dipolar solvent molecules. Thus, all three can contribute to the final polarized structure around the excited ion or dipole. In addition, there is an electronic polarization contribution from the solvent molecules that may also form a significant part of the total solvation energy. This contribution can be estimated from the polarizability of the solvent molecules. For example, the electrons of oxygen in water are polarizable and are known to contribute significantly to the total solvation. Therefore, the time scales of solvation dynamics can be rather wide. For example, the electronic and libration motions are very fast; however, translational and rotational motions can be significantly slower. An additional factor is the collective nature of the solvation dynamics. It can be very different from the dynamics that involves molecular-length scales. For example, dynamics at $k = 0$ is very different from the dynamics at $k = 2\pi/\sigma$, where σ is the molecular diameter. In terms of numbers, intermediate wave numbers range from 0.1 Å^{-1} to 10 Å^{-1}, roughly speaking. This is not the range where Navier–Stokes hydrodynamics is valid, as discussed in **Chapter 3**.

6.2.1 Measuring Ultrafast, Sub-100 fs Decay

The fluorescence Stokes shift studies, although (due to its simplicity) employed widely in experimental studies, suffer from the following limitations. (a) The short time limit is often not accessible because the excitation pulse has a duration comparable to the inverse of initial solvation rate. (b) It is not possible to separate the vibrational relaxation of the solute from the solvation energy progression because at short times the two processes can occur at the same time scale. (c) As the total fluorescence from the solution is sum total of the fluorescence from all the molecules of the system, this tool contains both homogeneous and inhomogeneous contributions. This can give misleading information about dynamics.

In order to overcome these difficulties, several sophisticated methods have been employed. Three-pulse photon echo technique has been developed to measure ultrafast solvation dynamics [7]. The key idea of photon echo spectroscopy is that subsequent to optical excitation in an inhomogeneous ensemble of absorbers, the phase of different absorbers becomes uncorrelated with time because the transition frequencies of different absorbers are different. A second pulse is then used to initiate rephasing. If the inhomogeneous width greatly exceeds the homogeneous width, an echo will be produced at a fixed time and the echo will be narrow in time.

Ideally, one would like 3PEPS technique to measure relaxation at different transition frequencies separately. One of course needs to make the assumption that these transition frequencies uniquely represent dynamics of a local region.

The key theoretical quantity involved in the photon echo experiment is the system bath coupling potential V_{SB}. Let us now partition V_{SB} into an average component $\langle V_{SB} \rangle$ and a fluctuating component δV_{SB}:

$$V_{SB}(t) = \langle V_{SB} \rangle + \delta V_{SB}(t). \tag{6.4}$$

Now both linear and nonlinear spectroscopy techniques are calculated from the line-broadening function defined by

$$g(t) = \frac{1}{\hbar^2} \int_0^t d\tau \int_0^\tau d\tau' \langle \delta V_{SB}(\tau') \delta V_{SB}(0) \rangle, \tag{6.5}$$

where $\delta V_{SB}(t)$ is the Heisenberg operator of δV_{SB} at high temperature. Thus, $g(t)$ is the same frequency modulation time-correlation function introduced by Kubo in his treatment of NMR line shape. We have discussed this later in **Chapter 12**. Function $g(t)$ can be expressed in terms of another key quantity of the nonlinear spectroscopy, which is the spectral density $\rho(\omega)$. This contains the information about the time scales and strengths of the system bath interaction. Assuming a Brownian Oscillator model, the line-shape function can be written as a function of the spectral density:

$$g(t) = -\frac{i\lambda t}{\hbar} + \int_0^\infty d\omega \rho(\omega) \coth(\hbar\beta\omega/2)(1 - \cos\omega t) + i\int_0^\infty d\omega \rho(\omega) \sin\omega t, \tag{6.6}$$

where λ is the reorganization energy defined as the first moment of the spectral density

$$\lambda = \hbar \int_0^\infty d\omega\, \omega\, \rho(\omega). \tag{6.7}$$

Echo signals are most often discussed in terms of the frequency fluctuation time-correlation function, $M(t)$, defined as

$$M(t) \equiv \frac{\langle \Delta\omega(0) \Delta\omega(t) \rangle}{\langle \Delta\omega^2 \rangle} = \frac{1}{\langle \Delta\omega^2 \rangle} \int_0^\infty d\omega \omega^2 \rho(\omega) \coth\left(\frac{\hbar\beta\omega}{2}\right) \cos\omega t, \tag{6.8}$$

where $\langle \Delta\omega^2 \rangle$, the mean-square fluctuation amplitude, is commonly referred to as the coupling strength

$$\langle \Delta\omega^2 \rangle = \int_0^\infty d\omega \omega^2 \rho(\omega) \coth\left(\frac{\hbar\beta\omega}{2}\right). \tag{6.9}$$

Now, the fluorescence Stokes shift function $S(t)$ defined in Eq. (6.1) can be written in terms of the spectral density as

$$S(t) = \frac{\hbar}{\lambda} \int_0^\infty d\omega \omega \rho(\omega) \cos(\omega t). \tag{6.10}$$

Note that Eqs. (6.8) and (6.10) are equivalent in the high-temperature limit. Note that $M(t)$ (given by Eq. (6.8)) is close to the second term of $g(t)$ (given by Eq. (6.6)).

In 3PEPS experiments, one measures the integrated 3PE signal in the phase-matching direction as a function of τ (delay between first and second pulses) for a population period T (delay between second and third pulses). The integrated signals are found to be quite symmetric and can be fitted to Gaussian functions. The shift in peak position $\tau^*(T)$, which specifies the delay of the echo signal from zero delay of the τ axis, is the primary variable of interest here. Measurement of τ^* for a series of different values of T, the population period, contributes what we have called a 3PEPS measurement. In the high-temperature limit, one can show that

$$\tau^*(T) = \frac{M(T)}{\sqrt{\pi\left(\frac{2\lambda}{\hbar^2 \beta} + f(T)\right)}}. \tag{6.11}$$

Equation (6.11) shows how the time-dependent echo peak shift is related to the Bohr frequency-correlation function defined in Eq. (6.8). The contribution of $f(T)$ can be ignored under the assumption that the value of $\frac{2\lambda}{\hbar^2 \beta}$ is much larger than $f(T)$ at high temperature.

Thus, in the high-temperature limit, photon echo experiments provide information about the frequency-fluctuation time-correlation function, $M(t)$, which is essentially the same as the equilibrium solvation time-correlation function $C_S(t)$ and in the high-temperature limit also the same as $S(t)$ obtained from the fluorescence Stokes shift experiments.

Apart from fluorescence up-conversion and 3PEPS techniques described above, there are several other techniques that have been used to study the solvation dynamics of dipolar liquids. Transient absorption spectroscopy and X-ray absorption spectroscopy are examples of such techniques. X-ray absorption spectroscopy has been used to probe the solvation shell around electronically excited atomic solutes, on the solvent shell rearrangements around iodide turning to iodine in water.

6.3 PHENOMENOLOGICAL THEORIES: CONTINUUM-MODEL DESCRIPTIONS

6.3.1 Homogeneous Dielectric Models

Homogeneous dielectric models are generalizations of the equilibrium solvation models of Born and Onsager to the time domain. In this model the solvent is replaced by a frequency- (ω) dependent dielectric continuum, with dielectric function, $\varepsilon(\omega)$, and the polar solute by a molecular cavity of some simple shape, mostly spherical [8]. The solvation energy is obtained by evaluating the reaction field of the polar solute molecule inside the molecular cavity. If the dipolar molecule is approximated by a spherical cavity with a point dipole at its center and the solvent by a dielectric continuum with a single relaxation time, then the continuum model predicts that the solvation time correlation function is a single exponential with a time constant given by

$$\tau_L^d = \left(\frac{2\varepsilon_\infty + \varepsilon_c}{2\varepsilon_0 + \varepsilon_c}\right)\tau_D, \tag{6.12}$$

where ε_0 and ε_∞ are, respectively, the static and the infinite-frequency dielectric constants, and τ_D is the Debye relaxation time of the dipolar solvent. ε_c is the dielectric constant of the molecular cavity. In the case of an ion, the continuum model predicts slightly faster solvation, and the time constant is given by

$$\tau_L = \left(\frac{\varepsilon_\infty}{\varepsilon_0}\right)\tau_D. \tag{6.13}$$

τ_L is the longitudinal polarization relaxation time of the unperturbed solvent. For typical values of ε_0, ε_∞, and ε_c, the difference between τ_L^D and τ_L is unimportant and we may simply state that homogeneous continuum models predict that the solvation energy correlation function should relax exponentially with a time constant given by τ_L. For water, $\tau_D = 8$ ps, $\varepsilon_0 = 80$, $\varepsilon_\infty = 5$, so τ_L becomes 0.5 ps. However, if ε_∞ is set to 2, then τ_L becomes 0.2 ps. As discussed below, the solvation dynamics in water exhibits ultrafast dynamics on a similar time scale except that the decay in this case is nonexponential. It is interesting to note that use of a multiexponential representation of $\varepsilon(\omega)$ does not lead to the experimentally observed results. One needs a molecular and non-Markovian theory to explain the observed results, as will be discussed.

As already discussed, in the continuum model, the solvent is approximated as a bulk structureless continuum characterized by a frequency-dependent dielectric constant, $\varepsilon(\omega)$. The solute probe molecule is, for simplicity, approximated as a spherical cavity with a radius a and frequency-independent dielectric constant ε_c.

The energy of interaction between the dipolar solute and the bulk solvent is given by

$$\Delta E(t) = -\frac{1}{3}(\varepsilon_c + 2)\mu(t) \cdot R(t), \qquad (6.14)$$

where $R(t)$ is the time-dependent reaction field due to the polarization of solvent on the solute and $\mu(t)$ is the dipole moment of the solute particle at time t. An expression for the frequency-dependent reaction field was derived by Nee and Zwanzig [9], and is given by

$$R(\omega) = r(\omega)\mu(\omega), \qquad (6.15)$$

where the frequency-dependent prefactor $r(\omega)$ is given by the following elegant expression:

$$r(\omega) = (8\pi\rho/3)\left\{[\varepsilon(\omega) - \varepsilon_\infty]/\varepsilon_\infty[2\varepsilon(\omega) + \varepsilon_\infty]\right\}. \qquad (6.16)$$

Note that the time dependence of $\mu(\omega)$ is due to the rotational motion of the solute probe which also makes a contribution to solvation dynamics if solvent relaxation is slow.

To proceed further we need to specify the frequency-dependent dielectric constant. Dielectric relaxation of many dipolar liquids can be described by the following simple form of the frequency-dependent dielectric constant:

$$\varepsilon(\omega) = \varepsilon_\infty + \frac{\varepsilon_0 - \varepsilon_\infty}{1 + i\omega\tau_D}, \qquad (6.17)$$

where ε_∞ is the infinite-frequency dielectric constant and ε_0 is the static dielectric constant. τ_D is called the Debye relaxation time. Both ε_∞ and τ_D are obtained by fitting the measured frequency-dependent dielectric function with the above form of $\varepsilon(\omega)$. One can obtain $r(t)$ by Laplace inversion to the time domain of Eq. (6.16). If we neglect solute rotation, then solvation time correlation follows the time dependence of $r(t)$ that, with expression (Eq. (6.17)) for $\varepsilon(\omega)$, is single exponential,

$$C_s(t) = e^{-t/\tau_L^d}, \qquad (6.18)$$

with $\tau_L^d = \dfrac{\varepsilon_c + 2\varepsilon_\infty}{\varepsilon_c + 2\varepsilon_0}\tau_D$, which is the same as Eq. (6.12).

In the presence of solute rotation, the rate of relaxation increases as

$$\frac{1}{\tau} = \frac{1}{\tau_L^d} + 2D_R, \qquad (6.19)$$

where D_R is the rotational diffusion constant of the solute.

Note that these simple expressions were derived under the assumptions that the molecular cavity is spherical and the solvent dielectric response is of simple Debye form. Both these assumptions were relaxed in a later study that investigated consequences of an ellipsoidal cavity and a non-Debye dielectric response. It was found that if the molecular cavity is ellipsoidal, the decay is bi-exponential even with a Debye form of $\varepsilon(\omega)$; the relaxation times, however, differ at most by 30–40% under extreme conditions, thus the solute shape effect is not important in solvation dynamics.

The effect of the non-Debye form was studied by using the Davidson–Cole form for $\varepsilon(\omega)$. It was found that the short-time dynamics were significantly affected even when the deviation from Debye behavior was small. In general, if the dielectric relaxation is multiexponential, the solvation time-correlation function will also be multiexponential.

Dielectric relaxation of many dipolar solvents is described by the sum of two terms—each describing one exponential relaxation. Such a form is called the Budo formula in literature. In this case, one uses an expression for the frequency-dependent dielectric constant of the polar solvent as,

$$\varepsilon(\omega) - \varepsilon_\infty = (\varepsilon_0 - \varepsilon_\infty)\left[\frac{g_1}{1 - i\omega\tau_{D1}} + \frac{g_2}{1 - i\omega\tau_{D2}}\right], \quad (6.20)$$

with $g_1 + g_2 = 1$. Here g_1 and g_2 are the relative contributions of the two dispersion terms to $\varepsilon(\omega)$ with relaxation times τ_{D1} and τ_{D2}. When dielectric relaxation is given by the Budo formula, solvation time-correlation function is also bi-exponential with time constants

$$\tau_{L1} = (S_+ + 2D_R)^{-1}, \quad (6.21)$$

$$\tau_{L1} = (S_- + 2D_R)^{-1}. \quad (6.22)$$

Here $S_\pm$ can be calculated using the dielectric relaxation times (τ_{D1} and τ_{D2}), relative contributions of the two dispersion terms (g_1 and g_2), ε_∞, ε_0, and ε_c values. The relative contributions of these two relaxation processes toward total solvation time-correlation function also depend on these parameters. D is the rotational diffusion constant of the probe. Experimentally, the faster component of dielectric relaxation τ_{D2} is found to be an order of magnitude smaller than τ_{D1} and g_2 is found to be between 0.1 and 0.2.

At 20°C, the dielectric relaxation data of propanol give $\tau_{D1} = 430$ ps, $\tau_{D2} = 21.9$ ps, $g_1 = 0.91$, and $g_2 = 0.09$. Using the other parameter for propanol and probe rotational diffusion $D_R = 5 \times 10^8$ s^{-1}, one obtains $\tau_{L1} = 13$ ps and $\tau_{L2} = 92$ ps. The solvation time-correlation function has a $1/e$ time of ≈ 33 ps.

Note that when we use either the Debye or the Budo form of the dielectric relaxation function $\varepsilon(\omega)$, we do not include an inertial component of rotational motion. The inertial component can be included by using either a Brownian oscillator model (of the type discussed above) or by using empirical forms that can be fitted to experiments or simulations. Readers may consult Ref. 18(a) for more details.

6.3.2 Inhomogeneous Dielectric Models

An ionic or a dipolar solute creates a distortion to the surrounding dipolar solvent. As a result, dielectric properties of the solvent in the immediate neighborhood of the solute can differ significantly from the bulk away from the solute. One expects both the static and dynamic effects to get modified. This can affect both the equilibrium solvation energy and the dynamics of dipolar solvation. The first theoretical study of the effect of polar solute on electrostatic properties of the solvent was carried out by no other than Peter Debye who modeled the surrounding solvent by a distance-dependent dielectric function $\varepsilon(r)$ such that $\varepsilon(r)$ has a small value at the surface of the solute and approaches the bulk value at large distances from the solute surface [10]. This model has been extended to treat both the equilibrium and dynamical aspects of dipolar solvation. The theoretical analysis closely follows the one used for position and distance independent dielectric constant of the homogeneous case.

The main effects of the introduction of the position-dependent dielectric constant into the continuum theory is the appearance of multiple time scales, the fastest of which is identical to τ_L (or τ_L^d for solute dipole) and the slowest one is close to but smaller than τ_D. If we further simplify the inhomogeneous model by representing the position-dependent dielectric constant by permitting $\varepsilon(r)$ to have discrete values within concentric spheres around the polar solute, then it is possible to obtain a simple analytical solution. See Fig. 6.3 for a schematic illustration of the model.

Because of the spherical symmetry of the model and the discrete nature of the dielectric function, it is easy to apply the quasistatic boundary value conditions at

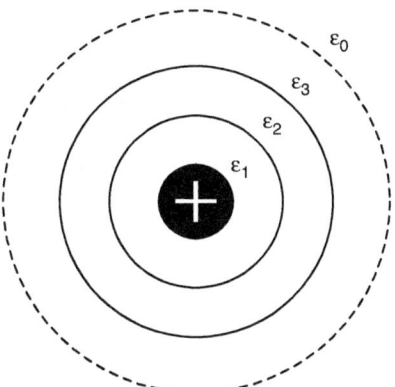

Figure 6.3 A schematic illustration of the inhomogeneous dielectric continuum model. The ion being solvated is at the center. The regions between different concentric circles (spheres in three dimensions) have different dielectric behavior.

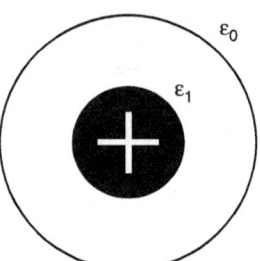

Figure 6.4 Illustration of the simple case with only two inhomogeneous dielectric regions.

each shell boundary. The rest of the procedure is similar to the one used in the homogeneous case. The final expression $R(t)$ for an ion is given by

$$R(t) = \sum_{k=1}^{N+1} a_k e^{-t/\tau_k}, \tag{6.23}$$

where

$$a_k = \left(\frac{1}{\varepsilon_\infty} - \frac{1}{\varepsilon_k}\right)\left(\frac{1}{r_{k-1}} - \frac{1}{r_k}\right), \text{ and } \tau_k = \left(\frac{\varepsilon_\infty}{\varepsilon_k}\right)\tau_D. \tag{6.24}$$

Here N is the number of concentric shells, ε_k is the dielectric constant of kth shell, and r_k is the radius. If the solute probe does not execute any translational or rotational motion, then $R(t)$ describes the time dependence of solvation energy.

Let us analyze the situation where the surrounding solvent can be represented by two inhomogeneous regions, that is, only one concentric shell. This situation is shown in Fig. 6.4.

In this case, the time dependence of the reaction field is given by

$$R(t) = a_1 e^{-t/\tau_1} + a_2 e^{-t/\tau_L}, \tag{6.25}$$

with

$$a_1 = \left(\frac{1}{\varepsilon_\infty} - \frac{1}{\varepsilon_1}\right)\left(\frac{1}{r_0} - \frac{1}{r_1}\right), \tau_1 = \frac{\varepsilon_\infty}{\varepsilon_1}\tau_D, \tag{6.26}$$

and

$$a_2 = \left(\frac{1}{\varepsilon_\infty} - \frac{1}{\varepsilon_0}\right)\frac{1}{r_1}, \quad \tau_L = \frac{\varepsilon_\infty}{\varepsilon_0}\tau_D. \tag{6.27}$$

If ε_1 is sufficiently small, that is, if the effect of the ion/dipole sufficiently lowers the dielectric response of the surrounding solvent, then we get a time constant that can be much larger than τ_L. That is, τ_1 can obey the condition $\tau_D > \tau_1 \gg \tau_L$.

Experiments and simulations have indeed found the existence of such time scales. Many experimental results can be fitted to a bi-exponential form with time constants close to τ_D and τ_L.

There is a deeper physical reason for the apparent success of the inhomogeneous continuum model. The dielectric response at small length scales (or, large wave numbers) is indeed described by a smaller value of the dielectric constant than the value in the bulk. This will naturally give rise to slower solvation dynamics. Length or wave-number dependence of the dielectric constant has been discussed in **Chapter 5**.

In most cases, if not all, the charge distribution in the excited state of the probe molecules is extended. Such a charge distribution cannot be treated either as a point dipole in the center of a sphere or by a point charge. A theoretical treatment based on a continuum-model description of the solvent to deal with such realistic cases was developed by Song, Marcus, and Chandler [11]. These authors used realistic charge distribution of the solute and frequency-dependent dielectric function-based description of the solvent to obtain excellent agreement with experimental results. This calculation still remains the ultimate continuum model in the theory of solvation dynamics.

Dielectric relaxation experiments essentially measure the total dipole moment time-correlation function $(C_M(t))$ in the frequency domain. The relationship between frequency-dependent dielectric function $\varepsilon(\omega)$ and total moment time-correlation function $(C_M(t))$ has been discussed in **Chapter 4**.

6.3.3 Dynamic Exchange Model

While the inhomogeneous continuum model does take into account the inhomogeneity in the solvent environment around the solute probe, it is limited by the lack of information about heterogeneity at a molecular-length scale. Such a situation arises in systems like micelles and reverse micelles, protein, and DNA hydration layer, not to mention water in the cellular environment about which our knowledge is still far from complete [12]. Another popular example is water under nano-confinement, like carbon nanotubes. One of the characteristics of water in complex systems is that the main contributor to the dynamics is the movement of individual water molecules at the surface. This movement can be in the form of either rotation or translation or may involve both. It may also involve breaking of a hydrogen bond with the polar/charged groups present on the surface of micelles, proteins, and DNA. The dynamics of breaking of such specific interactions can play an important role because water molecules bonded to surface, even by a weak and transient bond, cannot participate in the polar response involved in the solvation dynamics and/or dielectric relaxation of the solution [13].

A simple demonstration of such a phenomenological model is provided by the dynamic exchange model, which relates the polarization relaxation to a dynamic exchange between a "bound" and a "free" state at the surface of a hydration layer. The bound water molecule is not permanently bound, of course, but restricted by interactions with the surface atoms of the micelles/proteins/DNA. The main characteristics are that a bound molecule cannot rotate or translate like

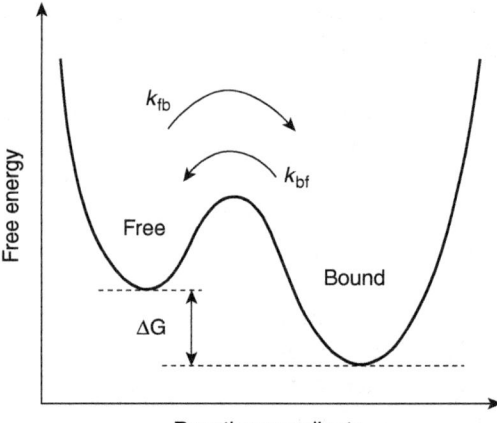

Figure 6.5 Schematic potential energy diagram showing the free water state, bound water state, and activated complex in dynamical equilibrium at a protein or micellar surface. This dynamic equilibrium is one of the processes in the hydration layers of different biological and macromolecular surfaces.

a free molecule. However, a bound water molecule can become free (for example, by breaking a hydrogen bond) and then contribute to polarization relaxation.

The dynamical equilibrium between the free and the bound water is characterized by equilibrium constant K_{bf}:

$$K_{bf} = \frac{k_{fb}}{k_{bf}} = e^{-\Delta G/k_B T}, \qquad (6.28)$$

where ΔG is the difference in free energy between bound and free state. k_{bf} and k_{fb} are the rate constants of the bound-to-free and free-to-bound transitions, respectively. A schematic illustration of the free energy surface is given in Fig. 6.5.

Orientations of free and bound water molecules are denoted by $\mathbf{\Omega}_f$ and $\mathbf{\Omega}_b$, respectively. Two time-dependent densities $\rho_f(\mathbf{\Omega}_f, t)$ and $\rho_b(\mathbf{\Omega}_b, t)$ denote the rotational dynamics of free and bound water molecules. These densities undergo change not only due to rotational diffusion but also due to interconversion between the two species. The dynamics can be described by two coupled reaction–diffusion equations:

$$\frac{\partial \rho_f(\mathbf{\Omega}_f, t)}{\partial t} = D_{Rf}^W \nabla_{\Omega_f}^2 \rho_f - k_{fb}\rho_f + k_{bf}\rho_b, \qquad (6.29)$$

$$\frac{\partial \rho_b(\mathbf{\Omega}_b, t)}{\partial t} = D_{Rb}^W \nabla_{\Omega_b}^2 \rho_b - k_{bf}\rho_b + k_{fb}\rho_f. \qquad (6.30)$$

where D_{Rx}^W is the rotational diffusion constant of x species $(x = f, b)$.

The coupled equations can be solved for the orientational correlation functions necessary to obtain the dielectric function. In the limit of large barrier from bound to free state, one finds a new slow time scale in the orientational relaxation of the water in the surface:

$$\tau_{slow} \approx k_{bf}^{-1}. \tag{6.31}$$

The value of the slow time constant (τ_{slow}) is determined by the strength of the localizing interaction that the water molecules have with the surface, and the amplitude of the slow component is determined by the number of such interaction centers on the surface. If the barrier is of the order of $5k_B T$ or so, then τ_{slow} is of the order of a few hundred picoseconds.

6.4 EXPERIMENTAL RESULTS: A CHRONOLOGICAL OVERVIEW

6.4.1 Discovery of Multiexponential Solvation Dynamics: Phase-I (1980–1990)

A large number of solvents were studied during this period using laser pulses with a few picoseconds time resolution. The main results of the experiments may be summarized as follows:

(1) The observed solvation times are largely insensitive to the probe size and shape, and depend primarily on the properties of the polar solvent studied. Linear response is probably valid.
(2) The solvation time correlation function, $S(t)$, was found to be nonexponential, sometimes with a large unresolved ultrafast sub-picosecond component. This nonexponentiality was found to prevail not only at short but also at long times ($t > \tau_L$) as shown in Fig. 6.6. This indicates the presence of a distribution of relaxation times and the $S(t)$ versus t curve could often be well represented by a stretched exponential form such as the Kohlrausch–Williams–Watts function, with a stretched exponential parameter, which has a value significantly less than unity.
(3) The average solvation time τ_s, defined as a time integral of $S(t)$, is generally larger than τ_L and usually lies between τ_L and τ_D. In some instances, solvation times were measured to be more than a order of magnitude longer than τ_L.
(4) Interestingly, the solvation time for LDS-750 in methanol and butanol was found to be smaller than τ_L. Solvent translational modes were believed to be responsible for the faster relaxation observed in these solvents.
(5) The observed deviation of τ_s from τ_L could be related to the static dielectric constant ε_0 (or $\varepsilon_0/\varepsilon_\infty$).
(6) Strongly nonexponential, and even exponential, relaxation with a very different time scale from τ_L^{ion} was observed for normal alcohols.

(7) Furthermore, the simulation studies of solvation dynamics in water showed the importance of the solvent structure in the dynamics of solvation.

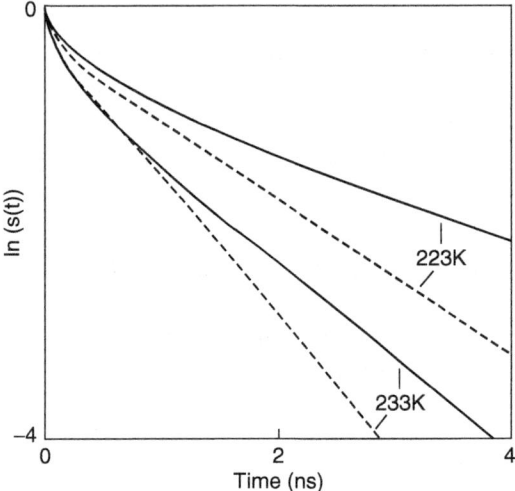

Figure 6.6 Comparison between observed (dashed line) and calculated (solid line) solvation time correlation functions in n-propanol. The calculated curves were obtained using the continuum-theory expressions and approximating the dielectric response by two Debye forms. Reprinted with permission from *J. Chem. Phys.* **86**, 6221 (1987). Copyright (1987) American Institute of Physics.

6.4.2 Discovery of Subpicosecond Ultrafast Solvation Dynamics: Phase-II (1990–2000)

While the emphasis in the early studies on solvation dynamics was on the nature and the utility of the solvation time-correlation function and whether it can be described by the continuum models and simple microscopic theories developed around the same time period, the focus changed naturally when ultrashort laser pulses with single-photon counting techniques became available. Perhaps the two most well-known studies of this period are the measurement of the solvation time-correlation function in acetonitrile and in water. Both studies revealed the presence of a dominant ultrafast, presumably Gaussian, component with time constant in the sub-100 fs range. In fact, the initial part of the solvation time-correlation function was not directly measured in these initial studies but was estimated by using a variety of methods, including the amount of Stokes shift that could not be measured, and also using computer simulations.

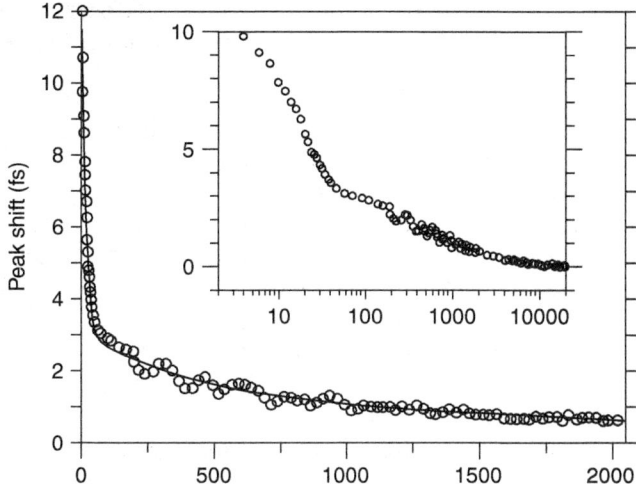

Figure 6.7 Three-pulse photon echo peak shift data of eosin in water (circles) are shown with a simple three exponential fit (solid line) as a guide to the eye. The fit includes a ∼17 fs (73%) fast component and two slower components, ∼400 fs (15%) and ∼2.7 ps (12%). The inset, on a log scale and without fit, shows that the peak shift decays to zero by $T \sim 15$ ps. Reprinted with permission from *J. Chem. Phys.* **110**, 5884 (1999). Copyright (1999) American Institute of Physics.

In Fig. 6.7, we show the measured solvation time-correlation function in water where the data were fitted to a sum of three exponentials. In Table 6.1, we present the time constants and the corresponding weights of the multiexponential fitting of the solvation time-correlation function for some of the important solvents:

$$S(t) = \sum_{i=1}^{n} a_i \exp(-t/\tau_i). \tag{6.32}$$

Here $\sum_{i=1}^{n} a_i = 1$. τ_i is the time constant of the ith component of the solvation time-correlation function and a_i is the corresponding weight. The average solvation time is defined as

$$\langle \tau \rangle = \sum_{i=1}^{n} a_i \tau_i. \tag{6.33}$$

6.4.3 Solvation Dynamics in Complex Systems: Phase-III (2000–)

Subsequent to the success of the solvation studies in water, attention turned toward complex chemical and biological systems [12]; some examples are micelles,

Table 6-1. MULTIEXPONENTIAL FITTING PARAMETERS AND CORRESPONDING AVERAGE SOLVATION TIMES FOR SEVERAL IMPORTANT COMMON SOLVENTS.

Solvent	a_1	τ_1 (ps)	a_2	τ_2 (ps)	a_3	τ_3 (ps)	a_4	τ_4 (ps)	$\langle\tau\rangle$ (ps)
Water	0.73	0.017	0.15	0.4	0.12	2.7			0.396
Acetonitrile	0.686	0.089	0.314	0.63					0.26
Methanol	0.101	0.030	0.340	0.28	0.298	3.20	0.268	15.3	5.0
Ethanol	0.085	0.030	0.230	0.39	0.182	5.03	0.502	29.6	16
Acetone	0.565	0.187	0.435	1.09					0.58
DMSO	0.5	0.214	0.408	2.29	0.092	10.7			2.0
DMF	0.508	0.217	0.453	1.70	0.039	29.1			2.0
Formamide	0.083	0.030	0.454	0.16	0.399	2.94	0.064	57.9	5.0
Chloroform	0.356	0.285	0.644	4.15					2.8
Benzene	0.366	0.234	0.6	1.89	0.034	24.7			2.1

reverse micelles, microemulsions, protein hydration layer, and grooves of DNA. Water molecules in such systems face a constrained atmosphere. As a result, the otherwise extended hydrogen bond network in the bulk water is compromised in such complex systems. Unfortunately, it is difficult to extract dynamic information about dynamics of water in such systems. Therefore, the further studies in the field of solvation dynamics focused on extracting the structural and dynamical characteristics of water in such constrained atmospheres.

The first such study was reported on the dynamics of water in a cyclodextrin (CDX) cavity. Solvation dynamics of a solute probe in a CDX solution revealed two components slower than those present in bulk water. The slowness was attributed to the presence of nearly immobile water within the CDX cavity. Subsequent studies revealed slow components in the solvation dynamics of solute probes in micelles, reverse micelles, and several other complex chemical systems. However, unambiguous interpretation of the observed slow components has not often been possible, because of the heterogeneity of solvating environment and the multiple occupation sites of the probe. In some cases, even the motion of the solute probe can be involved, which makes the use of solvation dynamics to study water environmental dynamics uncertain. In addition to the usual fluorescence Stokes shift studies, 2D-IR spectroscopy has also been used to understand the dynamics of water in such constrained systems. The interesting question of the relative importance of nanoconfinement versus interactions of water with the molecules of the confining surface has been addressed recently.

A large number of studies have been devoted to the study of the protein hydration layer in several different protein systems, employing TDFSS, dielectric relaxation, and magnetic resonance techniques. Theoretical and computational studies have also been devoted to this problem. Unfortunately, however, no consensus has yet evolved on the time scales of dynamic response of the hydration layer. While several studies report the presence of a significant slow component (of the time constant in the time scale range of 20–100 ps), other studies have claimed the absence of such slow components. A few experimental studies have even reported much longer time scales. Computer simulations have reported a change in the nature of hydrogen-bond-breaking dynamics, which has been attributed to the formation of a quasi-two-dimensional network of water molecules around the protein, with polar/charged amino acids acting as pivotal/binding sites of water molecules.

Simulation studies have also reported the observation of a slow component in the orientational dynamics of water in DNA grooves. As mentioned above, the issue of the existence of a slow component in the solvation dynamics in complex systems has largely remained controversial. There has not been any consensus on the amplitude and the time constant of this component and estimates vary from system-to-system and group-to-group.

6.5 MICROSCOPIC THEORIES

Several microscopic theories of nonequilibrium solvation dynamics have been developed and they do not differ greatly from each other [4, 5]. As in the continuum-model theories, the main thrust has been to obtain the time dependence of the polarization relaxation of the solvent in the absence of the electric field of the solute. So, at a simple level, the primary job at hand is to obtain a microscopic theory of natural polarization relaxation of the system. In the following, we briefly describe a simple approach that captures the essence of the outlines of the basic ideas and techniques involved.

6.5.1 Molecular Hydrodynamics Description

This approach employs the same theoretical method used to describe orientational relaxation, discussed in a previous chapter. In this approach, the basic variables are the position- ($\mathbf{r}$) and orientation- ($\mathbf{\Omega}$) dependent number ($\rho(\mathbf{r}, \mathbf{\Omega}, t)$), momentum ($g(\mathbf{r}, \mathbf{\Omega}, t)$) and angular momentum ($g_\Omega (\mathbf{r}, \mathbf{\Omega}, t)$) densities. These quantities are conserved and hence are regarded as slow variables. Using the general ideas of molecular hydrodynamics, a detailed theory of polarization relaxation has been developed in the last decade. This approach to solvation dynamics is based on the observation that the time-dependent solvent polarization of a dipolar liquid is related to the number density of the solvent by the following expression:

$$\mathbf{P}(\mathbf{r}, t) = \mu \int d\mathbf{\Omega} \hat{\alpha}(\mathbf{\Omega}) \rho(\mathbf{r}, \mathbf{\Omega}, t), \qquad (6.34)$$

where $\rho(\mathbf{r}, \mathbf{\Omega}, t)$ is the position- ($\mathbf{r}$), orientation- ($\mathbf{\Omega}$), and time- (t) dependent density of the solvent, $\mathbf{P}(\mathbf{r}, t)$ is the polarization vector, $\hat{\alpha}(\mathbf{\Omega})$ is a unit vector with orientation $\mathbf{\Omega}$, and μ is the magnitude of the dipole moment of the solvent molecules. The time-dependent solvation energy may be given by the following expression:

$$E_{\text{solv}}(t) = -\frac{1}{2} \int d\mathbf{r} \mathbf{D}(\mathbf{r}) \cdot \mathbf{P}(\mathbf{r}, t), \qquad (6.35)$$

where $\mathbf{D}(\mathbf{r})$ is the bare electric field (electric displacement vector) of the polar solute molecule whose solvation is being studied.

Therefore, the main quantity of interest is the position- and time-dependent polarization of the solvent. To obtain the time-dependent polarization vector, one needs a kinetic equation for $\rho\,(\mathbf{r},\boldsymbol{\Omega},t)$. These aspects have been discussed in Chapter 5 and need not be repeated here.

Historically, the molecular hydrodynamic description to treat ion solvation dynamics was first employed by Wolynes and coworkers and has subsequently been extended by many groups. There are several notable features of this approach that distinguish it from continuum-model-based theories. First, it takes into account the effects of intermolecular correlation on solvent dynamic response. At long length scales, the theory reduces to continuum-model description, but at molecular-length scales, the dynamic response is quite different. In the absence of the translational-mode contribution, molecular correlations lead to a substantial slowing of the dynamics. Interestingly, however, the translational mode accelerates decay precisely in the same length scale, thus partly removing the slow dynamics due to intermolecular correlations.

6.5.2 Polarization and Dielectric Relaxation of Pure Liquid

We have discussed this in detail in Chapter 5. We shall review briefly some salient aspects relevant to solvation dynamics. From Eq. (5.36) (linearized GSE), one can obtain an expression for polarization relaxation in a pure dipolar liquid if it is assumed that the direct correlation function is given by linearized equilibrium theories. Both longitudinal and transverse components were found to relax exponentially with time constants given by Eqs. (5.47) and (5.48).

Note that the polarization relaxation time and also the solvation dynamics in a liquid are determined by both the equilibrium and the dynamic properties of the solvent. This complex dependence often makes simple interpretation of solvation dynamics experiments difficult.

Solvation dynamics probes the longitudinal component. If the translational contribution is neglected, the $\mathbf{k} = 0$ mode relaxes faster, in accordance with Onsager's comment. If, however, the translational contribution is significant ($p' \geq 0.5$), polarization modes at intermediate wave vectors relax faster and the Onsager conjecture is no longer valid. The translational mechanism of polarization relaxation (where we consider relaxation of solvent structure around a positive ion) is depicted in Fig. 6.8.

6.5.2.1 Effects of Translational Diffusion in Solvation Dynamics

The molecules can relax either by rotating or by moving a small distance to attain the desired configuration. At small distances from the ion, which correspond, approximately, to intermediate wave vectors, the orientational relaxation is particularly slow, so in that region the translational contribution may be significant. But at large distances ($\mathbf{k} = 0$), the orientational relaxation is fast, so the translational contribution is unimportant. In this limit, Onsager's inverse snowball picture breaks down.

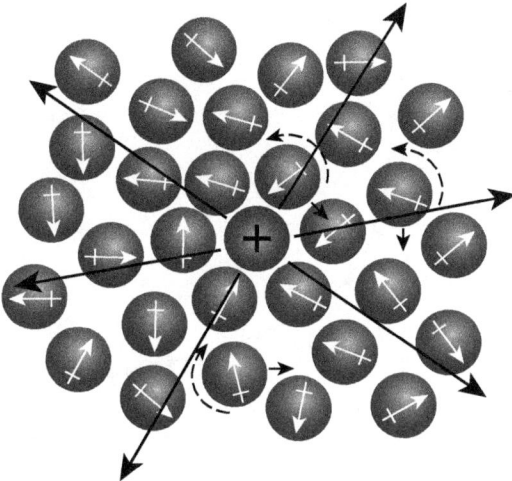

Figure 6.8 A pictorial description of the role of the translation diffusion in the solvation dynamics of a positive ion that is at the center of the figure. The solvent molecules can relax either by rotating (dashed arrows) at fixed position or by translating (solid arrows) a small amount in the required direction. The molecules close to the ion experience mostly the large wave vector field where orientational relaxation is slow, so the translational mechanism is important in this region. However, far from the ion, the translational motion of solvent molecules is not important.

This simple theory, as already pointed out, cannot describe short time, underdamped, and/or inertial limit of polarization relaxation. This requires a more general treatment and we briefly elucidate the main approach. It can be shown that under rather general conditions, the polarization fluctuation can be expressed in the following form:

$$P(k, z) = \frac{P(k, t=0)}{z + \Sigma(k, z)}, \qquad (6.36)$$

where $\Sigma(k, z)$ is the wave-number- (k) and frequency- (z) dependent generalized rate.

The time-dependent polarization is obtained from the above equation by inverse Laplace transform. This generalized rate contains contributions both from translational and rotational motion.

When this generalized expression is used to obtain the solvation time-correlation function, the agreement of the theory with both experiment and simulation is found to be quite satisfactory (see Fig. 6.9).

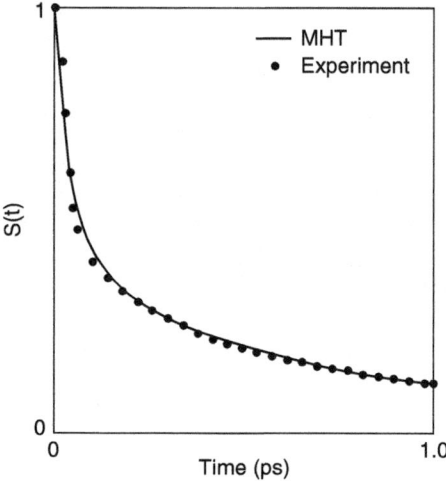

Figure 6.9 Comparison between the molecular hydrodynamics theory prediction (solid line) and the experimental results (solid circles) for the solvation time-correlation function, $S(t)$, plotted as a function of time for the solvation of an ionic solute in water. The theoretical calculations have been carried out with an ionic solute having size comparable to that of C-350, which is the solute used in experiment. Reprinted with permission from *J. Chem. Phys.* **102**, 1390 (1995). Copyright (1995) American Institute of Physics.

6.6 SIMPLE IDEALIZED MODELS

Both theoretical and simulation studies of solvation dynamics in real dipolar liquids are fraught with difficulties not only because of the long-range nature of the dipolar potential, but also because real molecules often are nonspherical and have internal degrees of freedom. For theory, these complications make evaluation of the angular pair and two-particle direct-correlation functions quite difficult. Simulations not only face the above difficulties, but an additional complication faced is an accurate description of the polarizability of the atoms and molecules. In theory, a part of this can be accounted for by using an experimental memory function, but for simulations, the effects of polarizability on dynamics remain a formidable problem.

Therefore, in order to obtain a measure of accuracy of the various approximations of the theories, several idealized models have been proposed and studied. We shall discuss two such models. First, we shall discuss a Brownian-dipolar-lattice (BDL) model, which was originally proposed by Lax to study equilibrium properties and has been generalized to treat dynamics by Zwanzig [15]. The second model we shall consider is the Stockmayer liquid (also a popular model). For both the models, both simulations and detailed studies have been carried out, and results are quite educative, and serve as good pedagogical examples.

6.6.1 Overdamped Solvation: Brownian Dipolar Lattice

In this model, the solvent molecules are point dipoles positionally fixed at the sites of a simple cubic lattice and free to execute Brownian rotation in the presence of surrounding dipoles. Thus, each dipole experiences two types of torque: a systematic torque due to interaction with other dipoles in the system and a random torque responsible for the Brownian motion. Both dielectric relaxation and solvation dynamics of this lattice have been studied in detail. Surprisingly, even such a simple model shows remarkably rich dynamics.

One crucial aspect studied in great detail is the validity of the continuum model. The BDL model is particularly suitable to explore this question because it has no inertial motion and also molecules are point dipoles and there is no translational motion in this case. It was found that the model reproduces the role of inter-dipolar correlation in giving rise to slow orientational relaxation. As shown in Fig. 6.10, the continuum model grossly overestimates the initial value of the torque–torque time-correlation function (TTTCF) and also underestimates the time constant of decay of the torque–torque time-correlation function. As a result, the integrated value may sometimes look correct but the underlying dynamics is incorrectly represented in the continuum model.

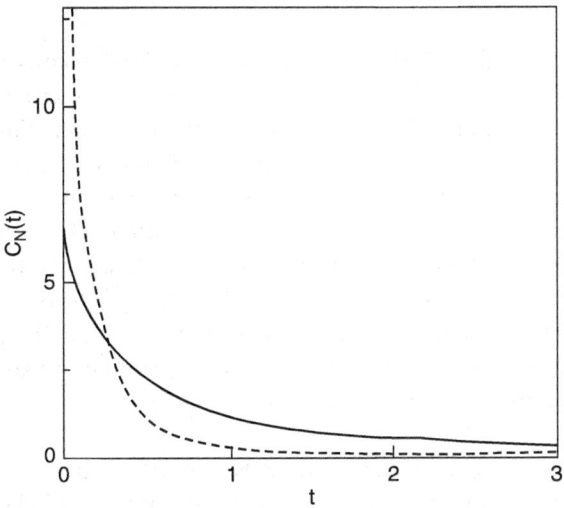

Figure 6.10 Comparisons between the torque–torque correlation function obtained from the simulation (solid curve) of the Zwanzig lattice model and from the Nee–Zwanzig continuum model (dashed curve) at $\eta = 0.3333$. Reprinted with permission from *J. Chem. Phys.* **97**, 3610 (1992). Copyright (1992) American Institute of Physics.

The study of solvation dynamics of a dipole also provided rather educative results. It was found that while the continuum model fails to capture the increased nonexponentiality of the decay with increase in the polarity of the system, the microscopic theories could capture the nonexponential solvation dynamics surprisingly well.

There are several other relevant issues that have been explored in the BDL model. First, it has been shown that the details of solute–solvent interactions do not play an important role in the solvation dynamics. Therefore, the linear "unperturbed" response of the solvent is sufficient to describe solvation. Second, in an indirect way, the role of solvent translational modes has been verified. Because the lattice models do not have solvent molecules moving around, the intermolecular correlations are more important, which leads to such a clear failure of the continuum-model-based theories.

6.6.2 Underdamped Solvation: Stockmayer Liquid

A Stockmayer liquid consists of a collection of spheres with point dipoles embedded at the center of each. These spheres interact by both Lennard-Jones (LJ) and dipolar interactions. In the absence of the point dipoles, the spheres face no rotational friction and the translational dynamics of such LJ spheres have been studied in detail. This system, therefore, provides a simple model where the only source of rotational friction is dipolar interactions. Another advantage is that we do have analytical expressions for the pair- and direct-correlation functions; therefore, we can compare theory and simulations almost on an equal footing.

The main results of the studies on Stockmayer liquid are summarized below:

1. The decay of the solvation time-correlation function consists of a large contribution, about 70–80% of the whole, from the inertial component. This is followed by a long, slow, exponential-like decay. Thus, the present model is complementary to the Brownian dipolar lattice model which considers only overdamped motion where inertial decay is absent.
2. The simulation results can be explained quantitatively by the molecular hydrodynamic theory without any adjustable parameter. Because of the simplicity of the model, one could explore the specific role of the probe solute translational motion in the solvation process, as this provides a stringent test of the microscopic theories. Quantitative agreement further supports the validity of the molecular theories.

The study of these two simple models proved to be of immense value in understanding various aspects of solvation dynamics and dielectric relaxation. We shall now consider more complex systems.

6.7 SOLVATION DYNAMICS IN WATER, ACETONITRILE, AND METHANOL REVISITED

Solvation dynamics (SD) in water, acetonitrile, and methanol deserve special mention. For water, the continuum model predicts that solvation dynamics should be extremely fast with a time constant between 0.2 and 0.5 ps. In an important paper, Jimenez et al. [17] reported the results of SD of the excited state of the dye, Coumarin 343 (C343), in water. Their results are shown in Fig. 6.11. The initial part of the solvent response of water is extremely fast (a few tens of femtoseconds) and constitutes more than 60% of the total solvation. The subsequent relaxation occurs in the picosecond time scale. The decay of the solvation time-correlation function, $S(t)$, is fitted to a function of the following form:

$$S(t) = a_g \exp\left(-\frac{1}{2}\omega_g^2 t^2\right) + a_1 \exp\left(-\frac{t}{\tau_2}\right) + a_2 \exp\left(-\frac{t}{\tau_3}\right), \quad (6.37)$$

where a_g, a_1, and a_2 are the relative weights of the initial Gaussian and the subsequent exponential decay processes, ω_g is the frequency of the Gaussian decay component, and τ_2 and τ_3 are relaxation time constants of the two exponential decay processes. The early simulation studies also predicted a very fast initial component with a Gaussian time constant less than 20 fs. Jimenez et al. experimentally detected a Gaussian component (frequency 38.5 ps^{-1} (28 fs), 48% of the total amplitude) and a slower bi-exponential decay with time constants of 126 (20%) and 880 (35%) fs, respectively. Several other experimental and simulation studies on SD of large dye molecules as well as electrons in water have demonstrated that the dynamics of solvation in water is indeed ultrafast and occurs in the femtosecond time scale. More recently, higher-order nonlinear optical measurements such as three-pulse photon echo peak shift measurements have been carried out to study the SD. Solvation dynamics study of dye molecule, eosin in water, revealed a substantial amplitude (about 60%) of aqueous solvation occurs within 30 fs. A three-exponential fit (up to 100 ps) to the data of eosin in water yields time constants of 17 fs (73%), 330 fs (15%), and 3 ps (12%) (see Fig. 6.7). This ultrafast solvation can be attributed to the high-frequency intermolecular vibrational/librational modes of water which are the hindered translational band at 180 cm^{-1} due to the hydrogen bond (HB) network and the 600 cm^{-1} band due to libration.

Molecular hydrodynamic theory and generalized continuum-model description provide a near quantitative agreement with experimental data. The excellent agreement at short times of the generalized continuum models with the experimental results suggests that the ultrafast component of SD is indeed collective in nature where a large number of water molecules participate. Note that this explanation is different from the inertial response of single water molecules. In the latter case, the collective polarization potential (the solvation potential) is not the driving force of the relaxation.

The continuum model predicts that in acetonitrile the longitudinal relaxation time is ~0.2 ps. However, solvation dynamics experiments on acetonitrile reveal an ultrafast component of time scale 70–90 fs. Total solvation time-correlation function has been fitted to a double exponential (0.089 ps (68%) and 0.63 ps (0.32%)) with average solvation time of 0.26 ps. Here the ultrafast solvation is powered largely by the small value of the rotational time constant (~400 fs

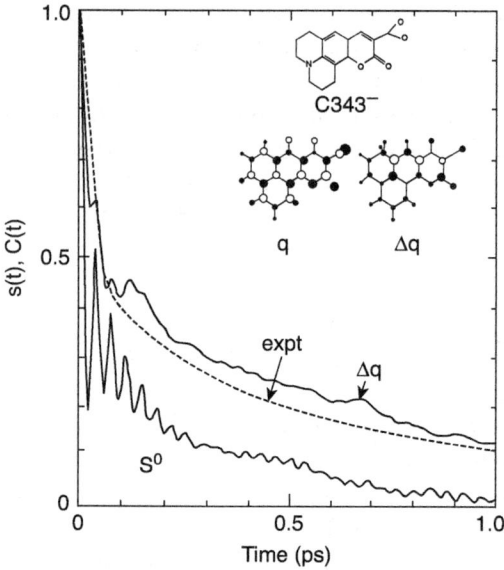

Figure 6.11 Experimental (dashed line, denoted by "expt") and simulated (solid line, denoted by "Δq") solvent response functions for C343 in water as obtained by Maroncelli, Fleming, and coworkers [17] are shown in this figure. Also shown is a simulation for a neutral atomic solute with the Lennard-Jones parameters of the water oxygen atom (denoted by S^0). The experimental data were fitted to Eq. (6.37) (using the constraint that the long time spectrum match the steady-state fluorescence spectrum) as a Gaussian component (frequency 38.5 ps^{-1}, 48% of total amplitude) and a sum of two exponential components: 126 fs (20%) and 880 fs (35%). Reprinted with permission from Macmillan Publishers Ltd: *Nature*, **369**, 471, (1994). Copyright (1994).

at 300 K). The relatively large value of the static dielectric constant again facilitates ultrafast solvation.

Solvation dynamics of methanol also shows an ultrafast component of time scale of 30 fs. Here one finds the contribution of ultrafast component toward the total solvation time-correlation function is only 15%. This can be explained in terms of a small-amplitude fast relaxation in the rotational relaxation of methanol. The rotation around C–OH bond is very fast which gives rise to very fast polar response.

6.7.1 The Sub-100 fs Ultrafast Component: Microscopic Origin

The extended MHT offers a microscopic explanation of the origin of the ultrafast response of dipolar liquids [4, 5]. The analysis suggests that the dipolar solvent must contain ultrafast components that can couple to the solvent polarization to observe ultrafast solvation. They may be librations, H-bond excitations, or fast single-particle orientations. The analysis also suggests that the force constant for polarization fluctuation must also be large. As will be discussed below, this is large if the static dielectric constant ε_0 of the solvent is large. Under the harmonic approximation, the following expression for the free energy expression for the polarization fluctuation of the system was derived:

$$F(\{P_L(q)\}) = \frac{(2\pi)^3}{2V} \int dq K_L(q) P_L^2(q), \qquad (6.38)$$

where $K_L(q)$ represents the wave-number-dependent force constant of the longitudinal component of the polarization fluctuation $P_L(q)$, and V is the volume of the system. The wave-number-dependent longitudinal dielectric function $\varepsilon_L(q)$ is given by

$$K_L(q) = \frac{2}{(2\pi)^2} \frac{\varepsilon_L(q)}{\varepsilon_L(q) - 1}, \qquad (6.39)$$

Now, $\varepsilon_L(q)$ of the dipolar liquid is large at small q and $K_L(q)$ is nearly equal to $2/(2\pi)^2$ at small q. At intermediate q ($q \approx 2\pi/\sigma$), $\varepsilon_L(q)$ is negative, with a small absolute value, much less that unity. Thus, $K_L(q)$ exhibits a pronounced softening in this region. $\varepsilon_L(q)$ becomes unity at large q, where $K_L(q)$ diverges. However, the contribution to the solvation energy from such a large q is negligible. Most of the contribution comes from small q, with a significant amount also coming from intermediate q. Therefore, for the present purpose, the force constant of polarization relaxation is the largest at small q.

As also noted by van der Zwan and Hynes [18], solvation dynamics of an ion can be viewed as a relaxation in a harmonic polarization potential where the curvature of the potential well is determined by the force constant $K_L(q)$. At small wave number, the curvature is steep, $K_L(q)$ is large, and thus the relaxation is fast. At intermediate wave numbers, on the other hand, softening of the force constant makes the relaxation slow (Fig. 6.12).

The effects of the force constant and the natural dynamics on the ultrafast response can be understood more convincingly if one considers the following generalized Langevin equation:

$$\ddot{P}_L(q) = K_L(q) P_L(q) - \int dt' \Gamma_R(q, t - t') \dot{P}_L(q, t') + R(t), \qquad (6.40)$$

which describes the relaxation of the longitudinal component of the solvent orientational polarization density. The double dot denotes the second-order time derivative. Γ_R is the rotational dispersive kernel and the effects of the fast orientations and high-frequency librations are embodied in this kernel. Equation (6.40) makes clear the effects of the force constant and the natural fast dynamics of the medium on the ultrafast polarization relaxation.

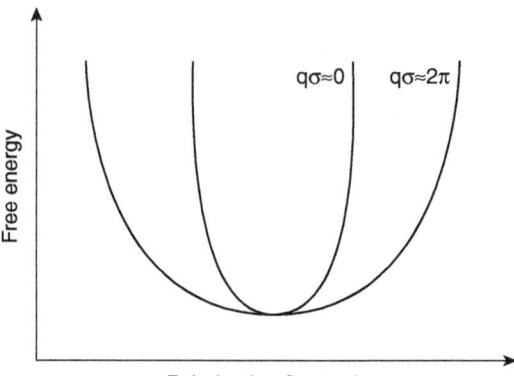

Figure 6.12 A schematic representation of the polarization fluctuation free-energy surfaces at two wave numbers at which solvent orientational polarization relaxation takes place at substantially different rates. The large value of force constant at $q \approx 0$ makes the surface steep and hence the relaxation becomes fast. At intermediate wave numbers ($q \approx 2\pi/\sigma$), the value of the force constant is small and therefore, the relaxation becomes slow.

Therefore, it is obvious that a large value of $K_L(q)$ alone is not sufficient to switch on the ultrafast relaxation. It has to couple to the high-frequency natural dynamics of the medium. For water, the dynamics is provided by the intermolecular vibrations (at ≈ 180 cm^{-1}) and the librational modes (at ≈ 600 cm^{-1}). For acetonitrile, however, the very fast single-particle orientational dynamics is partly responsible. For methanol, the ultrafast component (relatively smaller amplitude) derives contributions from fast single-particle orientation and libration.

6.8 EFFECTS OF SOLVATION ON CHEMICAL PROCESSES IN THE SOLUTION PHASE

We already have discussed that two important solvents, namely, water and acetonitrile, both exhibit a dominant ultrafast component, with a sub-100 fs time constant, in solvation dynamics, which was not anticipated at the time of discovery because the continuum model predicted much slower solvation. The question now naturally arises: what are the effects of this dominant ultrafast component in chemistry? The answer turned out to be quite interesting, as discussed below. In fact, the discovery of the ultrafast component in turn leads to the resolution of some longstanding puzzles in physical chemistry. We shall discuss three examples here where the solvation dynamics itself plays a dominant role. There are other examples also, but the following three will be quite illustrative.

6.8.1 Limiting Ionic Conductivity of Electrolyte Solutions: Control of a Slow Phenomenon by Ultrafast Dynamics

Ionic conductivity in water in the limit of infinite electrolyte dilution is determined by the interaction of the ion with the surrounding dipolar water molecules. The frictional force exerted by water molecules on a small ion (such as Li^+, Na^+, K^+, Cl^-) is determined largely by ion–dipole interaction. In many aspects, equilibrium and dynamical features of this dynamics are similar to solvation dynamics because both are governed by the same interactions. In fact, the ion–dipole part of the friction can be explained by expressions, both in dielectric continuum model [19, 20] and in molecular theories [21, 22], that are similar to those for solvation time-correlation function. Therefore, it can be anticipated that ultrafast solvation dynamics can play an important role in determining the limiting ionic conductivity [22].

The limiting ionic conductivity (LIC), usually denoted by Λ_0, exhibits several unusual features, the most important of which is a strong nonmonotonic dependence of Λ_0 on the atomic radius of the small rigid ions (see Fig. 6.13). The second important feature is that the maximum value of Λ_0 is quite large for a strongly polar liquid such as water. We discuss below how ultrafast solvation plays an important role in explaining both these aspects.

The theoretical analysis follows the following steps. The limiting ionic conductivity (Λ_0) is assumed to be given by Nernst equation, which relates Λ_0 to the self-diffusion constant. The latter is related to friction by Stokes–Einstein relation. One can now calculate the friction by using well-known statistical mechanical methods.

It is found that indeed the existence of ultrafast solvation reduces the friction on the ion drastically and brings the theory into good agreement with experimental results, as shown in Fig. 6.14.

6.8.2 Effects of Ultrafast Solvation in Electron-Transfer Reactions

Effects of ultrafast solvation dynamics on electron transfer reactions shall be discussed in detail in **Chapter 10**, but we should mention here that the discovery of ultrafast solvation helped to resolve a longstanding paradox in this area. Early theoretical studies that used a diffusive exponential decay of the solvation time-correlation function (with the time constant given by the longitudinal relaxation time, τ_L) predicted a strong dependence of the electron-transfer rate on the solvation time. Experimentally, however, mostly a weak dependence has been observed. If we now include a substantial ultrafast component in $S(t)$, then the dependence of ETR rate on the longitudinal relaxation time becomes considerably weaker, as is indeed found in experiments.

6.8.3 Nonequilibrium Solvation Effects in Chemical Reactions

Solvent polarity plays an important role in those solution-phase chemical reactions that involve ions or dipoles. Solvation free energy makes an important contribution

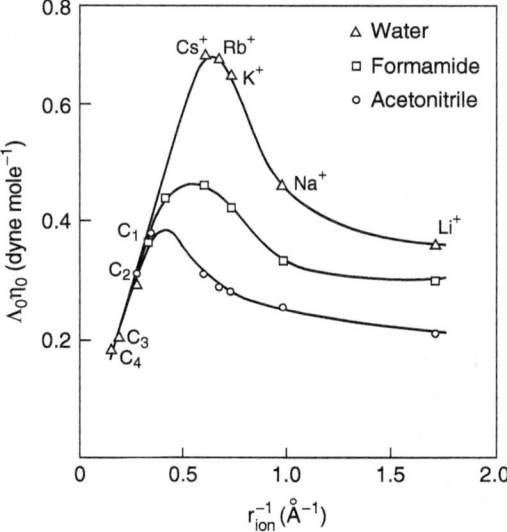

Figure 6.13 Experimental values of the Walden product ($\Lambda_0 \eta_0$) of rigid, monopositive ions in water (open triangles), acetonitrile (open circles), and formamide (open squares) at 298 K plotted as a function of the inverse of the crystallographic ionic radius, r_{ion}^{-1}. C_n denote alkyl ammonium ions. The solid line in each case is simply an aid to the eye. Reprinted with permission from *Acc. Chem. Res.* **31**, 181 (1998). Copyright (1998) American Chemical Society.

to the activation barrier of the reaction because of the solvation of reactants, products, and transition states owing to strong electrostatic interactions. If the charge distribution changes significantly along the reaction coordinate, one expects a considerable modification of the free-energy surface of the reaction depending on how fast or slow solvent can rearrange to solvate the change. This effect will be severe when the time scales of the intrinsic reaction and of the solvent relaxation are comparable. Examples of such reactions include proton transfers, ion pair interconversions, isomerizations, and S_N^2 reactions.

If a charge is displaced in a reaction, the surrounding dipoles will respond in a time-dependent way to this motion. At one extreme, the charge may move so rapidly that solvent dipoles are nearly frozen during the reactive event. At the other extreme, the charge may move so slowly that solvent dipoles have enough time to rearrange themselves around the charge so that at each stage of the reaction, equilibrium solvation is attained. However, several solution-phase reaction events fall in between these two extremes where an intermediate level of nonequilibrium solvation should apply. In any event, it is clear that the polar solvent dynamics are coupled to some degree to the dynamics involved in the intrinsic chemical charge-transport step. The solvent influences enter via the time-dependent friction

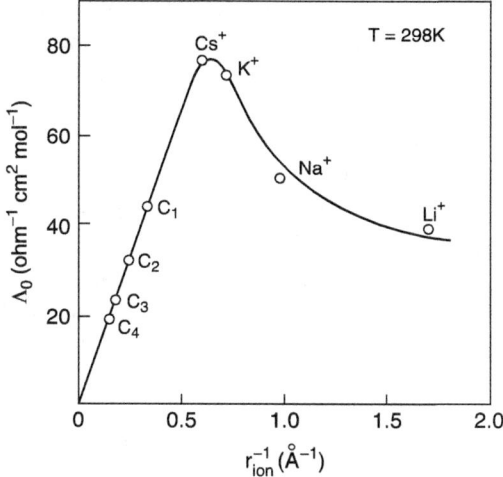

Figure 6.14 The comparison of the values of the limiting ionic conductivity Λ_0 in water of rigid monopositive ions with the prediction of the molecular theory that takes into account of the ultrafast sub-50 fs solvation dynamics in liquid water. Here Λ_0 is plotted as a function of the inverse ionic radius r_{ion}^{-1} in water at 298 K. The solid line represents the predictions of the microscopic theory. The open circles denote the experimental results. Reprinted with permission from *Acc. Chem. Res.* **31**, 181 (1998). Copyright (1998) American Chemical Society.

coefficient, $\zeta(t)$, which can be related to the solvent motion. Grote and Hynes [18(d)] derived the rate of a chemical reaction as

$$k = k^{TST} (\omega_r/\omega_b) \tag{6.41}$$

Here, the transition-state-theory rate constant k^{TST} describes the reaction if and when the solvent is fully equilibrated with the charge, so that solvent dynamics plays no role. The reactive frequency ω_r is a measure of the actual reactive motion, and this includes both the motion of the charge over the intrinsic barrier of frequency ω_b and the dynamics of solvent in response to this motion. Let us assume that the electrostatic solvent frequency is ω_s, which modifies the effective potential energy seen by the charge at the short times when the solvent polarization is not in equilibrium with charge. Let us consider two regimes in terms of ω_s.

6.8.3.1 STRONG SOLVENT FORCES

In this regime, electrostatic solvent frequency exceeds the barrier frequency, $\beta = \omega_s^2/\omega_b^2 > 1$.

(a) For small relaxation times, $\omega_b \tau_e$, the solvent responds rapidly to equilibrate the moving charge. Hence, the friction developed in the process will be very little as the charge accelerates down the barrier and $\kappa \approx 1$, i.e., k will be given by its transition state value k^{TST}.

(b) As $\omega_b \tau_e$ is increased, the solvent response increasingly lags behind the charge motion. The retarding friction then decreases k below k^{TST}. The process can be understood as follows. When the charge moves to the barrier top, it finds itself in an effective potential well of frequency $\omega_{\text{eff}}^2 = -\omega_b^2 + \omega_s^2 > 0$ and the solvent has no time to adjust. The motion in this potential well is oscillatory and the barrier is crossed and recrossed making the transmission coefficient less than unity. The rate constant becomes inversely proportional to the solvent relaxation time

$$\frac{k}{k^{TST}} \approx \frac{\omega_b}{\zeta_D} \propto \tau_e^{-1}, \tag{6.42}$$

so the approach to this diffusion-controlled limit is not as rapid as predicted by a simple Kramers' approach, in which the time-dependent details of the friction $\zeta_D(t)$ are ignored.

6.8.3.2 Weak Solvent Forces
In this regime, the charge-solvent force is weak and β is less than one.

1. For small values of $\omega_b \tau_e$, the rate can be expressed as in transition-state theory. The explanation in this regime is the same as in the case of strong solvent force regime.
2. On increasing $\omega_b \tau_e$, the transmission coefficient decreases. However, this decrease does not go to zero (as in strong regime) but rather to a finite limiting value:

$$\kappa_{\min} = \frac{\omega_{\text{eff}}}{\omega_b} = \left[(\omega_b^2 - \omega_s^2)/\omega_b^2\right]^{1/2}. \tag{6.43}$$

This interesting feature can be understood as follows. When $\tau_e \gg \omega_b^{-1}$, the charge leaving the chemical barrier top moves not on a potential barrier of bare frequency ω_b, but on a modified barrier of reduced frequency $\omega_{\text{eff}} = (\omega_b^2 - \omega_s^2)^{1/2}$. This reflects the fact that, as long as β is less than one, the effective potential still has negative curvature. Hence, the charge will proceed on to the product region undeterred by the nearly frozen solvent configuration and κ remains finite. This picture is in complete contrast to a simple Kramer's prediction, where the transmission coefficient continues to decrease to zero with increasing solvent relaxation times, in which the short-time aspects of the friction are ignored.

6.9 SOLVATION DYNAMICS IN SEVERAL RELATED SYSTEMS

In the following subsections we discuss recent developments in several important related areas where studies on solvation dynamics were motivated by the new results obtained in bulk.

6.9.1 Solvation in Aqueous Electrolyte Solutions

In the experimental studies of solvation dynamics in aqueous electrolyte solutions of varying ion concentration, it is found that the relaxation of ion–solution interaction energy can be separated into its solvent and ionic components. The solvent response was found to be much faster than the ionic relaxation. An interesting observation of these studies is an increase in the long-time solvation rate with an increase in the ion concentration. Computer simulation studies of solvation dynamics in ionic solutions revealed a fast Gaussian decay characterized by the Gaussian time constant, τ_G, which is followed by a very slow exponential-like decay. It has also been found that the Gaussian time constant is practically independent of ion concentration, whereas the long-time exponential time constant depends rather strongly on ion concentration and it decreases when the ion concentration is increased. Recently, a molecular theory of ion solvation dynamics in aqueous electrolyte solutions has been developed, which properly includes the molecularity of both solvent and ions. According to this theory, at short times the relaxation of $S(t)$ is Gaussian with a time constant that is essentially the same as in bulk water. The long-time decay is predicted to be exponential with a time constant, τ_M, which is inversely proportional to the conductivity (σ) of the solution. Thus, while the initial Gaussian decay depends only *weakly* on ion concentration, the long-time decay of the solvation time-correlation function decreases with ion concentration. Thus, the predicted ion-concentration dependence of solvation dynamics is consistent with the experimental results of solvation dynamics in solutions of varying ionic strength. Self-consistent mode coupling theory for ion atmosphere relaxation shows that the ion atmosphere relaxation is nonexponential. However, a study of solvation dynamics using this mode-coupling theory has not as yet been attempted.

6.9.2 Dynamics of Electron Solvation

Onsager's famous comment at the 1976 Banff conference on solvated electrons, which is famously known as *inverse snowball effect*, motivated the study of solvation around an electron. The quantum nature of the electron, its light mass and high polarizability, coupled with large polarizability of oxygen and the hydrogen bond network of water, make the issue of solvation dynamics of electron in water unconventional. Initial femtosecond spectroscopic studies showed that before a newly created electron gets solvated, it takes about 100 fs to get localized around a position. Computer simulation studies found the solvent response function to be bimodal with an initial Gaussian component of about 25 fs [23]. The remaining solvation is relatively slow with a time constant of 250 fs. The observed dynamics of hydrated electrons in water by using femtosecond pump-probe spectroscopy primarily reflected the p-state solvation with a time constant of initial ultrafast inertial solvation in the 30–80-fs range [24]. The current understanding of the physical process involved in electron solvation appears to be as follows. At first the excited electron undergoes displacements in the search of a pre-existing trap. This search is accompanied and facilitated by the orientational polarization of the water molecules away from the electron. This collective polarization gives rise to a spectral shift in the 30–80-fs range, just as in an ion in bulk water. This leads to the formation of an equilibrated and solvated p-state electron. The nonadiabatic transition to the s-state occurs on a much slower time scale. The initial part of the solvation of the newly created s-state is again fast, occurring on the sub-100-fs time scale.

6.9.3 Solvation Dynamics in Supercritical Fluids

The critical point of water is located at pressure (P_c) 22.1 MPa, temperature (T_c) of 847 K, and density (ρ_c) 0.32 gm/cm^3. By supercritical water (SCW) one usually means water at a high temperature (above 847 K) and relatively high density [25]. At such high temperatures, the extended hydrogen bond network of liquid water essentially disappears and water exhibits certain remarkable properties. The dielectric constant of SCW is about 6–8, making it similar to organic solvents in many respects. Thus, many organic solutes, such as benzene and toluene, are soluble in SCW, which makes it suitable for future use in material extraction and cleaning processes. Simulation studies of solvation dynamics in SCW indicated a biphasic decay of solvation energy, with an ultrafast decay, similar to the one observed for bulk water. This is surprising because here the density is low, and the extended hydrogen bond network is nonexistent, thereby eliminating the contributions from the libration and the intermolecular vibration modes. These results were subsequently corroborated by theory, which shows that the ultrafast component arises here from very fast rotational motion of small water molecules. Recent simulation studies find that the solvation dynamics in SC, CHF_3, and CO_2 are also biphasic in nature. The fast component of the total solvation energy here decays with a time constant of about 1 ps. The other component relaxes at a rate with time constant in the tens of picoseconds regime. A set of very recent experimental studies using a time-correlated single-photon-counting technique has, however, indicated that the slow component has a time constant of about 50–70 ps, which is much slower than that observed in the above simulation.

6.9.4 Nonpolar Solvation Dynamics

Although by solvation dynamics we usually mean dipolar solvation dynamics, and although most theoretical and experimental studies have also focused on dipolar solvation, solvation by nonpolar forces is also important in many physical situations [26]. Examples include solvation of nonpolar solutes in nonpolar solvents, such as alkanes in carbon tetrachloride. The magnitude of solvation energy of the solute involved in nonpolar solvation dynamics is less than that in polar solvation dynamics, but can still be of the order of a few $k_B T$'s.

Several experimental and theoretical studies have explored equilibrium and dynamic aspects of nonpolar solvation dynamics. The observed characteristic features of nonpolar solvation dynamics can be summarized as follows:

1. The amplitude of the nonpolar solvation and the Stokes shift due to it are less than 50% of the usual amplitudes observed in polar solvation dynamics. The main contributing interaction is the long-range dispersion forces like Lennard-Jones interaction.
2. Dynamics of nonpolar solvation, as measured by energy–energy time-correlation function, is again nonexponential, with fast and slow components.
3. Although dipolar solvation dynamics is dominated by the orientational motion of the solvent molecules, nonpolar solvation dynamics is dominated more by the translational motion. As a result, the slowest time scales in nonpolar solvation dynamics can be much slower than what is commonly found in dipolar solvents, other things (viscosity, size of the solvent molecules) being equal.

Although dipolar solvation dynamics is intimately connected with the dielectric relaxation of the solvent, nonpolar solvation dynamics is connected with the dynamic structure factor of the solvent. Although polar solvation dynamics plays an important role in charge-transfer reactions, nonpolar solvation dynamics has been connected with vibrational-energy relaxation. The rate of vibrational-energy relaxation and the nonpolar solvation dynamics in a nonpolar liquid are both dominated by the dynamical events only of binary interactions.

6.10 COMPUTER SIMULATION STUDIES: SIMPLE AND COMPLEX SYSTEMS

The experimental studies have in turn motivated a large number of computer simulation studies of solvation dynamics in both simple and complex systems, starting with the pioneering work of Maroncelli and Fleming [6(b)] who were perhaps the first to find the ultrafast solvation in bulk water. They observed the presence of a sub-50 fs time component in the decay of the solvation time-correlation function. Interestingly, this study also revealed the importance of the translational mechanism of polarization decay and analyzed the validity of the linear-response theory.

Because of the presence of molecular-length-scale heterogeneity, it is difficult to develop a simple analytical theory of solvation dynamics in self-organized assemblies (like micelles and reverse micelles) and in biological systems, like proteins and DNA. Molecular-dynamics simulations thus proved to be a valuable tool in this area, particularly because they can reveal molecular aspects of solvation not available otherwise.

Clearly, the main effect of a polar surface is to introduce a local distortion (or, frustration) in the extended hydrogen bond network of bulk water. This can give rise to both slower- and faster-than-bulk relaxation of water. Below we discuss the results of several important cases successfully studied by computer simulations.

6.10.1 Aqueous Micelles

The orientational dynamics of water molecules at the surface of micelles was found to be significantly slower than the bulk. A long time tail has also been found in the orientational dynamics of water molecules in aqueous SDS micelles. However, the translational diffusion of water molecules in the micellar surface was found to be slower by only 20%. Thus, the translational motion was found to be less affected than the rotational motion. Hydrogen-bond-lifetime dynamics of water molecules in aqueous micelles was also found to be significantly slower than in the bulk. The surface water molecules can be characterized depending on the number of hydrogen bonds they form with the polar head groups of micelle. For the CsPFO micelle, about 60% of the interfacial water molecules are singly hydrogen bonded with the micelle, while 33% form two such hydrogen bonds (HB). A small fraction of the molecules do not form any HB with the micelle. The dramatic slowing down of water dynamics along with the solvation dynamics has also been observed in simulations of CsFPO surfactant molecules in water. Recently, SD study of aqueous micellar solution of decyltrimethylammonium bromide (DTAB) revealed that the SD of bromide ions exhibits a slow component, which was about two orders of magnitude slower than that in bulk. The origin of the slow decay was found to be rather complex.

6.10.2 Water Pool in Reverse Micelles

Reverse micelles have been studied in great detail not only because it is an important system by itself but more importantly it presents a wonderful example of a confined liquid. Additionally, in this case the size of the system studied in simulations is comparable to that in experiments and this size can be varied over a rather wide range [27–29]. Perhaps the most interesting result that has come out from the studies is the existence of faster-than-bulk relaxation in the confined liquid. Such results have also been observed in experiments.

6.10.3 Protein Hydration Layer

The dynamics of water in the protein hydration layer has been a subject of considerable controversy. Several simulation studies have revealed significantly slow translational and rotational dynamics of water molecules in the hydration layer of proteins, while some have observed only a small retardation effect, with dynamics remaining close to those observed for bulk water. However, many times such difference arises because conclusions are arrived at from the study of different dynamical quantities. The residence time of water molecules in the layer has been found, in some simulations, to be sufficiently large, implying the presence of a rigid water network around the protein. Recent computer simulations have reported a change in the nature of hydrogen-bond-breaking dynamics, which has been attributed to the formation of a quasi-two-dimensional network of water molecules around the protein, with polar/charged amino acids acting as pivotal/binding sites of water molecules. The study of protein–water HB lifetime dynamics using classical atomistic molecular-dynamics simulations has shown that the structural relaxation of the protein–water HB is slower than that of the water–water HB. Slow protein–water HB dynamics may be correlated with the biological activity of the protein. A slow component of solvation dynamics that has been observed in the aqueous protein solution is an order of magnitude slower than that of the bulk. The origin of the slow component of solvation dynamics is attributed to the coupled motion of the protein side chain and water. Hence, the protein fluctuation plays an important role in hydration-layer-solvation dynamics.

6.10.4 DNA Groove Hydration Layer

Both translational and orientational dynamics of water molecules in the grooves of DNA have been found to be slower, especially in the minor grooves. The dynamics were found to be strongly dependent on the sequence and the AT minor groove waters were understandably slower. Hydrogen-bond-lifetime dynamics also becomes slower in the groove region and was found to be even slower for minor grooves. The solvation time-correlation function calculated in the aqueous B-DNA was found to have a distribution of time scales with a slow component of $\sim$250 ps. The slow component of the solvation time-correlation function was found to originate mainly due to the interaction of the nucleotides with dipolar water molecules and counterions. An interesting negative cross correlation was found between water and counterions, which makes an important contribution to the solvation dynamics at intermediate to longer times.

6.11 SUMMARY

Owing to continuous advances during the 1980s and 1990s, techniques to study solvation dynamics in liquids have become quite advanced. As a result, many new results have been obtained that were not anticipated. For example, the presence of a dominant sub-100 fs component in the solvation dynamics of water and acetonitrile came as a surprise. Theories developed earlier for slow solvation dynamics have been found to be quite successful, by using a generalized continuum model with an accurate frequency dependent dielectric function as the input. The agreement between solvation experiments on simple liquids and different versions of molecular theories has been quite satisfactory. The situation is different for complex systems. Here quantitative theory is still not available. In the latter class of problems, computer simulations have been quite successful. Thus, one now has a fairly reasonable understanding of the general features, even though still largely qualitative.

In the subsequent chapters we shall find use of solvation dynamics in various chemical processes.

REFERENCES

1. G. R. Fleming and P. G. Wolynes, *Physics Today* (Special Issue), 43, 36 (1990).
2. D. Bingemann and N. P. Ernsting, *J. Chem. Phys.* 102, 2691 (1995).
3. M. Maroncelli, J. McInnis, and G. R. Fleming, *Science* 243, 1674 (1989).
4. B. Bagchi, *Annu. Rev. Phys. Chem.* 40, 115 (1989).
5. B. Bagchi and A. Chandra, *Adv. Chem. Phys.* 80, 1 (1991).
6. (a) M. Maroncelli and G. R. Fleming, *J. Chem. Phys.* 86, 6221 (1987); (b) *J. Chem. Phys.* 89, 5044 (1988).
7. G. R. Fleming, T. Joo, and M. Cho, *Adv. Chem. Phys.* 101, 141 (1997).
8. B. Bagchi, D. W. Oxtoby, and G. R. Fleming, *Chem. Phys.* 86, 257 (1984).
9. T. Nee and R. Zwanzig, *J. Chem. Phys.* 52, 6353 (1970).
10. (a) B. Bagchi, E. W. Castner Jr., and G. R. Fleming, *J. Mol. Struct. Theor. Chem.* 194, 171 (1989); (b) E. W. Castner Jr., G. R. Fleming, B. Bagchi, and M. Maroncelli, *J. Chem. Phys.* 89, 3519 (1988).
11. X. Song, D. Chandler, and R. A. Marcus, *J. Phys. Chem.* 100, 11 954 (1996).
12. B. Bagchi, *Chem. Rev.* 105, 3197 (2005).
13. N. Nandi, K. Bhattacharyya, and B. Bagchi, *Chem. Rev.* 100, 2013 (2000).
14. M. J. Lang, X. J. Jordanides, X. Song, and G. R. Fleming, *J. Chem. Phys.* 110, 5884 (1999).
15. R. Zwanzig, *J. Chem. Phys.* 38, 2763 (1963).
16. H.-X. Zhou and B. Bagchi, *J. Chem. Phys.* 97, 3610 (1992).
17. R. Zimenez, G. R. Fleming, P. V. Kumar, and M. Maronecelli, *Nature* 369, 471 (1994).
18. (a) G. van der Zwan and J. T. Hynes, *J. Phys. Chem.*, 89, 4181 (1985); (b) *J. Chem. Phys.* 76, 2993 (1982); (c) *J. Chem. Phys.* 78, 4174 (1983); (d) *Chem. Phys.* 90, 21 (1984); (e) R. F. Grote and J. T. Hynes, *J. Chem. Phys.* 73, 2715 (1980).
19. R. Zwanzig, *J. Chem. Phys.* 52, 3625 (1970).
20. J. B. Hubbard and L. Onsager, *J. Chem. Phys.* 67, 4850 (1977).
21. P. Colonomos and P. G. Wolynes, *J. Chem. Phys.* 71, 2644 (1979).
22. B. Bagchi and R. Biswas, *Acc. Chem. Res.* 31, 181 (1998).
23. B. J. Schwartz and P. J. Rossky, *Phys. Rev. Lett.* 72, 3282 (1994).
24. K. Yokoyama, C. Silva, D. H. Son, P. K. Walhout and P. F. Barbara, *J. Phys. Chem. A* 102, 6957 (1998).
25. S. C. Tucker, *Chem. Rev.* 99, 391 (1999).
26. B. Bagchi, *J. Chem. Phys.* 100, 6658 (1994).
27. J. Faeder and B. M. Ladanyi, *J. Phys. Chem. B* 104, 1033 (2000).
28. D. E. Moilanen, E. E. Fenn, D. Wong and M. Fayer, *J. Chem. Phys.* 131, 014074 (2009).
29. R. Biswas and B. Bagchi, *J. Chem. Phys.* 133, 084509 (2010).

7

Activated Barrier-Crossing Dynamics in Liquids

7.1 INTRODUCTION

Many chemical, physical, and biological processes involve the crossing of a high-energy state, called the activation barrier, that separates the reactant from the product states. The traditional model of the crossing of such a barrier uses a one-dimensional bistable potential where a sizeable activation barrier separates the initial state (reactant) from the final state (product). See Fig. 7.1 for an illustration of the activated barrier-crossing process.

In such a picture, there are two important steps. First, the reactant must acquire sufficient energy to reach the activated state. In liquids, thermal equilibrium is expected to exist and the probability of the reactant having this energy is given by the Boltzmann law. When activation energy is large, this probability is small. However, even after reaching the barrier, the final step of barrier crossing could be difficult because the forward speed at the barrier top can be small and the reactant can be returned to the reactant side by collision with a solvent molecule. This can happen even after the trajectory has crossed somewhat into the product side. Because the force acting on the reactant (or the product) near the barrier top is small (being a maximum in the reaction potential energy surface), several such re-crossings can occur as a result of collisions (and re-collisions) with the surrounding solvent molecules. Such crossings and re-crossings of barrier top lower the rate of the reaction. Such effects are called the dynamic solvent effects. These dynamic solvent effects are determined not only by the rate of solvent relaxation, but also by the curvature or the frequency of the reaction-energy surface at the barrier top, denoted by ω_b and shown in Fig. 7.1.

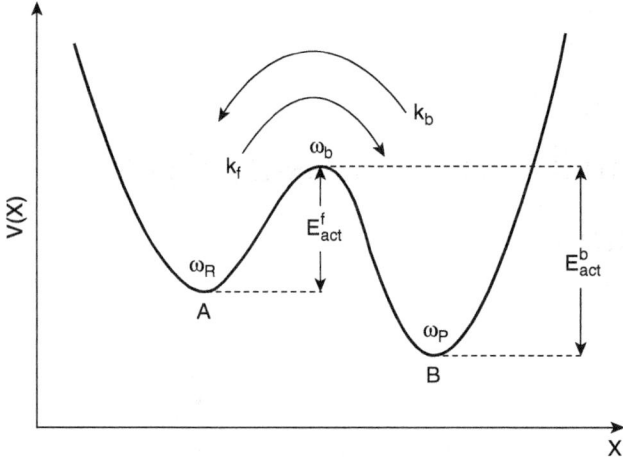

Figure 7.1 A bistable potential with two states reactant (A) and product (B). ω_R and ω_P are the frequencies of the reactant and the product well and ω_b is the frequency at the barrier. k_f and k_b are the rates for forward and backward reactions, respectively and E_{act}^f and E_{act}^b are the corresponding activation energies.

Recent studies of chemical reaction dynamics have focused on various aspects of barrier-crossing dynamics in simple and complex systems [1–4]. Aided by experiments with improved time resolution in laser spectroscopy, new theoretical techniques and theories, and computer simulations, tremendous progress has been made in the last four decades. During this period, several fundamental aspects of reaction dynamics have been addressed and much new understanding has evolved.

In Fig. 7.1 above, the reactant A goes to the product B, as denoted by

$$A \underset{k_b}{\overset{k_f}{\rightleftharpoons}} B. \tag{R1}$$

We traditionally write the rate equation as

$$\frac{d[A]}{dt} = -k_f[A] + k_b[B], \tag{7.1}$$

and a similar rate equation for product B. The above equation can then be solved easily to obtain the well-known expression for the exponential time dependence of the reactant concentration:

$$\frac{[A(t)] - [A]_{Eq}}{[A(0)] - [A]_{Eq}} = \exp(-t/\tau_{rxn}), \tag{7.2}$$

where $\tau_{rxn}^{-1} = k_f + k_b$. $[A]_{Eq}$ and $[B]_{Eq}$ are the equilibrium concentrations reached after sufficiently long time. τ_{rxn}^{-1} is the phenomenological rate constant measured in many chemical and biological processes that are modeled as activated barrier crossing as in Fig. 7.1. A useful relation among $[A]_{Eq}$, $[B]_{Eq}$, k_f, and k_b is obtained by using the detailed balance condition, which relates the two rate constants by the following expression:

$$K = \frac{k_f}{k_b} = \frac{[B]_{Eq}}{[A]_{Eq}}, \qquad (7.3)$$

where K is the equilibrium constant of the reaction, related to the free energy of the reaction by $\Delta G = -RT \ln K$.

In many chemical reactions one finds that such a simple rate law description is not totally valid, and one needs a time-dependent rate constant to describe the time dependence of $[A]$. In many bimolecular diffusion-controlled reactions, the validity of the rate law description is not obvious *a priori*. Many of these issues have been recognized and resolved in the last three decades. It is indeed remarkable that interest in this problem has remained unabated even after a hundred years of study partly because the ideas generated in this problem have found wide applications in understanding many different natural processes.

7.2 MICROSCOPIC ASPECTS

Questions that often arise in the mind of a newcomer to the area of chemical kinetics can be articulated as follows. How can we use deterministic, macroscopic chemical kinetics equations to describe reactions that are occurring at a molecular level where fluctuations and system size effects can be important? How critically do these rate laws depend on the initial conditions and on the decay dynamics of the relevant time-correlation functions? In the next two subsections we address these issues.

7.2.1 Stochastic Models: Understanding from Eigenvalue Analysis

A stochastic model relates the rate of a chemical reaction to the underlying transition probabilities between different microscopic states and is defined in the following way. Assume that each of the molecules is always in some well-defined microscopic state and these states are discrete and fall in two categories, reactant (say A) and product (say B), which are separated by a boundary in space of microscopic states as shown in Fig. 7.2. The transition from one state to another state, which one molecule continuously undergoes through collisions, is represented by instantaneous jumps from one to another state.

Mathematical formulation of the above stochastic model for the reactions shown by reaction (**R1**) can be given by the following master equation:

$$\frac{dn_i}{dt} = \sum_j \left(W_{ji} n_j - W_{ij} n_i \right), \qquad (7.4)$$

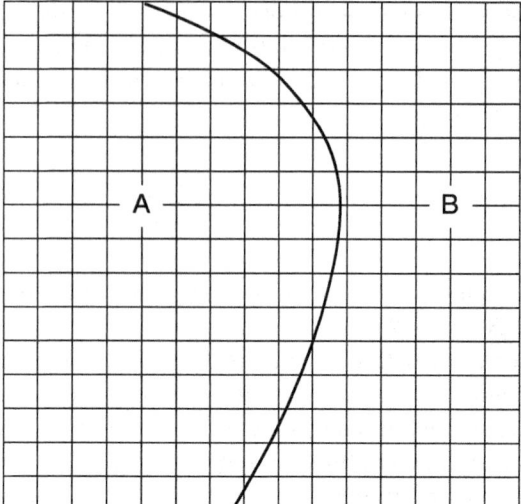

Figure 7.2 A schematic representation of the stochastic model. The grid points indicate different microscopic states and the line separates the reactant (A) and product (B) states.

where $n_i(t)$ is the number of molecules in state i and $n_j(t)$ is the number of molecules in state j at time t. W_{ij} is the transition probability per unit time from state i to state j. The sum of n_i overall reactant states is N_A and overall product states is N_B.

One of the first successful applications of a stochastic model to describe a chemical reaction (bond dissociation) was made by Montroll and Shuler [5] in an elegant analysis where the reactant states are the eigenstates of a harmonic oscillator and the bond-breaking point is equivalent to the arrival of the system at a high eigenstate. The dynamics of the reacting system is described by Eq. (7.4) with an absorbing boundary at a given eigenstate.

The solution of Eq. (7.4) is given by [6]

$$n_i(t) = n_i(\infty) + \sum_l c_{il} \exp(-\lambda_l t). \tag{7.5}$$

λ_l is the relaxation rate associated with lth mode of relaxation, which is a function of the transition probability W_{ij}.

Mathematically one can construct λ_l from W_{ij} by finding the eigenvalues of the matrix of coefficients on the right-hand side of the coupled Eq. (7.4). One of these eigenvalues is necessarily zero, and is associated with the constant term $n_i(\infty)$ in Eq. (7.5); the negatives of the remaining eigenvalues (which are themselves negative) are relaxation rates λ_l. A typical eigenvalue spectrum is shown in Fig. 7.3.

In Fig. 7.3 the relaxation rates are indexed in such a way that the smallest nonzero is called λ_1 and the next smallest is λ_2, and so on. At short times, many modes of relaxation contribute; however, when elapsed time t is very large, $(\lambda_2 - \lambda_1) t \gg 1$,

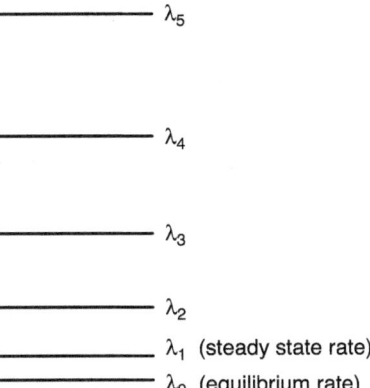

Figure 7.3 A typical eigenvalue spectrum of the transition matrix W is shown.

then every term beyond $l = 1$ in Eq. (7.5) will have negligible contributions compared to $l = 1$. So, for larger t (when the transient faster modes of relaxation are irrelevant, i.e., have negligible contribution) the relaxation rate is solely governed by the slowest mode of relaxation. After transient relaxation, the population in each state i varies with time as

$$n_i(t) = n_i(\infty) + c_{i1} \exp(-\lambda_1 t), \qquad (7.6)$$

or,

$$-\frac{dn_i}{dt} = \lambda_1 \left[n_i(t) - n_i(\infty) \right]. \qquad (7.7)$$

N_A and N_B will vary with respect to time according to

$$-\frac{dN_A}{dt} = \lambda_1 \left[N_A(t) - N_A(\infty) \right], \qquad (7.8)$$

$$-\frac{dN_B}{dt} = \lambda_1 \left[N_B(t) - N_B(\infty) \right]. \qquad (7.9)$$

Because the total number of molecules $N_A + N_B$ is constant, Eqs. (7.8) and (7.9) are equivalent to

$$-\frac{dN_A}{dt} = \frac{dN_B}{dt} = k_f N_A - k_b N_B, \qquad (7.10)$$

where k_f and k_b are constants that can be identified as the phenomenological rate constants of the forward and reverse reactions. Their sum, $k_f + k_b$, is λ_1 and their ratio, k_f/k_b, is the equilibrium constant $K = N_B(\infty)/N_A(\infty)$. After the elapse of the transient period $(1/(\lambda_2 - \lambda_1))$ it is the $1/\lambda_1$ which sets the characteristic time scale of the chemical reaction (relaxation of reactant and product states described by reaction **R1**).

7.2.2 Validity of a Rate-Law Description: Role of Macroscopic Fluctuations

One finds experimentally that the rate constant of a given chemical reaction may depend on the concentration, pressure, and temperature of the system, *but not on the volume or any extensive property of the system*. For a bimolecular reaction, this independence of the system size is not obvious *a priori*. This problem is related to the question of system size independence of a transport property and a microscopic understanding requires understanding of the effects of fluctuations on the reaction rate.

The following analysis (due to Zwanzig/van Kampen) shows why fluctuation effects can be ignored in the limit of large volume. Let us consider the following simple bimolecular reaction taking place in a container of volume V:

$$A + B \underset{k_b}{\overset{k_f}{\rightleftharpoons}} A + A. \tag{R2}$$

Let us assume the number of A molecules is m and the number of B molecules is n. Because the reaction consists of interconversion between B and A, the total number $N = m + n$ is conserved. A chemical state of the system is specified by $[m, n]$. We further assume that transitions are made only to neighboring states. For a forward reaction the transition rate is simply the forward rate constant k_f times the number of A molecules times the concentration of B,

$$w_f(m, n \to m+1, n-1) = k_f m \frac{n}{V}, \tag{7.11}$$

where the origin of the volume term in the denominator lies in the probability that an A molecule finds a B molecule, which is n/V.

Similarly for the backward reaction we can write

$$w_b(m, n \to m-1, n+1) = k_b m \frac{m}{V}. \tag{7.12}$$

Note that the right-hand side contains only m as no B molecule is involved. We are now in a position to construct a master equation of motion for the time-dependent joint probability $P_{m,n}(t)$ of states having m and n number of molecules of A and B at time t. The master equation is given by

$$\frac{dP_{m,n}}{dt} = -w_f(m, n \to m+1, n-1)P_m - w_b(m, n \to m-1, n+1)P_m$$
$$+ w_f(m-1, n+1 \to m, n)P_{m-1} + w_b(m+1, n-1 \to m, n)P_{m+1}. \tag{7.13}$$

Because the total number of molecules ($N = m + n$) is conserved, only one index is sufficient to describe the state of the system.

The preceding master equation is not very useful for analytical study. We therefore need to transform this equation into a form that is amenable to analysis, such as a Fokker–Planck equation. A systematic method to do this transformation is to employ Van Kampen's system-size-expansion technique [7] to expand the master equation (Eq. (7.13)) into a corresponding Fokker–Planck equation. Let us introduce the volume of the system as system-size-expansion parameter. The expansion parameter must be a measure of the size of the fluctuation or jump length. Now we define a new intensive variable corresponding to m (which is concentration in this case) as

$$C = \frac{m}{V}, \quad C_0 = \frac{N}{V}, \quad \rho(C, t) = V P_{m,n}(t). \tag{7.14}$$

Note that we have not defined any extra concentration variable for n as $n = N - m$.

The transition probabilities naturally scale as

$$w_f = V W_f(C) \text{ and } w_b = V W_b(C),$$

where $W_f(C) = k_f C(C_0 - C)$ and $W_b(C) = k_b C^2$ are the transition probabilities in terms of concentration variable.

The *Kramers–Moyal expansion* of the master equation in terms of the intensive variable C is

$$\frac{\partial \rho}{\partial t} = \sum_{k=1}^{\infty} \frac{(-1)^k}{k!} \frac{1}{V^{k-1}} \left[\frac{\partial}{\partial C}\right]^k a_k(C) \rho(C), \tag{7.15}$$

where $a_k(C) = W_f(C) + (-1)^k W_b(C)$ are the Kramers–Moyal moments.

Now, truncation of the above expansion beyond second order gives rise to the following Fokker–Planck equation

$$\frac{\partial \rho}{\partial t} = -\frac{\partial}{\partial C} a_1(C) \rho(C) + \frac{1}{2V} \frac{\partial^2}{\partial C^2} a_2(C) \rho(C), \tag{7.16}$$

where $a_1(C) = k_f C(C_0 - C) - k_b C^2$ and $a_2(C) = k_f C(C_0 - C) + k_b C^2$.

In the limit of infinite volume, the diffusion term disappears, and the above Fokker–Planck equation reduces to

$$\frac{\partial \rho}{\partial t} = -\frac{\partial}{\partial C} a_1(C) \rho(C). \tag{7.17}$$

The above first-order differential equation is similar to a (noise-free) Liouville equation corresponding to the deterministic equation of motion

$$\frac{dC}{dt} = a_1(C) = k_f C(C_0 - C) - k_b C^2, \tag{7.18}$$

and describes the macroscopic evolution of a system. Note that the above deterministic equation is chemical kinetics rate equation corresponding to the reaction **R2**.

In the case of finite volume V, the diffusion term cannot be neglected, and, in contrast to the earlier case (infinite-volume limit), an initially sharp distribution will broaden in time. Then there will be fluctuations about the mean concentration.

The above analysis shows the role of macroscopic fluctuations on rate-law description and clarifies the conditions under which the deterministic chemical kinetics equations with a few rates can be used to describe the chemical reactions.

7.2.3 Time-Correlation-Function Approach: Separation of Transient Behavior from Rate Law

There is an elegant time-correlation formalism based on linear-response theory. In chemical reaction dynamics this approach was initiated by Yamamoto [8] and was further developed by Chandler [9]. This approach relates the rate of decay of the population of the reactant during a chemical reaction to the regression of the spontaneous reactant population fluctuation at equilibrium.

Let us define $\delta[A(t)]$ as the departure from the equilibrium population $[A]_{Eq}$ as $\delta[A(t)] = [A(t)] - [A]_{Eq}$. Linear-response theory then states that the time-correlation function of the concentration fluctuation should decay with the same time constant as in a chemical reaction:

$$\frac{\langle \delta[A(0)]\delta[A(t)]\rangle}{\langle \delta[A(0)]^2\rangle} = \exp(-t/\tau_{rxn}). \tag{7.19}$$

If we now define a dynamical variable $N_A(t)$ whose average gives the time-dependent concentration of species A, then linear-response theory provides the following interesting relation:

$$\frac{\langle \delta N_A(0)\delta N_A(t)\rangle}{\langle (\delta N_A(0))^2\rangle} = \exp(-t/\tau_{rxn}). \tag{7.20}$$

Remember that the time-correlation function on the left side is determined by the natural (intrinsic) microscopic dynamics of the system at equilibrium. The above equation is remarkable because it relates a macroscopic relaxation time with a microscopic correlation function. Note that the above equation cannot be valid at short times because the time-correlation function must decay as a Gaussian function at very short times. The exponential decay can occur at long times as shown in Fig. 7.4. This nonexponential decay at short times is usually referred to as a "transient behavior" where the phenomenological rate-law description of chemical kinetics is not valid. In Fig. 7.4, τ_{mol} represents the time scale for transient behavior to relax.

The preceding time-correlation-function representation of the phenomenological rate has been used to develop a very practically useful method to calculate the rate with the help of computer simulations. This method starts with a definition of the dynamical variable $N_A(t)$. One usually chooses the activated state at $X = X_b$ (see Fig. 7.1) as the divider between the reactant and the product when there is only one reaction coordinate. Then $N_A(t)$ can be defined operationally as $N_A(t) = H_A[X(t)]$, where H_A is a Heaviside function such that $H_A[X] = 1$, for $X < X_b$, and zero otherwise. The above definition of $N_A(t)$ leads to the following values of the averages:

$$\langle H_A \rangle = X_A,$$

$$\langle H_A^2 \rangle = \langle H_A \rangle = X_A,$$

$$\langle (\delta H_A)^2 \rangle = \langle H_A^2 \rangle - \langle H_A \rangle^2 = X_A X_B,$$

$$\frac{\langle \delta N_A(0) \delta N_A(t) \rangle}{\langle (\delta N_A(0))^2 \rangle} = (X_A X_B)^{-1} \left[\langle H_A(0) H_A(t) \rangle - X_A^2 \right], \quad (7.21)$$

where X_A, X_B are the mole fractions of A and B species, respectively, $X_A + X_B = 1$. In order to find an expression of the rate constant, we need to take a time derivative of the time-correlation function of $\delta N_A(t)$. This should give rise to a time-dependent rate constant. Time derivative of the Heaviside function $H_A(t)$ gives rise to a velocity along the reaction coordinate multiplied by a delta function of X. The final expression of the time-dependent rate for an intermediate time scale $\tau_{rxn} \gg t \gg \tau_{mol}$ is given by

$$k_f(t) = \frac{1}{X_A} \langle v(0) \delta[X(0) - X_b] H_B[X(t)] \rangle. \quad (7.22)$$

This elegant expression was presented by Chandler and has the following meaning. The time-dependent rate is determined by finding a reactant at time $t = 0$ at the activated state $X = X_b$. The contribution of this reactant to rate is nonzero if it is found on the product side at all later time t. Thus, those reactants that cross the activated state but come back to the reactant side do not contribute to the rate. A characteristic behavior of $k_f(t)$ is shown in Fig. 7.4. An important limiting case of the above general expression is the value of the rate at $t = 0$. This rate neglects the return of the reactants at future times. This is the transition-state theory limit and is given by

$$k_f^{TST} = \frac{1}{X_A} \langle v(0) H[v(0)] \delta[X(0) - X_b] \rangle, \quad (7.23)$$

where $H[v(0)]$ is a Heaviside function selecting only the positive values of initial velocity.

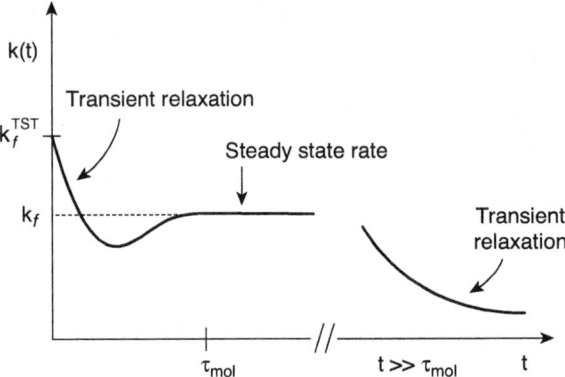

Figure 7.4 Characteristic behavior of the time dependent rate given by Eq. 7.22 with $\lambda = k_f + k_b$ is shown. The steady state plateau gives the long time rate.

If we assume that the reactant potential energy surface is harmonic with frequency ω_R, then the above averaging can easily be evaluated to obtain the following expression for the rate:

$$k_f^{TST} = \frac{\omega_R}{2\pi} \exp(-E_{act}^f/k_B T). \tag{7.24}$$

The above discussion is valid when the reaction coordinate is one dimensional. While such a one-dimensional description is sufficient for many situations, at least for semiquantitative purposes, it breaks down in many complex situations where the reaction coordinate cannot be approximated as one dimensional.

7.3 TRANSITION-STATE THEORY

If the dynamic solvent effects are neglected, then the rate of the barrier crossing is given by the well-known transition-state theory (TST). The elegant expression of the TST rate given by Eq. (7.23) can be used to derive a well-known expression of the transition-state rate first derived by Wigner and Eyring. In the preceding equation, the averages require summing over all the eigenstates of the reactant system with Boltzmann weight, along with normalization in the denominator. Thus, the denominator will consist of canonical partition function of reactant A. The average of the quantity in the bracket can be evaluated by following the method of Eyring as follows. In the transition state, we can separate the degrees of freedom into two bunches. One of them is the reactive motion along the reaction coordinate, as denoted by the Heaviside function in Eq. (7.23). The second contains all the other degrees of freedom. The second one therefore can be represented by the partition function of the transition state of the system without the reactive motion. We denote it by $Q^\ddagger$. Now, we evaluate the velocity term. If we regard this velocity term as originating from a vibration with frequency ν, then averaging over this unstable vibration gives the famous Eyring prefactor $k_B T/h$. The combination of all these factors gives rise to the following expression for the TST rate in its most popular (and oft-quoted) form:

$$k_f^{TST} = \frac{k_B T}{h} \frac{Q^{\ddagger}}{Q_A} \exp\left(-E_{act}^f/k_B T\right). \tag{7.25}$$

As already mentioned, $Q^{\ddagger}$ and Q_A are the partition functions of the activated state (without the unstable vibrational mode) and of the reactant.

However, this elegant form is often not useful in the condensed phases, as evaluation of partition functions is prohibitively difficult. Using the thermodynamic relation, $\Delta G^{\ddagger} = -k_B T \ln\left(k_{eq}^{\ddagger}\right)$, where $k_{eq}^{\ddagger}$ is the equilibrium constant between the reactant and the transition state and is given as $k_{eq}^{\ddagger} = \frac{Q^{\ddagger}}{Q_A} \exp\left(-E_{act}^f/k_B T\right)$, the above equation can be converted into a more useful form:

$$k_f^{TST} = \frac{k_B T}{h} \exp(-\Delta G^{\ddagger}/k_B T), \tag{7.26}$$

where $\Delta G^{\ddagger}$ is the free energy difference between the activated state and the reactant, which can be written as

$$\Delta G^{\ddagger} = \Delta H^{\ddagger} - T \Delta S^{\ddagger}, \tag{7.27}$$

where $\Delta H^{\ddagger}$ and $\Delta S^{\ddagger}$ are the enthalpy of activation and entropy of activation, respectively. It is a standard procedure to relate the enthalpy of activation to the experimentally observed activation energy. The role of the entropy of activation is more subtle. In many reactions, the activated state is more compact and $\Delta S^{\ddagger}$ is negative. This is often used as the reason to explain the observed decrease of the rate from the value given by $k_f^{TST} = \frac{k_B T}{h} \exp(-E_{act}^f/k_B T)$. A major advantage of the thermodynamic TST expression is that one can include effects of solute–solvent interactions through $\Delta H^{\ddagger}$ and $\Delta S^{\ddagger}$.

The validity of the TST can be limited in the condensed phases where the reactive motion is often coupled strongly to the solvent, as has already been discussed. The reduction of the reaction rate by the solvent frictional forces experienced during the barrier crossing can be discussed by using the well-known Kramers' theory that provides a simple dependence of rate on friction (ζ) and predicts that, in the limit of high friction, the rate should vary inversely with friction. If one further assumes that this friction is proportional to the zero-frequency shear viscosity (η) (via the Stokes–Einstein relation with a proper boundary condition), then one finds a very simple relation connecting the rate to the solvent viscosity.

7.4 FRICTIONAL EFFECTS ON BARRIER-CROSSING RATE IN SOLUTION: KRAMERS' THEORY

For those reactions which involve spatial movement (such as rotation or translation) of molecules or groups of atoms as the rate-determining process, the TST can overestimate the rate by even an order of magnitude. More importantly, it cannot describe the dependence of the rate on the solvent viscosity. This dependence is often easy to measure and is a popular method to understand the reaction mechanism.

The effects of the solvent frictional forces can be easily understood from the expression for the time-dependent rate constant $k_f(t)$, given by Eq. (7.22). If the trajectory of a reactant is followed with time, then one would find many re-crossings. Each of the crossings of the reactant from the reactant to the product side will be erroneously considered as a reaction event in TST. Thus, TST will overestimate the rate. This aspect can also be understood by plotting the time-dependent rate $k_f(t)$ against time, shown in Fig. 7.4.

The theory of the effects of friction on the rate of chemical reaction has drawn a great deal of attention from physicists and chemists alike. In order to study the effects of frictional forces on the rate of a chemical reaction in solution, Kramers' modeled the reactive motion as the passage of a Brownian particle over a one-dimensional potential barrier. Kramers' assumed that the motion along the reaction coordinate is given by an ordinary Langevin equation. The reaction potential energy was assumed to be piecewise harmonic:

$$V(X) = \frac{1}{2}\mu\omega_R^2 X^2, \tag{7.28}$$

and near the barrier

$$V(X) = E_{act}^f - \frac{1}{2}\mu\omega_b^2(X - X_b)^2. \tag{7.29}$$

Here ω_R and ω_b are the harmonic frequencies of the reactant well and the barrier top and μ is the effective mass. The ordinary Langevin equation for motion in the reactant well is, therefore, given by

$$\mu\frac{dv}{dt} = \mu\omega_R^2 X - \zeta v(t) + R(t). \tag{7.30}$$

The random force term $R(t)$ is related to the friction term ζ by the first fluctuation–dissipation theorem. Subsequent steps for the calculation of the rate is nontrivial because Eq. (7.30) is a stochastic equation. In the first step this equation is converted into a Fokker–Planck equation for the phase–space probability density. The Fokker–Planck equation is a second-order partial differential equation, which has two solutions. First is the equilibrium solution, given by the Boltzmann law. A second steady-state solution can be obtained which gives the rate of flow of the current from the reactant to the product side. Kramers' expression for the rate is given by

$$k_f^K = \frac{\omega_R}{2\pi\omega_b}\left[\left(\frac{\zeta^2}{4} + \omega_b^2\right)^{\frac{1}{2}} - \frac{\zeta}{2}\right]\exp\left(-E_{act}^f/k_B T\right). \tag{7.31}$$

This expression has several interesting limiting behaviors. When the value of the friction coefficient is much larger than that of the barrier frequency, that is, ζ/ω_b is much larger than unity, then rate is predicted to be inversely proportional to the friction. This limit is called the Smoluchowski limit and the Smoluchowski rate is given by the following well-known expression:

$$k_f \equiv k_f^{SL} = \frac{\omega_R \omega_b}{2\pi \zeta} \exp(-E_{act}^f/k_B T). \tag{7.32}$$

In the opposite limit of very low friction, that is ζ/ω_b is very small, Kramers' rate expression reduces to the transition-state result:

$$k_f^{TST} = \frac{\omega_R}{2\pi} \exp\left(-E_{act}^f/k_B T\right). \tag{7.33}$$

The above two limits can be understood from simple physical reasoning. If the barrier is sharp, then the particle spends too short a time on the barrier top to feel the frictional forces and we obtain the transition-state result. On the other hand, if the barrier is flat, then the motion is diffusive on the barrier top and we get inverse friction dependence of the rate.

The experimental relevance of the Smoluchowski limit comes from the proportionality between the friction and the viscosity, as predicted by the Stokes law. As remarked earlier, if the reactive motion along the reaction coordinate involves spatial movement of molecules, then the friction coefficient ζ is the translational friction that is proportional to solvent viscosity (η). Smoluchowski rate predicts a rate that decreases as $1/\eta$.

7.4.1 Low-Friction Limit

The limit of zero friction, however, is trickier. Kramers' expression and the TST expression break down in this limit, for the following interesting reason. When the viscosity becomes very small, the rate-determining step is the accumulation of energy necessary for the reactant to climb to the barrier top. For a reaction characterized by a one-dimensional reaction energy surface, the sole source of this energy is collisions with the surrounding solvent molecules. As friction becomes very small, this source of energy also decreases. The rate thus becomes proportional to friction at very small values of the friction parameter as shown in Fig. 7.5. Kramers himself obtained an approximate expression for the rate k_f in this low-friction, energy-controlled regime, which is given by

$$k_f(\zeta \to 0) \cong \frac{E_{act}^f}{k_B T} \zeta \exp\left(-E_{act}^f/k_B T\right). \tag{7.34}$$

It is clear by comparing the expressions of the rate in the limits of zero friction and large friction that the rate must go over a maximum (or turn over) as the value of the friction is increased from very small values. This turn over behavior

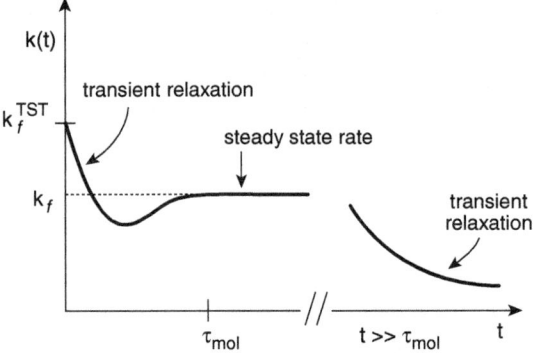

Figure 7.5 A schematic illustration of the dependence of transmission coefficient, defined as k_f/k_f^{TST}, on friction, ζ. Note the opposite dependencies at low and high values of the friction.

has been a subject of intense debate. Unfortunately, Kramers' expression in the low-frequency limit is not sufficiently accurate to give an estimate of the value of the friction at which this turn over occurs.

Starting with the pioneering work of Skinner and Wolynes [10], several attempts have been made to derive, albeit by approximation, expressions that are valid over the entire friction range. Skinner and Wolynes obtained an expansion of the rate in powers of friction coefficient (ζ), given by

$$\kappa = \frac{k_f}{k_f^{TST}} = \frac{2\pi\zeta}{\omega_b}\left[1 + \frac{\frac{2\pi\zeta}{\omega_b}}{2} + \frac{\left(\frac{2\pi\zeta}{\omega_b}\right)^2}{2\pi}\right]^{-1}. \qquad (7.35)$$

The above function shows a crossover behavior at intermediate value of friction (Fig. 7.5) and Kramers' behavior in large friction limit.

7.4.2 Limitations of Kramers' Theory

Kramers' theory provides us with a simple and elegant expression for the frictional dependence of rate but this theory has some limitations that make it difficult to apply to some realistic cases. The limitations are following:

(A) Kramers' treatment is one dimensional. In many reactions the reactive mode is coupled to other nonreactive modes. (For instance examples of nonreactive modes may be the vibrational motions in directions perpendicular to the reactive motions.) The multidimensional nature of the potential surface may have an important effect on barrier crossing, especially at low viscosity.

(B) The assumption that the solvent forces on the reactive motion are uncorrelated in time is not valid for sharp barriers where ω_b^{-1} is smaller than bath correlation time (at high viscosities). In this limit, the reactant does not remain long enough at the barrier top to probe all the solvent motions. So, the zero-frequency friction *overestimates* the solvent drag on the reactive motion, and one must consider the frequency-dependent friction.

(C) The Langevin description is applicable only to those cases where the reacting molecules are much larger than the solvent molecules.

(D) The influence of the solvent on the reactant potential energy surface is ignored.

7.4.3 Comparison of Kramers' Theory with Experiments

While the rate of a reaction can be measured by various techniques, it is sometimes hard to obtain an accurate viscosity dependence of the rate that is required to test the predictions of the theories. This is because it is not possible experimentally to vary viscosity of a given solvent without varying the reaction energy surface. For example, if we change the temperature, we not only change viscosity, but also the dielectric constant of the solvent. Similarly, a change of pressure also changes the thermodynamic properties of the solvent. In order to circumvent this difficulty, one uses a two-step method. First, one uses a series of homologous solvents to construct "isoviscosity" plots where temperature is varied at constant viscosity. For example, the viscosity of butane at a given temperature is equal to hexane at a higher temperature. These isoviscosity plots are used to extract the intrinsic activation energy of the reaction, with the assumption that across a homologous series, the activation energy does not vary greatly. One can then select a few solvents and vary viscosity by varying temperature. By this method, one can generate a large number of data at various viscosities and temperatures, and study the viscosity dependence of rate.

Extensive studies of this kind have been carried out, primarily using isomerization dynamics of fluorescent dyes, like diphenyl butadiene (DPB), DODCI, stilbene, and several others [11–14]. The results were analyzed using Kramers' theory. In order to fit the data to Kramers' theory (Eq. (7.31)), we need an estimation of the barrier frequency ω_b. Although this was used as a fitting parameter, there is a correlation between the activation energy and the barrier frequency (ω_b should increase with activation energy/barrier height), which serves as a consistency check on the analysis. The outcome of this analysis was highly interesting and to a great extent changed our understanding of barrier-crossing dynamics. We summarize the basic results below:

1. For reactions characterized by low barrier height and low barrier frequency ($\omega_b \leq 10^{12}$ s^{-1}), the observed viscosity dependence of the rate is found to obey Kramers' theory prediction quite well, especially at low viscosities. At high viscosities, small deviations from the Smoluchowski limit are observed. Examples are provided by DPB in alcohols.

2. In the opposite limit of a high barrier and a large barrier frequency (an example is DODCI in alkanes), Kramers' theory fails to describe the observed viscosity dependence, particularly at large viscosities. At small viscosities, Kramers' theory is still found to describe the trend, but the viscosity dependence at the large viscosity end is found to be much weaker than the Kramers' prediction.
3. In most cases, the observed rate can be fitted to a power-law dependence of viscosity $k = A\eta^{-\alpha}$ with the exponent α is found to have a value between zero and unity.
4. While experimental results appear to be consistent within a homologous series (such as alkanes or alcohols), they change drastically when the nature of the solvent is changed (such as alkanes to alcohols).

7.4.4 Comparison of Kramers' Theory with Computer Simulations

As mentioned above, experiments suffer from the difficulties that viscosity or friction cannot be changed without affecting the reaction energy surface. In such a situation computer simulations provide an ideal testing ground for the stochastic theories already discussed. Simulation results of isomerization dynamics by Montgomery et al. [15] revealed that the inertial effects play a significant role in producing non-TST trajectories and the theory of isomerization dynamics based on the Smoluchowski limit (where inertial variables are neglected) may overestimate the rate. Visscher [16] studied the escape from a monostable potential using Fokker–Planck equation (including inertial terms) numerically. Visscher's calculations illustrate non-monotonic behavior with good agreement with Kramer's formalism but the particular form of potential used prohibits the study of the effects of re-crossing of trajectories having positive initial velocities.

7.5 MEMORY EFFECTS IN CHEMICAL REACTIONS: GROTE–HYNES GENERALIZATION OF KRAMERS' THEORY

Kramers' theory attempts to account for the solvent frictional forces on activated processes in solution by assuming an ordinary Langevin equation for motion along the reaction coordinate where the random force is totally uncorrelated in time. Because the random force at two different times is uncorrelated, the friction is frequency independent. For some chemically activated processes in solution, there is a clear separation in the time scale between the correlation time of the interacting solvent molecules and the rate of the motion of the reactant molecule. So, the assumption of white noise holds and Kramers' theory gives a reasonable description of the viscosity dependence of the rate constant. However, there are situations where the use of frequency-independent friction can lead to erroneous results. For example, when the motion near the top of the barrier takes place on a picosecond or subpicosecond time scale, the solvent forces at two different instants can become correlated. For reactions in slow, viscous liquids, the use of Kramers' theory can lead to inaccurate results. Because these liquids have very large value of viscosity, a large fraction of viscosity comes from slow motions. Such slow motions cannot couple to barrier-crossing dynamics. Kramers' theory highly overestimates the effect of friction on rate as shown in Fig. 7.6. Experimental evidence of this rate decoupling from solvent viscosity comes from the fractional dependence of the rate on solvent viscosity.

Grote and Hynes [3] generalized Kramers' theory by removing the assumption of white noise. One very interesting result of their theory is that the rate constant depends on the friction at a frequency comparable to that of the barrier, ω_b. Because for many reactions the barrier frequency, ω_b can be quite high, this friction can be much smaller than the zero-frequency friction. Thus, Kramers' theory can vastly overestimate the effects of friction on those chemical reactions, which involve sharp, high-frequency barriers.

In order to include non-Markovian effects, Grote and Hynes assumed the following generalized Langevin equation (GLE) for the dynamics along the reaction coordinate

$$\mu \frac{dv}{dt} = F(X) - \int_0^t d\tau \zeta(\tau) v(t-\tau) + R(t), \quad (7.36)$$

where the reaction coordinate connects the two minimum points at the two isomeric forms (*cis* and *trans*) on the ground-state potential surface through the transition-state barrier between them. μ is the effective mass, v is the velocity along the reaction coordinate, and $F(X)$ is the systematic force arising from the potential in the barrier region. $\zeta(t)$ is the time-dependent friction, and $R(t)$ is the random force from solvent assumed to be Gaussian. $F(X)$ is assumed to arise from a static potential, which is an inverted parabola, so that

$$F(X) = \mu \omega_b^2 X. \quad (7.37)$$

In general, both $\zeta(t)$ and $R(t)$ appearing in Eq. (7.36) arise from microscopic motions of heat-bath (solvent) modes interacting with the reaction coordinate. For isomerization reactions in solvents, both of them arise from the microscopic motions of the solvent molecules interacting with the isomerizing moiety. Therefore, they must be related to each other. They are related by the following relation (known as fluctuation–dissipation theorem):

$$\zeta(t) = \beta \langle R(0)R(t) \rangle, \quad (7.38)$$

where β is the inverse of the Boltzmann constant (k_B) times the absolute temperature (T) and $\langle \ldots \rangle$ represents the statistical average over heat-bath modes at temperature T. By using the probability distribution from the generalized Fokker–Planck equation, Grote and Hynes obtained, after some rather lengthy analysis, the following simple and elegant expression for the rate constant, k^{GH}:

$$k_f^{GH} = k_f^{TST}(\lambda_r/\omega_b), \quad (7.39)$$

where k_f^{TST} is the transition-state rate constant given by

$$k_f^{TST} = \frac{\omega_R}{2\pi} \exp(-E_{act}^f/k_B T). \tag{7.40}$$

Here ω_R is the frequency of motion in the reactant well, and E_{act}^f is the activation energy needed for the reactant molecule to surmount the barrier. The frequency with which the reactant molecule passes by diffusive Brownian motions through the barrier region is λ_r and is given by the following self-consistent relation:

$$\lambda_r = \frac{\omega_b^2}{\lambda_r + \hat{\zeta}(\lambda_r)}, \tag{7.41}$$

where $\hat{\zeta}(\lambda_r)$ is the Laplace transform of the time-dependent friction

$$\hat{\zeta}(\lambda_r) = \int_0^\infty dt\, e^{-\lambda_r t} \zeta(t). \tag{7.42}$$

An alternative definition of λ_r obtained in terms of the velocity autocorrelation function

$$\lambda_r = \omega_b^2 \int_0^\infty dt \, \exp(-\lambda_r t) \langle v^2 \rangle^{-1} \langle v(0)v(t) \rangle. \tag{7.43}$$

The time correlation should be calculated in the absence of a barrier.

The Grote–Hynes approach is quite involved. Fortunately, Eq. (7.39) can also be derived in simpler ways by following the treatments put forward by Pollak [17] and separately by Munakata [18]. We discuss Pollak's approach below.

Pollak's approach draws insight from an earlier work of Zwanzig (see Refs. 1 and 2 in **Chapter 2**) who showed that the equation of motion of a solute particle linearly coupled to a system of harmonic oscillators can be exactly described by a generalized Langevin equation, with an analytic expression for the frequency-dependent friction in terms of the density of states of the oscillator and the coupling constants. This approach can then be used to build a description based on a Hamiltonian, which is satisfactory. At the end it leads exactly to the same expression derived by Grote and Hynes.

Instead of deriving the GLE, Pollak modeled the GLE via a harmonic bath. The total Hamiltonian can be written as

$$H = H_{sys} + H_{bath},$$

$$= \frac{p^2}{2M} + V(X) + H_{bath}(q_1, \ldots q_N, p_1, \ldots p_N; X), \tag{7.44}$$

where H_{bath} is the heat bath Hamiltonian described as

$$H_{bath} = \sum_{j=1}^{N} \left(\frac{p_j^2}{2m_j} + \frac{m_j}{2} \left[\omega_j q_j + \frac{C_j}{m_j \omega_j} X \right]^2 \right). \tag{7.45}$$

Here (p_j, q_j) are the momenta and coordinates of the jth bath oscillator whose mass and frequency are m_j, ω_j, respectively. C_j couples the bath oscillator to the system and measures the strength of the coupling of the system to the jth bath oscillator. Integrating the equations of motion for the combined Hamiltonian, $H_{sys} + H_{bath}$, over all bath variables $\{q_1, \ldots, q_N\}$, one finds the following generalized Langevin equation:

$$M \frac{d^2 X}{dt^2} = -\frac{dV}{dX} - M \int_0^t \zeta(t-\tau) v(\tau) d\tau + R(t). \tag{7.46}$$

The random force is given in terms of the initial conditions of the bath variables $\{q_j^0, p_j^0\}$ as

$$R(t) = \sum_{j=1}^{N} C_j \left[\left[q_j(0) + \frac{C_j}{m_j \omega_j^2} X(0) \right] \cos(\omega_j t) + \frac{p_j(0)}{m_j} \frac{\sin(\omega_j t)}{\omega_j} \right]. \tag{7.47}$$

Random and frictional forces are associated by the following fluctuation–dissipation theorem

$$\langle R(t) R(s) \rangle = k_B T M \zeta(t-s). \tag{7.48}$$

The time-dependent friction $\zeta(t)$ is

$$\zeta(t) = \frac{1}{M} \sum_{j=1}^{N} \frac{C_j^2}{m_j \omega_j^2} \cos(\omega_j t). \tag{7.49}$$

The time-dependent friction can be expressed in terms of the spectral density

$$\zeta(t) = \frac{2}{\pi} \int_{-\infty}^{\infty} d\omega \frac{J(\omega)}{\omega} \cos(\omega t), \tag{7.50}$$

where $J(\omega)$ is the spectral density of the bath, defined as

$$J(\omega) = \frac{\pi}{2} \sum_{j=1}^{N} \frac{C_j^2}{m_j \omega_j} \delta(\omega - \omega_j). \tag{7.51}$$

(i) Normal-mode analysis

Let us assume that the potential $V(X)$ can be approximated at the barrier $X = X_b$ as

$$V(X) \approx E_{act}^f - \frac{1}{2} M \omega_b^2 (X - X_b)^2. \qquad (7.52)$$

This harmonic approximation implies that in the vicinity of the barrier the total Hamiltonian can be written as a sum of $N + 1$ independent harmonic oscillators. This is achieved by first transforming to mass weighted coordinates

$$X' = \sqrt{M} X, \quad q'_j = \sqrt{m_j} q_j,$$

and then diagonalizing the $(N + 1) \times (N + 1)$ force constant matrix $\mathbf{K}$ defined by the second derivatives of the total potential energy surface evaluated at the saddle point

$$X' = X_b, \quad q'_j = -\frac{C_j}{\sqrt{m_j M}} \frac{1}{\omega_j^2} X_b.$$

Let $\{\lambda^2\}$ denote the eigenvalues, then the solution of the secular equation $\det(\mathbf{K} - \lambda^2 \mathbf{I}) = 0$ can be written as

$$\left[\prod_{i=1}^{N} (\omega_i^2 - \lambda^2) \right] [\omega_b^2 (\Gamma^2 - 1) - \lambda^2] = \sum_{i=1}^{N} \frac{C_i^2}{m_i M} \prod_{\substack{j \neq i \\ j=1,N}} (\omega_j^2 - \lambda^2), \qquad (7.53)$$

where $\Gamma^2 = \frac{1}{M \omega_b^2} \sum_{j=1}^{N} \frac{C_j^2}{m_j \omega_j^2}$.

The reactive frequency along the barrier will correspond to the negative eigenvalue of the secular equation, given in terms of the Laplace transform of the time dependent friction (Eq. (7.49)) as [17]

$$\lambda_r = \frac{\omega_b^2}{\lambda_r + (1/M) \hat{\zeta}(\lambda_r)}. \qquad (7.54)$$

(ii) The rate of escape

The TST rate is given as the ratio of the partition function of all bath modes at the barrier to the partition function of the total system in the reactant well:

$$k_f = \left[\prod_{i=1}^{N} \left(\frac{\omega_i}{\lambda_i} \right) \right] \frac{\omega_R}{2\pi} \exp\left(-\beta E_{act}^f\right), \qquad (7.55)$$

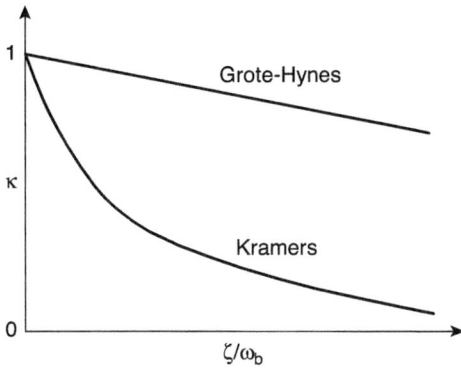

Figure 7.6 A comparative picture of the expected (and also calculated) friction dependence of the transmission coefficients obtained from the Kramers' and Grote–Hynes' approaches.

where $\lambda_i^2 (i = 1, N)$ are positive eigenvalues of the force constant matrix **K**. Also one can easily show from Eq. (7.53) that [17]

$$\prod_{i=1}^{N} \left(\frac{\omega_i}{\lambda_i} \right) = \frac{\lambda_r}{\omega_b}. \qquad (7.56)$$

Equation (7.55) can be written as

$$k_f = \frac{\lambda_r}{\omega_b} \frac{\omega_R}{2\pi} \exp\left(-\beta E_{act}^f\right). \qquad (7.57)$$

Equation (7.57) predicts the transition state result for very weak friction ($\lambda_r \sim \omega_b$), and the Kramers' result for low barrier frequency (i.e., $\omega_b \to 0$) so that $\hat{\zeta}(\lambda_r)$ can be replaced by $\hat{\zeta}(0)$. If the barrier frequency is large ($\omega_b \geq 10^{13} s^{-1}$) and the friction is not negligible ($\hat{\zeta}(0)/\mu \sim \omega_b$), then the situation is not so straightforward. In this regime, which often turns out to be the relevant one experimentally, the effective friction, $\hat{\zeta}(\lambda_r)$, can be quite small even if the zero frequency (i.e., the macroscopic) friction (proportional to viscosity) is very large. The non-Markovian effects can play a very important role in this intermediate regime.

One can also write the Grote–Hynes relation (Eq. (7.41)) as

$$\lambda_r^2 + \lambda_r \cdot \hat{\zeta}(\lambda_r) = \omega_b^2. \qquad (7.58)$$

The largest positive root of the above quadratic equation is

$$\lambda_r = -\frac{\hat{\zeta}(\lambda_r)}{2} + \left(\frac{\hat{\zeta}^2(\lambda_r)}{4} + \omega_b^2\right)^{1/2}. \tag{7.59}$$

From Eqs. (7.39), (7.40), and (7.59) we can write

$$k_f^{GH} = \left(\frac{\left(\frac{\hat{\zeta}^2(\lambda_r)}{4} + \omega_b^2\right)^{1/2} - \frac{\hat{\zeta}(\lambda_r)}{2}}{\omega_b}\right) \frac{\omega_R}{2\pi} \exp\left(-\beta E_{act}^f\right). \tag{7.60}$$

This is a landmark result in rate theory. Kramers' and transition-state rate constants are special cases of Eq. (7.60).

In order to apply the Grote–Hynes formula to realistic cases, we need a reliable expression for the frequency-dependent friction in terms of known quantities, especially as a function of viscosity.

7.5.1 Frequency Dependence of Friction: General Aspects

Many reactions in solution, for example, *cis* → *trans* isomerization in stilbene, involve large-amplitude motion of a bulky group (isomerizing moiety) twisting around a molecular axis. Therefore, this twisting motion will certainly experience the effect of frictional forces exerted by the surrounding solvent molecules. The friction experienced by the moving moiety consists of two parts: one contribution comes from the translational drag by the environment, which we shall call the translational friction, ζ, while the second could be a pure rotational friction. Generally the rotational friction in most cases is smaller than its translational counterpart unless the bulky twisting group is either very large or carries a charge. Therefore, the translational friction is the principal quantity that regulates the diffusive Brownian motion of the tagged moiety near the barrier region. Two approaches have been developed to estimate the frequency dependence of friction to be used in Grote–Hynes formula. These are discussed below.

7.5.1.1 FREQUENCY-DEPENDENT FRICTION FROM HYDRODYNAMICS

A hydrodynamic expression for the frequency-dependent friction can be obtained by solving the generalized Navier–Stokes equation. This calculation was carried out by Zwanzig and Bixon [19(a)]. The resulting expression involves, among other things the frequency-dependent shear and bulk viscosity. The final expression (with the correction of Metiu et al. [19(b)]) is given by

$$\hat{\zeta}(z) = \left(\frac{4\pi}{3}\right) \eta_s(z) R X^2 \left[2(X+1)P + (1+Y)Q\right], \tag{7.61}$$

where

$$X = (z\rho_0/\eta_s)^{\frac{1}{2}} R, \tag{7.62}$$

$$Y = z\left[c^2 + \frac{z\eta_l}{\rho_0}\right]^{-\frac{1}{2}} R, \tag{7.63}$$

$$P = \frac{3}{\Delta}(3 + 3Y + Y^2), \tag{7.64}$$

$$Q = \frac{3}{\Delta}\left[3 + 3X + X^2 + \frac{X^2(1+X)}{2 + \beta/\eta_s}\right], \tag{7.65}$$

$$\Delta = 2X^2[3 + 3Y + Y^2] + Y^2[3 + 3X + X^2] + \frac{3X^2(1+X)(2 + 2Y + Y^2)}{2 + \beta/\eta_s}. \tag{7.66}$$

η_s is the frequency-dependent shear viscosity, R is the radius of sphere, ρ_0 is the solvent density, c is the velocity of sound, and β is the slip parameter, which is zero for pure slip and infinity for stick boundary conditions. The longitudinal viscosity $\eta_l(z)$ is related to shear viscosity $\eta_s(z)$ and bulk viscosity $\eta_v(z)$ by

$$\eta_l(z) = \frac{4}{3}\eta_s(z) + \eta_v(z). \tag{7.67}$$

Expressions for the frequency-dependent viscosities $\eta_s(z)$ and $\eta_v(z)$ are assumed to be simple Maxwell forms

$$\eta_s(z) = \frac{\eta_s^0}{1 + z\tau_s} \text{ and } \eta_v(z) = \frac{\eta_v^0}{1 + z\tau_v},$$

where η_s^0 and η_v^0 are the zero-frequency shear and bulk viscosities. The time constants τ_s and τ_v are the relaxation times for shear and bulk modes of solvent.

A schematic plot for the frequency dependence of the hydrodynamic friction is shown in Fig. 7.7 for various values of shear viscosity. The frequency dependence becomes stronger as the shear viscosity increases. This implies that, if we use zero-frequency viscosity, then we vastly overestimate its effect, the effective friction at high viscosities is much smaller than its zero-frequency value. This is precisely the reason why the decrease of the rate slows down at high viscosities. Physically, this means that many of the low-frequency motions that contribute to $\zeta(z = 0)$ do not affect the reactive motion across the barrier if the barrier frequency is sufficiently high. In the long-chain alkanes, examples of low-frequency motions are those rotations around the backbone that involve cooperative motions of several backbone atoms. These kinds of motion will not respond if the liquid is driven at high frequency.

The above discussion of frequency-dependent friction based on the generalized hydrodynamic model becomes questionable when the barrier frequency is

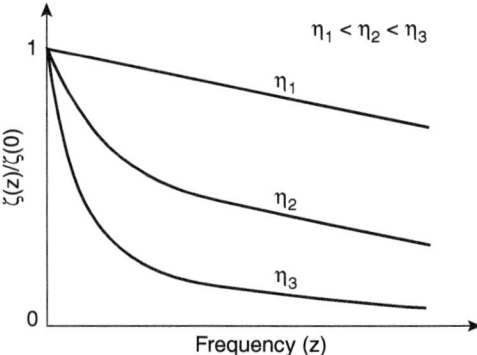

Figure 7.7 Values of $\zeta(z)/\zeta(0)$ are plotted against frequency z for various values of the zero-frequency shear viscosity. Note the sharper fall of the scaled friction at larger viscosity.

very high. The friction ζ on the tagged moiety in slow liquids is primarily determined by the molecular-length-scale processes and the rate is, therefore, determined primarily by the *local* dynamics. The mode-coupling theory includes these aspects rigorously.

7.5.1.2 Frequency-Dependent Friction from Mode-Coupling Theory

A generalized hydrodynamic description of friction becomes inadequate at short times or large frequencies. In chemical reactions, the barrier frequency is often in the excess of 10^{12} s^{-1}, or so. At such large frequencies, even the viscoelastic response will be dominated by such processes as collision and inertia. Generalized hydrodynamics is grossly inadequate when such processes become dominant. However, we also need a self-consistent description at longer times because the viscosity of the solution is often determined by slow processes. Therefore, we have the difficult task of having a unified description valid at both short and long times. Fortunately, mode-coupling theory provides a reliable description in both the limits.

According to this theory, the frequency-dependent friction, $\zeta(z)$, is given by the following simple expression [20]:

$$\frac{1}{\hat{\zeta}(z)} = \frac{1}{\hat{\zeta}_B(z) + \hat{\zeta}_{\rho\rho}(z)} + \hat{R}_{tt}(z), \qquad (7.68)$$

where $\hat{\zeta}_B(z)$ is the bare (collisional) part of the total friction, while $\hat{\zeta}_{\rho\rho}(z)$ is that component of the total friction that comes from the coupling of the reactive motion to the solvent density (ρ) fluctuation. $\hat{R}_{tt}(z)$ arises from the coupling to the transverse current mode of the solvent. Note here that if one ignores both ζ_B and $\zeta_{\rho\rho}$ and retains only R_{tt}, then one recovers a generalized hydrodynamic-type description for the friction experienced by a tagged molecule. However, in dense, slow liquids, the contributions from ζ_B and $\zeta_{\rho\rho}$ are far more important than R_{tt}, especially when the response is studied at high frequencies. The bare or the binary or the short-time friction, $\hat{\zeta}_B(z)$ is determined by equilibrium quantities. For example, for a hard-sphere fluid, this can be approximated by the well-known Enskog friction (ζ_E), which is given by

$$\zeta_E = \frac{8}{3m}(\pi m k_B T)^{1/2}\rho\sigma^2 g(\sigma), \qquad (7.69)$$

where $g(\rho)$ is the value of radial distribution function at contact. In writing the above equation, both the solvent and the moving parts of the reactant molecules are assumed as hard spheres of same diameter (σ) and mass (m). It turns out that the Enskog expression is not very reliable, for the following reason. Reactions with high barrier frequency ω_b probe dynamics which are in the collision-frequency range in liquids. Thus, these reactions probe the details of the repulsive part of the potential and the hard-sphere model gives too high a value at high frequency. Fortunately, it is easy to overcome this problem because one can derive accurate short-time expansions for the time-dependent friction. The simplest is to assume a Gaussian form for $\zeta(t)$:

$$\zeta(t) = \Omega_0^2 \exp\left(-(t/\tau_B)^2\right), \qquad (7.70)$$

where τ_B is the time constant associated with the above Gaussian relaxation and Ω_0 is the Einstein frequency of the liquid. The value of the friction at zero time is given by the following expression:

$$\zeta(t=0) = \Omega_0^2 = \frac{4\pi\rho}{3m}\int dr\, r^2 g_{12}(r) \nabla_r^2 v_{12}(r), \qquad (7.71)$$

where $g_{12}(r)$ is the solute–solvent radial distribution function, $v_{12}(r)$ is the interaction potential, ρ is the density of the solvent and m is the mass of the solute particle.

The collective contribution, $\hat{\zeta}_{\rho\rho}(z)$, is obtained from the mode-coupling theory and is given by the following simple and elegant expression:

$$\hat{\zeta}_{\rho\rho}(z) = \frac{\rho k_B T}{6\pi^2 m}\int_0^\infty dt\, \exp(-zt) \int_0^\infty dq\, q^4 F_s(q,t)[c(q)]^2 S(q,t), \qquad (7.72)$$

where $c(q)$ is the direct correlation function as a function of q. This function provides the coupling between the reactant and the solvent molecules. $F_s(q,t)$ is called the self-dynamic structure factor and it describes the self-motion of the solute (here reactant). In a homogeneous liquid, it is measured by incoherent neutron scattering. $S(q,t)$ describes the liquid (here the solvent)

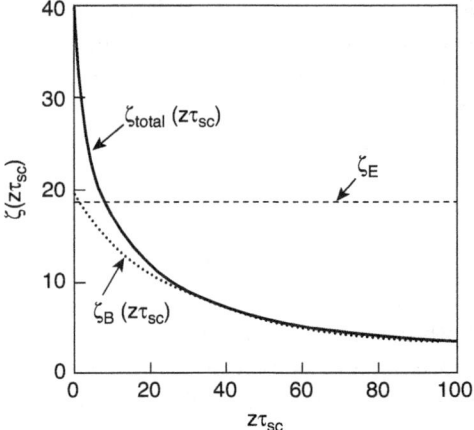

Figure 7.8 The frequency-dependent total friction, $\hat{\zeta}(z\tau_{sc})$ (solid line), and the binary friction, $\hat{\zeta}_B(z\tau_{sc})$ (dotted line), are plotted as a function of the scales Laplace frequency, $z\tau_{sc}$. For comparison, the Enskog friction, $\hat{\zeta}_E$, is also shown (large dashed line). Reprinted with permission from *J. Chem. Phys.* **110**, 7365 (1999). Copyright (1999) American Institute of Physics.

dynamic structure factor. Away from the glass-transition regime, these two can be approximated by simple expressions of the following forms:

$$F_s(q, t) = \exp[-Dq^2 t], \, S(q, t) = S(q) \exp\left[-Dq^2 t/S(q)\right], \quad (7.73)$$

where D is the diffusion coefficient of a solvent molecule related to the translational friction, ζ, by the relation, $D = k_B T/\mu\zeta$ and $S(q)$ is the static structure factor.

In Fig. 7.8 the total frequency-dependent friction, $\hat{\zeta}(z)$, is plotted against the Laplace frequency (z). In the same figure Enskog friction, ζ_E, and the binary contribution, $\zeta_B(z)$ are also shown.

Figure 7.8 makes it clear that neither the hard sphere model nor the hydrodynamics provide an accurate description of the frequency-dependent friction. The details of the interaction potential between the reactant and the solvent do matter in determining the rate of barrier-crossing dynamics.

7.5.2 Comparison of Grote–Hynes Theory with Experiments and Computer Simulations

The solution of Grote–Hynes formula (Eq. (7.39)) based on the hydrodynamic expression of frequency-dependent friction reproduced the qualitative features of viscosity dependence observed in experiments of Fleming and co-workers [13, 14]. Rothenberger et al. [12] fitted their experimental results on photoisomerization of *trans*-stilbene in liquid *n*-alkanes with Kramers' and Grote–Hynes' theories. They obtained better agreement with the latter theory using the values $\omega_b = 1.5 \times 10^{12} s^{-1}$ and $\omega_R = 9.6 \times 10^{13} s^{-1}$, where ω_R is the harmonic potential of reactant well. The value of ω_b is very low, which is hard to imagine when the barrier height is sufficiently large, as in DPB or in DODCI. The Grote–Hynes theory predicts an unphysical potential surface. The reason for the unphysical value of ω_b given by the theory is still not well understood. Among many factors that may be responsible for the failure of theoretical calculations in predicting reasonable value for ω_b the following three seem most likely:

(A) The one-dimensional picture of stochastic theories may not be applicable to the isomerization reactions.
(B) The hydrodynamic expression for the frequency dependent friction is not reliable.
(C) There is a hidden relationship between the intrinsic barrier height and the barrier frequency due to geometric constraint on isomerization processes and this is neglected in the fitting of Grote–Hynes theory to experimental data.

Experimental confirmation of the Grote–Hynes theory is, however, difficult because there are many competing contributions to rate which may mask the subtle effects predicted by the theory. Computer simulation studies can overcome some of these difficulties. In the computer simulation study of isomerization dynamics of a model reaction, Statman and Robinson [21] found that the friction along the reaction coordinate can be a factor of 2 less than that for the bulk solvent. Another important result of the simulation was that entropy of activation makes a nontrivial contribution to the rate constant. In this case the entropic contribution seems to arise from a change of hydrodynamic volume upon isomerization.

7.6 VARIATIONAL TRANSITION-STATE THEORY

Transition-state theory (TST) is the most widely used tool for calculating rate constants of chemical reactions occurring in the gas and condensed phases. As discussed above, one of the fundamental assumptions of TST is that the reactive trajectories originating from the reactant well cross the dividing surface only once (forward flux over dividing surface) and proceed to the product. Thus, the TST rate will be exact for the systems in which every reactive trajectory crosses the transition state only once. However, in high-dimensional or high-friction systems, the trajectories will re-cross the dividing surface many times. In these cases the TST provides an upper bound to the exact reaction rate. This is the basis of classical variational TST in which the *position of the dividing surface is (variationally) optimized to minimize the reaction rate* [22].

The transition states for microcanonical or canonical ensembles correspond to a minimum sum of states or a maximum free energy of activation, respectively. In the canonical variational transition-state theory (CVTST), the rate constant is obtained by minimizing the canonical TST generalized rate k^{CTST} constant $\left(k^{CTST}\right)$ along the reaction path at temperature T,

$$k^{CVTST}(T) = \min_{X}\left\{k^{CTST}(T,X)\right\} = k^{CTST}\left(T,X^{\neq}\right), \quad (7.74)$$

where X denotes the distance along the reaction coordinate and $X^{\neq}$ is the value of X that minimizes the rate constant along the reaction path at temperature T.

In the microcanonical variational transition-state theory (MVTST) the dividing surface is varied in order to minimize the rate calculated for a fixed energy (microcanonical TST rate, k^{MTST}) and is given by

$$k^{MVTST}(E) = \min_{X}\left\{k^{MTST}(E,X)\right\} = k^{MTST}\left(E,X^{\neq}\right). \quad (7.75)$$

The canonical rates can be calculated by integrating the rate expressions obtained in a microcanonical treatment over the energy, taking into account the statistical distribution over energy states (Boltzmann distribution).

7.7 MULTIDIMENSIONAL REACTION SURFACE

In many chemical and biological processes, the reaction free- (or potential-) energy surface is multidimensional. A well-known example is provided by protein folding, which, in an idealized picture, is the collapse and reorganization of an extended (fully denatured) amino acid chain. This process should involve at least two reaction coordinates—the radius of gyration for size and the fraction of native contacts. Descriptions of reactions involving more than one reaction coordinate are considerably more complex. One major difficulty is finding the path along which the reaction rate is maximum. In one dimension obviously this difficulty is not present. A contour diagram of two-dimensional PES is shown in Fig. 7.9.

One natural extension of one-dimensional transition-state theory is the multidimensional transition-state theory. In higher dimensions the saddle point is characterized by $N-1$ real modes of vibration. According to this theory the rate constant may be expressed as

$$k_f^{TST} = \frac{\prod_{i=1}^{N} \nu_i}{\prod_{i=1}^{N-1} \nu_i^b} \exp\left(-\beta E_{act}^f\right). \quad (7.76)$$

Here ν_i and ν_i^b are the vibrational frequencies in the initial well and at the saddle point, respectively. N is the total number of vibrational degrees of freedom. But a problem with the above equation is that it is a straightforward generalization of the equilibrium theory. As we know, the equilibrium theory highly overestimates the rate, so one must go beyond the equilibrium theory in higher dimensions also.

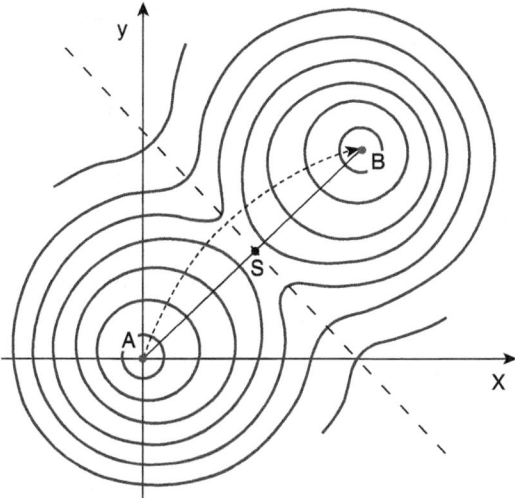

Figure 7.9 Equipotential contours in two dimensions are shown. Points A and B are the potential minima for the reactant and the product, respectively, and S is the saddle point. The dashed line shows a possible reactive pathway. In general there could be several paths all crossing near the saddle from the reactant to the product.

7.7.1 Multidimensional Kramers' Theory

This approach was developed by Landauer and Swanson [23], who extended the Kramers' theory to the many-dimensional case.

In the overdamped limit, the TST overestimates the rate, because the motion along the reaction coordinate is highly diffusive and the re-crossings are more prominent. In the case where the two potential-energy minima are located symmetrically about a plane through the saddle point, Landauer and Swanson obtained the following expression for the rate constant in the overdamped limit (details of the calculation are given in the **Appendix**):

$$k_f = 2 \left(\frac{\omega_s}{\eta} \right) k_f^{TST}, \tag{7.77}$$

where ω_s is the imaginary frequency at the saddle point, η is the viscosity, and k_f^{TST} is given by Eq. (7.76). Note that the reduction factor $\left(\frac{\omega_s}{\eta} \right)$ is the same as the one Kramers' found for the one-dimensional case.

In underdamped limit the TST rate also overestimates the rate. In this limit the particle that has crossed the rate may bounce back because the dissipation of excess kinetic energy is slow. In the underdamped limit, Landauer and Swanson have argued that a very extreme underdamping is required to produce a deviation from the transition-state theory. The reason is that the time available for the dissipation of excess kinetic energy is more in the multidimensional case than the one-dimensional case. In the one-dimensional case in the absence of damping the escape time is the oscillation time in the well. However, in the multidimensional case a particle must bounce back in a very particular way to re-cross the saddle point.

The major drawback of this theoretical approach is that it is based on the idea of two well-defined potential wells and the separation of time scale between the intrawell motions and interwell motions. However, many complex systems, such as proteins, evolve through multitudes of minima and maxima. In these systems the two-well concept and the separation of time scales are not valid. A more elaborated discussion of these complex systems is given in Chapter 9.

7.8 TRANSITION PATH SAMPLING

In many rare events/reactions, consisting of changes across many degrees of freedom, it is often a very difficult task to even find the most probable path of the event, let alone the free-energy barrier. In many of such cases, the difficulty arises from the fact that there is actually no unique path, and one needs to consider contributions from a large number of paths to compute the rate. A typical example is protein folding, where the folding funnel recently introduced envisages many paths from the unfolded to the folded states (see Chapter 9). Here both the unfolded state and the partly folded state have many minima, while the folded state is often unique. Finding the probable paths with significant weight to the transformation rate is nontrivial. In Fig. 7.10 we show a complex potential surface. In this case there is no unique path, and one needs to consider many paths and sum their contributions to obtain the rate.

Transition-path sampling does not require *a priori* knowledge of the reaction coordinate (or reaction mechanism). This significant advantage reflects its importance for the study of complex multidimensional systems. In the transition-path sampling, one finds the probability of a reactive path by piecing together small segments, which are propagated from the reactant basin (A) to the product basin (B). The propagation can be carried out either by Monte Carlo or by molecular dynamics. One can derive a neat expression for the reactive path probability by using the same set of treatments we used earlier to derive the time-correlation-function formalism. This expression is given by

$$P_{AB}(x(\tau)) = h_A(x_0) P(x(\tau)) h_B(x_\tau) / Z_{AB}(\tau), \tag{7.78}$$

where $x(\tau)$ denotes a trajectory connecting A and B regions, $h_A(x)$ equals unity if the argument x lies within basin of A and zero otherwise. A similar definition holds for $h_B(x)$. $P(x(\tau))$ is the statistical weight of a particular trajectory $x(\tau)$ and for a Markovian process $P(x(\tau))$ is given by

$$P(x(\tau)) = \rho(x_0) \prod_{i=0}^{\tau/\Delta t - 1} p(x_{i\Delta t} \to x_{(i+1)\Delta t}), \tag{7.79}$$

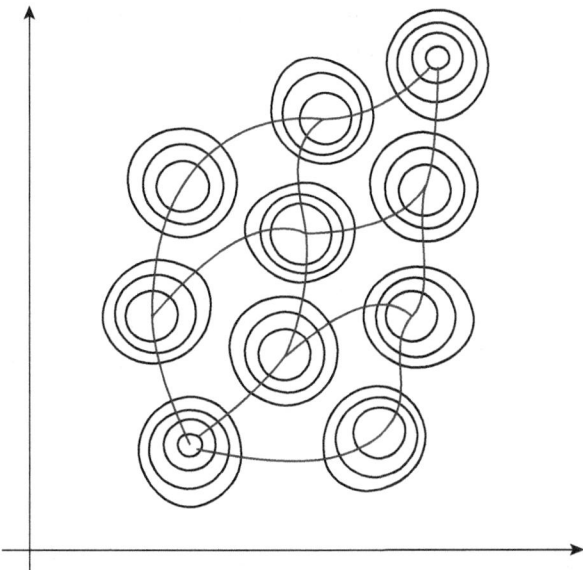

Figure 7.10 Schematic illustration of different pathways connecting different potential energy minima, shown by contours, in a two-dimensional potential energy surface.

where $\rho(x_0)$ denotes the distribution of initial starting points. $p\left(x_t \to x_{(t+\Delta t)}\right)$ is the transition probability from state x_t to $x_{(t+\Delta t)}$ in time interval Δt. $Z_{AB}(\tau)$ is a normalizing constant given by

$$Z_{AB}(\tau) = \int Dx(\tau) h_A(x_0) P(x(\tau)) h_B(x_\tau). \qquad (7.80)$$

Here $Dx(\tau)$ indicates the integration over all paths, $x(\tau)$. Therefore, $P_{AB}(x(\tau))$ gives the reactive or restricted path that takes one from the reactive basin to the product basin. The set of reactive paths is called the transition-path ensemble. The selection of the initial starting path is an important aspect of this technique. This first path can be an artificial one, similar to the selection of a solid as the starting configuration of a liquid in a Monte Carlo or molecular dynamics simulation. However, once a path is constructed one can bias this path toward more efficient reactive paths. One not only needs to find the ensembles of reactive paths but also the weight of each path contributing to the reaction. One needs to obey the principle of detailed balance for the transitions along the path.

In transition-path sampling one selects the reactive trajectories from a set of trajectories generated by the intrinsic dynamics of the system (in the absence of any external bias), so these trajectories are true dynamical trajectories and can be used to calculate reaction rates.

At equilibrium, the transition between stable states A and B are random and rare and the rate of transition is related to the time correlation of populations of stable states A and B:

$$C(t) = \frac{\langle h_A(x_0) h_B(x_t) \rangle}{\langle h_A(x_0) \rangle}, \qquad (7.81)$$

where $\langle .... \rangle$ is the equilibrium ensemble average. As already discussed, according to linear-response theory the dynamics of equilibrium fluctuations of the population is related to the relaxation from a nonequilibrium state. The time-dependent rate is

$$k(t) = \frac{dC(t)}{dt}. \qquad (7.82)$$

In the context of transition-path sampling, $C(t)$ in the above equation can be written in terms of integration over trajectories as

$$C(t) = \frac{\int Dx(t) h_A(x_0) P(x(t)) h_B(x_t)}{\int Dx(t) h_A(x_0) P(x(t))}, \qquad (7.83)$$

where the numerator is the partition function for the ensemble of trajectories that originates from region A and terminates in region B at time t. The denominator is the partition function for trajectories that originate from region A; however, there is no restriction on the end point. A typical behavior of $k(t)$ is shown in Fig. 7.4, and the plateau value gives the forward reaction rate, $k_{A \to B}$.

There is a very nice and exhaustive review written by Dellago, Bolhuis, and Geissler [24], which provides insight into the technique and application into this technique. We refer the reader to this review.

7.9 QUANTUM TRANSITION-STATE THEORY

The relatively simple rate expressions derived above, either for the transition-state theory or the Kramers' theory is valid only in the classical limit when the temperature is high and/or the mass of the reactant is large. Thus, the above expressions are applicable for a *cis–trans* isomerization in liquid at 300 K where a bulky phenyl group moves through the liquid from one position to the other. However, in many important chemical reactions such as exchange of hydrogen from one position of a molecule to another, or in the study of isotope effects on chemical reactions, the quantum nature of the reaction is of paramount importance and the classical theories are invalid. An important new phenomenon that enters into the process is the tunneling. Many proton-transfer reactions proceed primarily through tunneling. There is now an elaborate theory to study such quantum effects.

If the tunneling effects are not important, one can use the TST expression of rate in terms of the quantum partition function

$$k_f^{TST} = \frac{k_B T}{h} \frac{Q^{\ddagger}}{Q_A Q_B} exp\left(-E_{act}^f/k_B T\right), \quad (7.84)$$

where Q_A and Q_B are the quantum partition functions for the reactant and $Q^{\ddagger}$ is the quantum partition function of the transition state. An ad hoc modification of the above expression is to replace the activation energy E_{act}^f by $E_{act}^f - \hbar\omega_R/2$, where the subtraction takes into account the reduction in the activation energy due to the zero point vibrational energy to take into account reactant well vibration with frequency ω. Because the reactant well frequency depends on the mass by the standard relation ($\omega_R = \sqrt{k/m}$), where k is the force constant of vibration and m is the mass of the reactant, this reduction depends on mass m.

The above simple procedure can provide an explanation of some of the kinetic isotope effects, such as the decrease of rate when deuterium (D) is exchanged in place of hydrogen (H). The ratio of the rates, k_H/k_D, can often be quite large, signaling presence of such (trivial) quantum effects.

7.10 SUMMARY

As in many areas of molecular relaxation, this field has also been driven by many novel experimental results obtained by ultrafast laser spectroscopy. The new results that became available from the early 1980s motivated the use of new theories and computer simulations. Availability of population dynamics at short times made us look into the time-dependent (or, transient) behavior of rate description.

In this chapter we have discussed the conditions under which the rate of a reaction can be defined unambiguously. For a rate description to be useful, a chemical reaction must achieve a steady state early. We have described an elegant eigenvalue analysis due to Widom which can in principle quantify the range of validity of the rate equation description. Time-correlation-function formalism has been used to derive a useful description of the rate constant.

A large amount of experimental and theoretical effort has been directed at understanding the viscosity dependence of an activated reaction. The initial studies by Smoluchowski and Kramers have been extended by Grote and Hynes to include the effects of time-dependent response of the solvent. Since viscosity of a liquid often derives contributions from slow modes, like slow structural relaxation involving many particles, a fast chemical reaction might not couple to such modes and in turn is not affected by these modes. Thus, a fast chemical reaction couples only to the short-time response of the liquid. Grote–Hynes theory includes such sophisticated effects through a description which is termed non-Markovian. Such non-Markovian (or, memory) effects need the use of a generalized hydrodynamic description at the simplest level, and a mode-coupling theory-based approach of friction at an advanced level. Both have been discussed here.

Another important issue in the activated chemical reaction is the dimensionality of the reaction. In many chemical reactions, the reaction path involves more than one dimension. For such complex reactions, the expression of the rate also becomes complex, as described above, and also in the **Appendix**.

APPENDIX

Multidimensional Kramers' Theory

This approach was developed by Landauer and Swanson [23], and is an extension of Kramers' theory to the many-dimensional case. Here we will discuss the case when the two potential minima have the same value and are symmetrical about a hyperplane $z = 0$. This case can be easily generalized for the unsymmetrical potential well.

The probability density, ρ, in the N-dimensional phase space can be decomposed as the time-independent thermal equilibrium part (ρ_t) and the deviation from the equilibrium (ρ_d)

$$\rho = \rho_t + \rho_d. \quad (A.1)$$

ρ_d decays with time as

$$\rho_d = \sum_i \rho_{di} \exp(-t/\tau_i). \quad (A.2)$$

Here it is assumed that there is a clear separation of time scale between the motion within the potential well and between the two wells, that is, the barrier is sufficiently large to allow the separation of time scale between the two wells. Our primary interest here is to find the population difference between the two wells and the time constant, τ_i, associated with the relaxation of population difference.

The population difference (Δ) between the two wells is given by

$$\Delta = \int_A \rho_d d\Gamma - \int_B \rho_d d\Gamma = \int_A \rho_d d\Gamma - \left(\int_{\text{all space}} \rho_d d\Gamma - \int_A \rho_d d\Gamma \right) = 2 \int_A \rho_d d\Gamma. \quad (A.3)$$

(Note that $\int_{\text{all space}} \rho_d d\Gamma = 0$.)

The number of ensembles crossing the saddle point per unit time (**J**) is related to Δ and can be given as

$$2\mathbf{J} = -\frac{d\Delta}{dt} = \frac{\Delta}{\tau}, \quad (A.4)$$

where τ is the time constant associated with the population difference. Our next aim is to find Δ and **J** to obtain the time constant τ in overdamped and underdamped cases.

(a) Overdamped Case

As we know in the overdamped case the equation of motion of probability density of the ensemble under the influence of external potential V obeys the following Smoluchowski equation:

$$\frac{\partial \rho}{\partial t} = -\nabla \cdot \mathbf{J}, \tag{A.5}$$

where

$$\mathbf{J} = -D\left(\nabla \rho + \beta \nabla V \rho\right). \tag{A.6}$$

For the sake of simplicity we assume that the diffusion constant D is independent of position. Near the saddle point the potential V is approximated as $V = V_s - \frac{1}{2}\gamma z^2 + \frac{1}{2}\sum_{i=1}^{N-1} \gamma_i x_i^2$, and near the reactant well A as $V = V_A + \frac{1}{2}\sum_{i=1}^{N} \delta_i x_i^2$.

At equilibrium ρ will satisfy the Boltzmann distribution, $\rho = c \exp(-\beta V)$, where c is the normalization constant. In the nonequilibrium case one can write

$$\rho = \alpha \exp(-\beta V), \tag{A.7}$$

where variation of α indicates the extent of deviation from equilibrium. Now we can write Eq. (A.6) as

$$\mathbf{J} = -D(\nabla \alpha) \exp(-\beta V). \tag{A.8}$$

From Eq. (A.8) it is clear that a large deviation from equilibrium will occur only at a saddle point because $\exp(-\beta V)$ is smallest at the saddle point, because we are only interested in the case where the two potential minima have the same value and the potential is symmetrical about a plane through the saddle point. Let z be the coordinate perpendicular to the symmetry plane, then we can integrate Eq. (A.8)

$$\alpha(z) = -D^{-1} \int_0^z j_z \times \exp\left[\beta\left(V_s - \frac{1}{2}\gamma z'^2 + \frac{1}{2}\sum_{i=1}^{N-1} \gamma_i x_i^2\right)\right] dz'. \tag{A.9}$$

The population difference is given as

$$\Delta = 2\int_A dx_1 \ldots dx_N \rho = 2\alpha(A) \exp(-\beta V_A) \prod_{i=1}^{N} (2\pi k_B T/\delta_i)^{\frac{1}{2}}, \tag{A.10}$$

where $\alpha(A) = -D^{-1} \int_0^A j_z \times \exp\left[\beta\left(V_s - \frac{1}{2}\gamma z^2\right)\right] dz$ (j_z is the current at the saddle point and is given by $j_z = -D(\partial \alpha/\partial z)_{z=0} \exp(-\beta V_S)$).

Now the other quantity needed for the calculation of τ is the total current J crossing the saddle point. The current density in the symmetry plane containing the saddle point is

$$j_z = -D \left(\frac{\partial \alpha}{\partial z}\right)_{z=0} \exp\left[\beta\left(-V_s - \frac{1}{2}\sum_{i=1}^{N-1} \gamma_i x_i^2\right)\right]. \quad (A.11)$$

We now follow the following steps:

1. Integration over $N-1$ transverse coordinates gives total current (J)

$$J = -D \left(\frac{\partial \alpha}{\partial z}\right)_{z=0} \exp(-\beta V_s) \prod_{i=1}^{N-1} \left(\frac{2\pi k_B T}{\gamma_i}\right)^{\frac{1}{2}}. \quad (A.12)$$

2. We use Eqs. (A.4), (A.10), and (A.12) to obtain the rate

$$\frac{1}{\tau} = 2De^{-\beta(V_s - V_A)} \times \prod_{i=1}^{N} \left(\frac{2\pi k_B T}{\gamma_i}\right)^{\frac{1}{2}} \Bigg/ \left[\left(\frac{2\pi k_B T}{\gamma}\right)^{\frac{1}{2}} \prod_{i=1}^{N} \left(\frac{2\pi k_B T}{\delta_i}\right)^{\frac{1}{2}}\right]. \quad (A.13)$$

On more simplification we obtain

$$\frac{1}{\tau} = 2D\beta m \left(\frac{\gamma}{m}\right)^{\frac{1}{2}} \left[\left(\prod_{i=1}^{N} \nu_i \Bigg/ \prod_{i=1}^{N-1} \nu_i'\right) e^{-\beta(V_s - V_A)}\right], \quad (A.14)$$

where $(\delta_i/m)^{\frac{1}{2}} = 2\pi\nu_i$, and $(\gamma_i/m)^{\frac{1}{2}} = 2\pi\nu_i'$.

As we know that the multidimensional TST rate is $\left(\prod_{i=1}^{N} \nu_i \Big/ \prod_{i=1}^{N-1} \nu_i'\right) e^{-\beta(V_s - V_A)}$, the preceding expression of rate can be written in terms of TST rate as

$$\frac{1}{\tau} = 2D\beta m \left(\frac{\gamma}{m}\right)^{\frac{1}{2}} k_f^{TST}. \quad (A.15)$$

The quantity $D\beta m$ is related to viscosity (η) by the Stokes–Einstein relation and the quantity $(\gamma/m)^{1/2}$ is an angular frequency at the saddle point. This equation can be written as

$$\frac{1}{\tau} = 2\left(\frac{\omega_s}{\eta}\right) k_f^{TST}. \quad (A.16)$$

The factor $\left(\dfrac{\omega_s}{\eta}\right)$ is the same as in the Kramers' expression of rate for the one-dimensional overdamped case.

(b) The Underdamped Case

We have already discussed the underdamped reaction for one dimension. In this section we extend the previous treatment to many dimensions.

Let us first find the mean energy in excess of the saddle point energy, V_s, with which particles in thermal equilibrium cross the saddle point. Consider the particle with excess total energy between ε and $\varepsilon + d\varepsilon$, located at a position in the symmetry plane ($z = 0$) at potential $V_s + \delta V$. The kinetic energy of the particle is then $\varepsilon - \delta V$ and their total momentum is $(2m(\varepsilon - \delta V))^{\frac{1}{2}}$. Only one component ($z$-component) of this total momentum gives rise the flux along the z-direction.

The flux per unit energy across the saddle point is

$$\phi(\varepsilon)d\varepsilon = d\varepsilon \int p^{N-1} dx_1 \ldots\ldots dx_{N-1}$$

$$= d\varepsilon \int \left[2m \left(\varepsilon - \sum_{i=1}^{N-1} \gamma_i x_i^2 \right) \right]^{(N-1)/2} dx_1 \ldots\ldots dx_{N-1}. \quad (A.17)$$

On changing the variables as $t_i^2 = \gamma_i x_i^2$ and $t_i = \sqrt{\varepsilon} v_i$ we can write the above equation as

$$\phi(\varepsilon)d\varepsilon = c\varepsilon^{N-1} d\varepsilon \int \left[2m \left(1 - \sum_{i=1}^{N-1} v_i^2 \right) \right]^{(N-1)/2} dv_1 \ldots\ldots dv_{N-1}, \quad (A.18)$$

where c is a constant independent of ε. The flux per unit energy range, across the saddle point, would therefore be proportional to ε^{N-1}.

In thermal equilibrium the flux will be proportional to $\varepsilon^{N-1} \exp(-\beta\varepsilon)$. The mean excess energy of the flux across the saddle point will be

$$\langle \varepsilon \rangle = \int \varepsilon^N \exp(-\beta\varepsilon) d\varepsilon \bigg/ \int \varepsilon^{N-1} \exp(-\beta\varepsilon) d\varepsilon = Nk_B T. \quad (A.19)$$

The energy loss required, $k_B T$, is the same as in the one-dimensional case. The time available, however, for this loss of about $k_B T$ is likely to depend strongly on dimensionality and will be much larger in the many-dimensional case.

REFERENCES

1. P. Hanggi, P. Talkner, and Michal Borkovec, *Rev. Mod. Phys.* **62**, 251 (1990).
2. H. A. Kramers, *Physica*, **7**, 284 (1984).
3. R. F. Grote and J. T. Hynes, *J. Chem. Phys.* **73**, 2715 (1980).
4. J. Wang and P. G. Wolynes, *Phys. Rev. Lett.* **74**, 4317 (1975).
5. E. W. Montroll and K. E. Shuler, *Adv. Chem. Phys.* **1**, 361 (1958).
6. B. Widom, *Science*, **148**, 1555 (1965).

7. N. G. van Kampen, *Stochastic Processes in Physics and Chemistry*, North Holland Personal Library (1983).
8. T. Yamamoto, *J. Chem. Phys.* **33**, 281 (1960).
9. D. Chandler, *J. Chem. Phys.* **68**, 2959 (1978).
10. J. L. Skinner and P. G. Wolynes, *J. Chem. Phys.* **69**, 2143 (1978).
11. C. Rulliere, *Chem. Phys. Lett.* **43**, 303 (1976).
12. G. R. Rothenberger, D. K. Negus, and R. M. Hochstrasser, *J. Chem. Phys.* **79**, 5360 (1983).
13. S. P. Velsko and G. R. Fleming, *J. Chem. Phys.* **76**, 3553 (1982).
14. S. P. Velsko and G. R. Fleming, *Chem. Phys.* **65**, 59 (1982).
15. J. Montgomery, D. Chandler, and B. Berne, *J. Chem. Phys.* **70**, 4056 (1979).
16. P. B. Visscher, *Phys. Rev. B* **13**, 3272 (1976).
17. E. Pollak, *J. Chem. Phys.* **85**, 865 (1986).
18. T. Munakata, *Prog. Theor. Phys.* **75**, 747 (1986).
19. (a) R. Zwanzig and M. Bixon, *Phys. Rev. A* **2**, 2005 (1970); (b) H. Metiu, D. W. Oxtoby, and K. Freed, *Phys. Rev. A* **15**, 361 (1977).
20. R. K. Murarka, Bhattacharyya, R. Biswas, and B. Bagchi, *J. Chem. Phys.* **110**, 7365 (1999).
21. G. W. Robinson, W. A. Jalenak, and D. Statman, *Chem. Phys. Lett.* **110**, 135 (1984).
22. P. Pechukas, *Ann. Rev. Phys. Chem.* **32**, 159 (1981).
23. R. Landauer and J. A. Swanson, *Phys. Rev.* **121**, 1668 (1961).
24. C. Dellago, P. G. Bolhuis, and P. L. Geissler, *Adv. Chem. Phys.* **123**, 1–78 (2002).

8

Barrierless Reactions in Solution

8.1 INTRODUCTION

Many important chemical and biological reactions do not face a sizable activation barrier in their reactive motion along the reaction coordinate from the reactant to the formation of the product. Dynamics of such barrierless (sometimes referred to as "activationless") reactions naturally differ considerably from those chemical reactions where the reactant has to climb a high activation barrier (E_{act}) to reach the product. To begin with, these barrierless reactions can be very fast, often having time constants in the range of a few hundred femtoseconds (fs). The resistance to the motion along the reaction coordinate may come from the viscous forces only. It is interesting to note that historically (because of very high rates of these reactions) they were among the first to be studied whenever a new laser system with shorter time resolution was available.

Typical reaction-potential-energy surfaces for barrierless reactions are depicted in Fig. 8.2. The most important example is clearly the *vision transduction process*, which involves a critical initiation step during a barrierless *cis* to *trans* transformation of a retinal chromophore, rhodopsin [1]. The 11-*cis*-retinal chromophore lies in a pocket of the protein and isomerizes to all-*trans* retinal chromophore when light is absorbed (see Fig. 8.1). The isomerization of the retinal chromophore leads to a change of the shape of rhodopsin that triggers a cascade of reactions which leads to a nerve impulse that is transmitted to the brain by the optical nerve. [See *Human Eye, http://www2.mrc-lmb.cam.ac.uk/groups/GS/eye.html.*] Other important examples include isomerization of stilbene and diphenyl butadiene in alcohol solvents and of triphenyl methane dyes (crystal violet and ethyl violet) in lower alcohols. All these reactions are known to follow a barrierless reaction pathway.

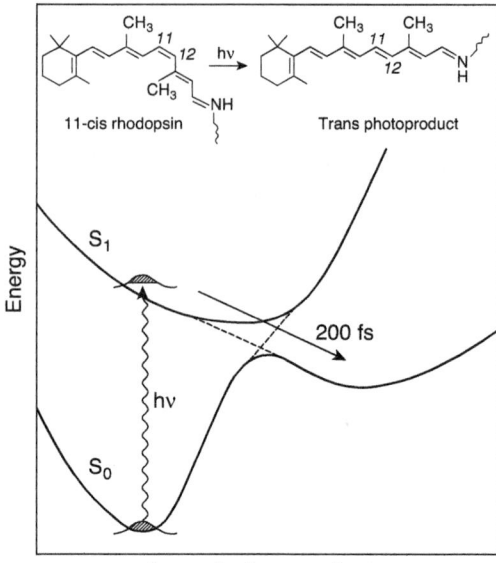

Figure 8.1 Schematic ground-state and excited-state potential energy surface for the 11-*cis* → 11-*trans* isomerization in rhodopsin. This is the optical process involved in vision transduction.

As already mentioned, barrierless reactions are generally characterized by high reaction rate, exhibit temperature dependence distinctly different from that of high barrier reactions, and are often strongly coupled to solvent viscosity [2–6]. Actually, viscosity dependence has often been used to characterize the mechanism of these reactions. Another important feature is that these reactions may depend on the initial conditions, thus not only the wavelength of the excitation light but also the frequency of both the excited and ground-state potential surfaces may play important roles in controlling the relaxation dynamics. Because of their high reaction rates, a detailed study of barrierless reactions had to wait for the development of ultrafast laser spectroscopy. For example, one recent study of rhodopsin isomerization (using a 10 fs probe pulse) showed that the reaction time is in the few hundreds of femtosecond range and is much faster than the previously reported value of 30 ps. Also by a femtosecond time-resolved spectroscopic study it is shown that the excited-state relaxation of crystal violet is completed within 500 fs and reported anomalous viscosity dependence. Similar enhanced rates due to the higher resolution of recent femtosecond investigations have been noted in the isomerization of *cis*-stilbene and in tetraphenylethylene.

A typical reaction-potential-energy surface for a barrierless reaction is depicted in Fig. 8.2. The reaction is excited from the ground state to the position denoted by X_0 on the excited-state surface. The subsequent relaxation brings the reactant down toward the potential energy minimum where the efficient sink X_s is also located.

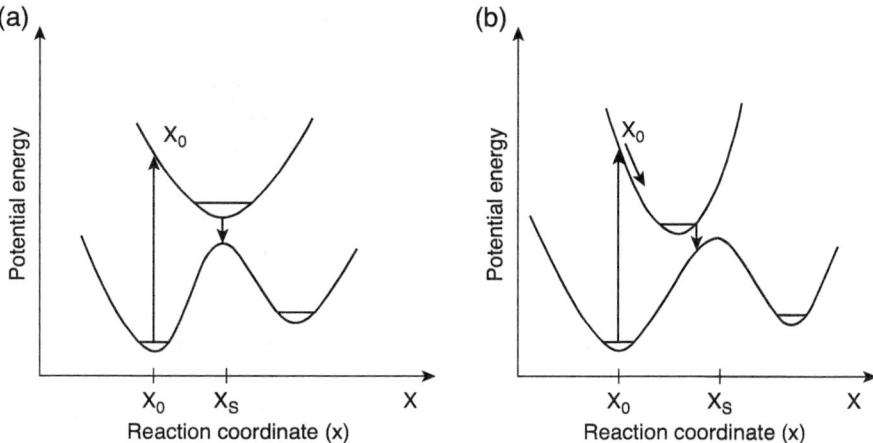

Figure 8.2 Schematic illustration of potential energy surfaces in which a nonradiative relaxation occurs from the excited-state surface from (a) a symmetric sink (decay channel) available at zero barrier and (b) a displaced sink with a small internal barrier. Reprinted with permission from *J. Phys. Chem.*, **94**, 9, (1990). Copyright (1990) American Chemical Society.

This initial relaxation may be resisted by the solvent friction for those cases where relaxation involves large amplitude motion of a bulky molecular group, as is believed to be the case in *cis–trans* isomerization, for example, in stilbene, or in triphenyl methane (TPM) dyes. Triphenyl methane dyes have played an important role in the study of barrierless reactions [3, 4, 6]. Although there is some controversy regarding the precise nature of the reaction coordinates, it is generally believed that it involves rotation of the phenyl rings around the bond between the phenyl group and the central carbon atom. It is worthwhile to note the following experimental results on electronic relaxation of TPM dyes that have motivated many theoretical studies.

The experimental observable in most of the reactions is the reactant survival probability, which is referred to here as $P_e(t)$. The salient features of the observed reaction kinetics are summarized below.

1. The time dependence of the reactant population (that is, of the excited-state survival probability) in solution is often strongly viscosity dependent. This usually indicates that the reactive motion is a diffusive motion along the reaction coordinate.
2. Interestingly, the decay is usually a single exponential at low viscosity, but becomes multiexponential as the viscosity of the solvent is increased. The change in the form of the decay curve is an important feature of the radiationless relaxation in the absence of an activation barrier.
3. The fluorescence quantum yield, ϕ_f, shows a fractional viscosity dependence. Following the initial suggestion by Förster and Hoffmann [2] that the fluorescence quantum yield should exhibit an inverse 2/3 power

dependence on viscosity (η), that is, $\varphi_f \approx \eta^{-2/3}$, several studies actually found adequate fits to this form over a limited viscosity range.

4. Barrierless reactions have been studied by using several different spectroscopic techniques, namely, fluorescence decay, excited-state absorption, and ground-state recovery. These techniques may lead to results at variance with each other, with fluorescence decay giving the fastest and ground-state recovery the slowest decay. A dependence on the wavelength of the excitation has also been observed.

There have been many theoretical studies to explain the general features observed in experiments. In the absence of an activation barrier, there is no clear separation of time scales between the motion in the reactive region and in the rest of the potential surface. Therefore, a steady-state solution (similar to the one used by Kramers') is no longer possible and one must solve for the full time-dependent probability distribution of the reactant on the reaction-potential-energy surface. As the decay of the reactant population ($P_e(t)$) is often not exponential over the sizable part of the decay, a unique reaction rate in the phenomenological sense may not exist in these reactions. It is common to define several different rate constants to characterize the decay. First is the long-time rate, k_L, defined as

$$k_L = - \lim_{t \to \infty} \frac{d}{dt} \ln P_e(t). \tag{8.1}$$

The second rate is the average rate, k_I, defined by the following expression:

$$k_I^{-1} = \int_0^\infty dt P_e(t). \tag{8.2}$$

Sometimes a third rate, k_1, is defined as the first moment of the survival probability:

$$k_1^{-1} = \int_0^\infty dt \, t P_e(t). \tag{8.3}$$

Before we start discussing other issues, let us point out that diverse processes in natural and biological sciences are modeled as barrierless reactions. Two notable examples are barrierless electron transfer reactions [7] and ligand binding in proteins [8]. Much of the theoretical and computational studies discussed below also apply to these two important processes. There is yet another class of activationless or barrierless processes that are dominated by entropic bottleneck. We shall discuss them in the next chapter (**Chapter 9**).

8.2 STANDARD MODEL OF BARRIERLESS REACTIONS

Although the experimental observable is the time-dependent reactant survival probability, $P_e(t)$, that can be observed experimentally, the fundamental theoretical quantity here is $P(X, t)$, which gives the position (X)-dependent population distribution on the reactant surface, where X gives the location of the particle along the reaction coordinate. The survival probability $P_e(t)$ is obtained from $P(X, t)$ by averaging over the entire reaction coordinate X:

$$P_e(t) = \int dX\, P(X, t). \tag{8.4}$$

$P_e(t)$ is directly amenable to experiments by various optical techniques, such as fluorescence quenching, ground-state recovery, or excited-state absorption.

Almost all the theories developed to address barrierless reactions are based on a simple stochastic description of the motion of the reactant on the reaction-potential-energy surface, and the reaction is described phenomenologically by a position- (or coordinate-) dependent sink term along the reaction coordinate. As already mentioned, such a description has been used in a wide variety of different problems ranging from protein dynamics [8] to electron-transfer reactions [7]. We shall term this model as the *"Standard Model"* of barrierless reactions. This model is described by the following equation:

$$\frac{\partial P(X, t)}{\partial t} = LP(X, t) - S(X)P(X, t), \tag{8.5}$$

where X is the reaction coordinate (which may depend on more than one dimension) and L is a stochastic operator, either Fokker–Planck or the Smoluchowski operator. The latter is safe to use only when momentum relaxation is much faster than position relaxation. The second term on the right-hand side is usually referred to as a sink term, which gives rise to the decay of reactant population at the reaction coordinate value X. In many cases this position dependence is rather strong, which makes this equation quite difficult to solve. The reaction potential-energy-surface enters in the specification of the stochastic operator L.

Many simple models have been proposed to understand these astonishingly fast reactions. Some of these models can be solved exactly, while others need numerical solution. In the following we discuss some of the currently popular models.

8.2.1 Exactly Solvable Models for Photochemical Reactions

When the time evolution operator L is a Smoluchowski operator, the standard model simplifies to the following expression:

$$\frac{\partial P(X, t)}{\partial t} = D\frac{\partial^2 P(X, t)}{\partial X^2} + B\frac{\partial}{\partial X}XP(X, t) - k_{nr}S(X)P(X, t) - k_r P(X, t), \tag{8.6}$$

where $D = k_B T/\zeta$ and $B = \omega^2 \mu/\zeta$, ω being the frequency of motion on the harmonic surface, ζ is the relevant friction coefficient, μ is the reduced mass of the

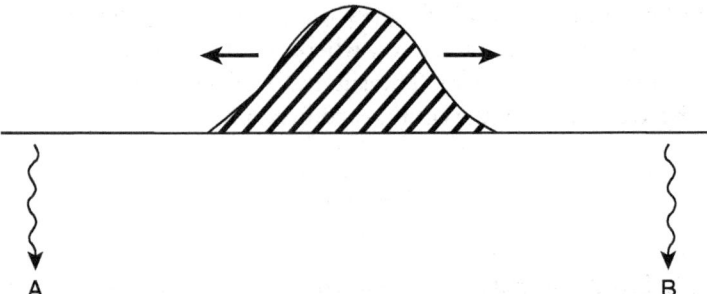

Figure 8.3 The Oster–Nishijima model, where A and B are the reactive centers.

reactive motion, and T is the temperature. k_{nr} is the magnitude of the radiationless rate at the origin and k_r is the radiative rate constant, independent of position. The radiative rate is present in many photochemical reactions, but could be absent in many cases. Equation (8.6) has been solved exactly for several specific cases. When the reaction zone is so narrow that it can be approximated by a point on the reaction coordinate, and at the same time the rate of decay from this narrow sink is so much faster than the rate of population relaxation along the reaction coordinate that we can replace the rate by an absorbing barrier (point of no return), then Eq. (8.6) is amenable to simple but elegant analytical solution, as discussed below.

8.2.1.1 Oster–Nishijima Model

In this model [5], the reaction energy surface is flat (that is, frequency ω is equal to zero) against the reaction coordinate and the reactive motion is a one-dimensional diffusion process. Thus, the diffusion coefficient D provides the only time scale in the problem. In many applications, D is the rotational diffusion coefficient. The reaction is assumed to occur at the two ends of the surface. A schematic drawing of the surface is shown in Fig. 8.3.

Mathematically, this model can be represented by the following sets of initial boundary values that the diffusion equation must satisfy:

$$P(X, t = 0) = P_0(X), \tag{8.7}$$

$$P(X = 0, t) = 0, \tag{8.8}$$

$$P(X = a, t) = 0. \tag{8.9}$$

The reaction surface is such that there is one absorbing barrier at $X = 0$ and another at $X = a$. This problem can be solved by the method of repeated reflections and the solution is given as [6]

$$P(X, t) = \int_0^a dX' P_0(X') P(X, t|X'), \tag{8.10}$$

where

$$P(X, t|X') = \exp(-k_r t)[4\pi Dt]^{-1/2}$$
$$\times \sum_{n=-\infty}^{\infty} \left\{ \exp\left[-\frac{(X-X'+2na)^2}{4Dt}\right] - \exp\left[-\frac{(X+X'+2na)^2}{4Dt}\right] \right\}, \quad (8.11)$$

which can be transformed into a convenient and simple form by making use of the Poisson summation formula. The final expression for the excited survival probability is quite elegant and is given by

$$P_e(t) = \frac{4}{\pi} \exp(-k_r t) \sum_{n=0}^{\infty} \frac{1}{2n+1} \times \exp\left[-\frac{(2n+1)^2 \pi^2}{a^2} Dt\right]$$
$$\times \int_0^a dX' P_0(X') \sin \frac{(2n+1)\pi X'}{a}. \quad (8.12)$$

The Oster–Nishijima model predicts an exponential decay after a small transient decay, which is the time required for the distribution to reach the reaction zones by diffusion. Integral in Eq. (8.2) can be evaluated for average rate and the resulting expression for delta function initial condition, $\delta(X - X_0)$, is $k_I^{-1} = X_0(a - X_0)/2D$.

This solution remains unchanged when the reactive motion is rotation on a fixed axis because the equation of motion of the probability distribution remains the same, which could be the case for triphenyl methane dyes.

8.2.1.2 Staircase Model

This model is supposed to mimic barrierless reaction on a highly anharmonic surface. In some respects this model is similar to the Oster–Nishijima model. The reaction energy surface is flat in this case also (and the reactive motion is pure diffusion), but the reaction is assumed to occur only at one end of the surface. A schematic drawing of the surface is shown in Fig. 8.4.

The sink is located near the crossing of two zero surfaces. Relaxation on this surface is modeled by placing an absorbing barrier at position C and a reflecting barrier at A, which is also chosen to be the origin of the coordinate system.

Mathematically, this model can be represented by the following sets of initial boundary values that the diffusion equation must satisfy:

$$P(X, t = 0) = P_0(X), \quad (8.13)$$

$$\frac{\partial P(X, t)}{\partial X}\bigg|_{X=0} = 0, \quad (8.14)$$

$$P(X = a, t) = 0. \quad (8.15)$$

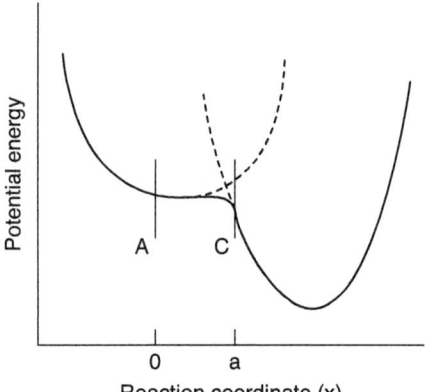

Figure 8.4 Schematic illustration of the Staircase model. Reprinted with permission from *J. Phys. Chem.*, **94**, 9, (1990). Copyright (1990) American Chemical Society.

The solution can again be represented by Eq. (8.10), where $P(X', t)$ is given by the following somewhat different expression [6]:

$$P(X,t|X') = \exp(-k_r t)(4\pi Dt)^{-1/2} \times \sum_{n=-\infty}^{\infty} (-1)^n \left\{ \exp\left[-\frac{(X-X'+2na)^2}{4Dt}\right] \right.$$
$$\left. + \exp\left[-\frac{(X+X'+2na)^2}{4Dt}\right] \right\}. \tag{8.16}$$

The final expression for the survival probability is obtained by carrying out the integration in Eq. (8.4):

$$P_e(t) = \frac{4}{\pi} \exp(-k_r t) \sum_{n=0}^{\infty} \frac{(-1)^n}{2n+1} \exp\left[-\frac{(2n+1)^2 \pi^2}{4a^2} Dt\right]$$
$$\times \int_0^a dX' \cos\frac{(2n+1)\pi X'}{2a} P_0(X'). \tag{8.17}$$

Though the above equation looks a little complicated, the integral in Eq. (8.2) can be evaluated analytically for delta function initial condition and the resulting expression is $k_I^{-1} = (a + X_0)(a - X_0)/2D$.

This model predicts decay a lot slower than that of the Oster–Nishijima model.

8.2.1.3 Pinhole Sink Model

This model is a bit more general where the motion along the reaction coordinate is governed by the force from the potential due to the free-energy surface (approximated as harmonic) and the viscous drag of the solvent [6]. All these competing factors can be modeled by a modified Smoluchowski Eq. (8.6). In this case the reaction window is assumed to be situated at the origin and is again an absorption barrier. The initial boundary conditions are

$$P(X, t = 0) = P_0(X), \qquad (8.18)$$

$$P(X = \pm\infty, t) = 0, \qquad (8.19)$$

$$P(X = 0, t) = 0. \qquad (8.20)$$

With these initial-boundary conditions, one obtains the following exact solution for the time dependence of the excited state population, $P_e(t)$ [6]:

$$P_e(t) = \exp(-k_r t) \times \int_0^\infty dX' \left[P_0(X') + P_0(-X') \right] \operatorname{erf} F(X', t), \qquad (8.21)$$

where

$$F(X, t) = \left[\frac{B}{2D(1 - e^{-2Bt})} \right]^{1/2} X e^{-Bt}. \qquad (8.22)$$

The error function, *erf a*, is defined by

$$\operatorname{erf} a = \frac{2}{\sqrt{\pi}} \int_0^a dy\, e^{-y^2}. \qquad (8.23)$$

If the initial population distribution along X on the excited state is given by the Boltzmann distribution on a harmonic surface, that is, $P_0(X)$ is a Gaussian distribution centered at the origin of the excited surface, then we obtain the following elegant expression for the excited-state probability:

$$P_e(t) = \frac{2}{\pi} \sin^{-1}(\exp(-Bt)). \qquad (8.24)$$

The above equation predicts non-exponential decay at all viscosities except at long times where the decay becomes single exponential. Since B is inversely proportional to the viscosity, the decay in long time limit is governed by t/η.

This elegant $\sin^{-1}$ or arcsin law finds application in other problem of chemical kinetics. However, in most photochemical reactions, the initial distribution is not the equilibrium one and one needs to use the more general expression. Figure 8.5 displays the distribution function $P(X, t)$ at different times for pinhole sink.

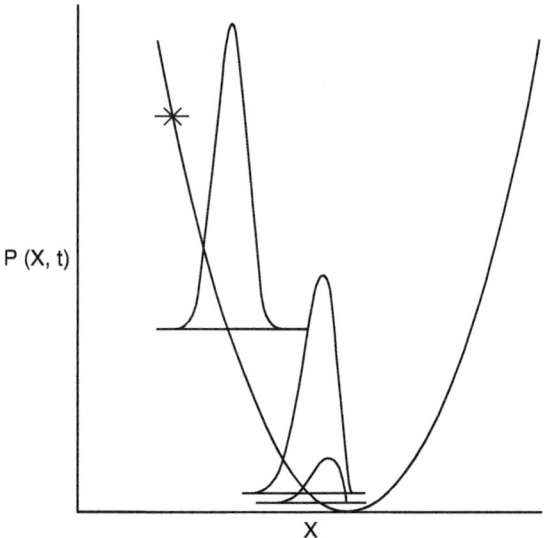

Figure 8.5 Distribution function $P(X, t)$ at three different times for pinhole sink. The * marks the initial excitation at X_0. Reprinted with permission from *J. Chem. Phys.* **78**, 7375 (1983). Copyright (1983) American Institute of Physics.

8.2.2 Approximate Solutions of Realistic Models

For many experimental situations, the limiting models described previously are not applicable, especially because the rate of decay even from a narrow sink is often not infinitely large. As a result, a second time scale due to the decay rate from the reaction zone enters the description. Now the theoretical models cannot be solved analytically. We describe below a few such models that have been solved successfully by numerical methods.

At this stage it is interesting to note a historical fact. As already mentioned, barrierless processes appear in many situations. They have mostly been modeled theoretically by what we call the Standard Model and expressed by Eq. (8.5). However, the essentially same equation has been solved numerically by employing at least four different methods! The results obtained are similar.

8.2.2.1 Delta Function Sink
The delta function sink is similar to the pinhole sink in the sense that the reaction occurs from a point sink on a harmonic surface, except that the rate at the reaction window is finite in this case:

$$S(X) = \delta(X - X_s), \qquad (8.25)$$

where X_s is the position of the sink. Even this model cannot be solved fully analytically, unlike that in the cases where the rate at the reaction window is infinite.

However, this can be solved by using the Green's function technique. The Green's function for diffusive motion on a one-dimensional harmonic surface is well known (see Chapter 2 for a discussion) and is given by the following expression:

$$G_0(X, t|X_0) = \sqrt{\frac{B}{2\pi D(1 - \Delta^2(t))}} \exp\left[-\frac{B(X - X_0\Delta(t))^2}{2D(1 - \Delta^2(t))}\right], \quad (8.26)$$

where $\Delta(t)$ is the position–position time correlation, $\langle X(0)X(t) \rangle$. In the overdamped (or Smoluchowski) limit, this function decreases monotonically:

$$\Delta(t) = \exp\left(-\frac{\mu\omega^2}{\zeta}t\right) = \exp(-Bt). \quad (8.27)$$

Thus, $\Delta(t)$ decreases faster as the frequency increases, while the solution in the case of delta function sink is given in a closed form, not as a function of time, but rather as a function of frequency z—the two functions are related by the usual Laplace transformation

$$P(X, z) = \int_0^\infty dt\, e^{-zt} P(X, t). \quad (8.28)$$

The final solution is given by

$$P(X, z) = G_0(X, z) - S(X_s)P(X_s, z). \quad (8.29)$$

Time dependent survival probability can be obtained numerically by Laplace inversion. The delta function sink is certainly more realistic than the pinhole sink, although the latter is solvable analytically.

8.2.2.2 Gaussian Sink

The limitation of all the above models is that the width of the sink is assumed to be zero. A more generalized sink function is the Gaussian sink, given by the following expression [6]:

$$S(X) = S(X_s) \exp\left(-(X - X_s)^2/\sigma^2\right). \quad (8.30)$$

where $S(X_s)$ is the rate at $X = X_s$. The Gaussian sink is realistic because the rate of nonradiative decay between two electronic surfaces is known to obey a negative exponential dependence on the energy gap between the two surfaces. Now, if the two surfaces are of the kind shown in Fig. 8.2 and if one assumes that both are harmonic near the origin with nearly the same frequency, then the energy gap between them is indeed a quadratic function of the reaction coordinate.

The Smoluchowski equation with the above sink cannot be solved analytically. However, one can use a technique, which combines the Green's function technique

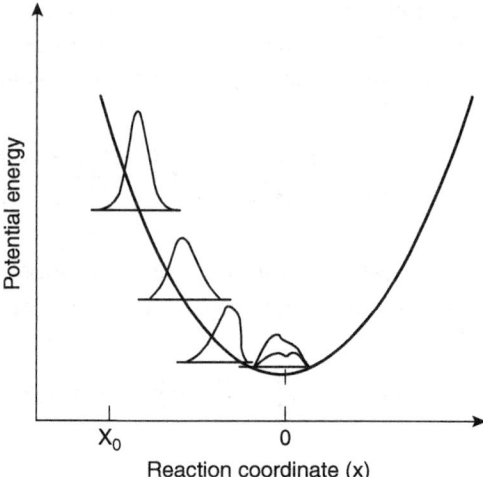

Figure 8.6 Distribution function $P(X, t)$ at five different times for a narrow Gaussian sink at the origin. The initial excitation is at X_0. Reprinted with permission from *J. Chem. Phys.* **78**, 7375 (1983). Copyright (1983) American Institute of Physics.

with a clever discretization of the sink function, to solve this equation. In this method, the sink is first discretized as follows:

$$S(X) = \sum_s k_s \delta(X - X_s). \tag{8.31}$$

The Green's function solution for a delta function source at $X = X'$ can be obtained and is a generalization of Eq. (8.29):

$$P(X, z) = G_0(X, z) - \sum_s k_s P(X_s, z). \tag{8.32}$$

This leads to a system of equations for $P(X_s, z)$ that can be solved by the matrix method.

The distribution function $P(X, t)$ at different times for a narrow Gaussian sink at the origin is shown in Fig. 8.6. Gaussian sink model can be derived from the energy gap law of non-radiative rates and assuming harmonic surfaces for both, the ground and the excited states.

8.3 INERTIAL EFFECTS IN BARRIERLESS REACTIONS: VISCOSITY TURNOVER OF RATE

Although we have used the Smoluchowski equation in the above discussions to describe the motion along the reaction coordinate, these results are not applicable at very low viscosities where inertial effects become important. In fact, one finds interesting nonmonotonic viscosity dependence in the viscosity dependence of rate at low viscosity, similar to the one observed in the Kramers' theory.

To describe the inertial effects on barrierless reactions we have to solve the full phase space Fokker–Planck equation. A straightforward generalization of Eq. (8.6) is to consider a full phase space Fokker–Planck equation of the following form:

$$\frac{\partial P(X, V, t)}{\partial t} = -V\frac{\partial P}{\partial X} + \frac{\omega^2 X}{\mu}\frac{\partial P}{\partial V} + \frac{\zeta}{\mu}\frac{\partial}{\partial V}\left(V + \frac{k_B T}{\mu}\frac{\partial}{\partial V}\right)P - k_{nr}S(X)P - k_r P, \tag{8.33}$$

where V is the velocity of the solute particle. It is hard to obtain an analytical solution of the above equation with an arbitrary position-dependent sink term. This is usually solved numerically by generating a set of stochastic trajectories. The statistical properties of these trajectories are representative of the phase space distribution, which is the solution to the Fokker–Planck equation. Averaging over many stochastic trajectories is therefore equivalent to solution of the partial differential equation. The results are expressed in terms of the scaled friction parameter $\beta = \zeta/\mu$, where ζ is the friction on the reactive motion and μ is the reduced mass.

Figure 8.7 displays a typical decay curve for a pinhole sink obtained at low viscosities. After an initial transient time, the logarithm of the excited-state

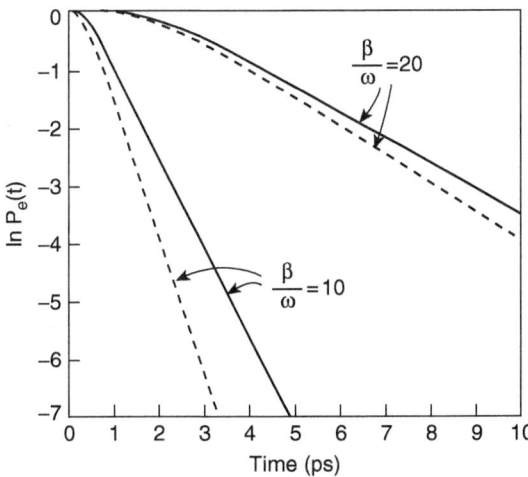

Figure 8.7 The decay of the excited-state population for the pinhole sink case for two different values of the friction parameter β. The SE result of the Bagchi, Fleming, Oxtoby model [6] is given by dashed lines. Reprinted from *Chem. Phys. Lett.* **99**, B. Bagchi, S. Singer, and D. W. Oxtoby, Non-monotonic dependence of electronic relaxation rate on solvent viscosity, p. 285 (1983). Copyright (1983) with permission from Elsevier.

population density, $P_e(t)$, decreases linearly with time. Figure 8.7 also shows the population decay predicted by Smoluchowski equation (SE) of the Bagchi, Fleming, Oxtoby (BFO) model [6]. For high β, the results of the SE and the present calculation agree quantitatively. However, as β decreases, SE starts deviating from trajectory calculation results as the SE consistently overestimates the rate of decay. The deviation becomes significant when β becomes smaller than ω, the frequency of the well.

When the sink function $S(X)$ is Gaussian centered at the minimum of the harmonic potential surface, then the trajectory calculation generates potential decay curves as shown in Fig. 8.8. The population decay predicted by BFO's solution of the SE is also shown in the figure. At large values of friction parameter β ($\beta \gg 1$), the SE results are in quantitative agreement with trajectory calculations. But as β becomes smaller ($\beta \approx \omega$), the population decay predicted by SE is again too rapid. This is due to the fact that SE neglects the oscillatory behavior of the solute particle in the harmonic well. In the Smoluchowski description the solute particle continuously approaches the equilibrium distribution, which is peaked at the origin where the decay probability is a maximum. In reality, in the low-friction

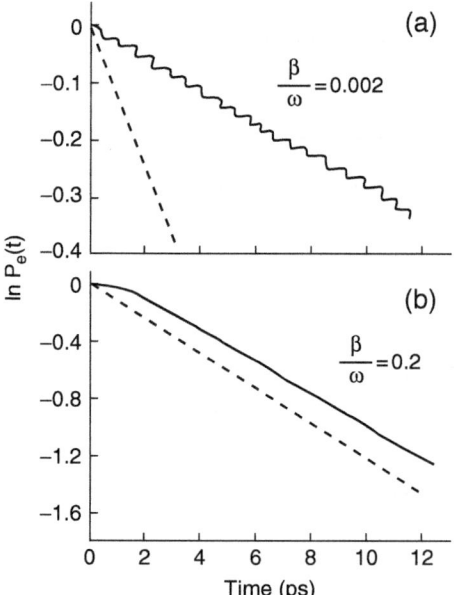

Figure 8.8 The decay of excited-state population for a Gaussian sink at very low (a) and at intermediate (b) values of β. The SE result is given by dashed line. Reprinted from *Chem. Phys. Lett.* **99**, B. Bagchi, S. Singer, and D. W. Oxtoby, Non-monotonic dependence of electronic relaxation rate on solvent viscosity, p. 285 (1983). Copyright (1983) with permission from Elsevier.

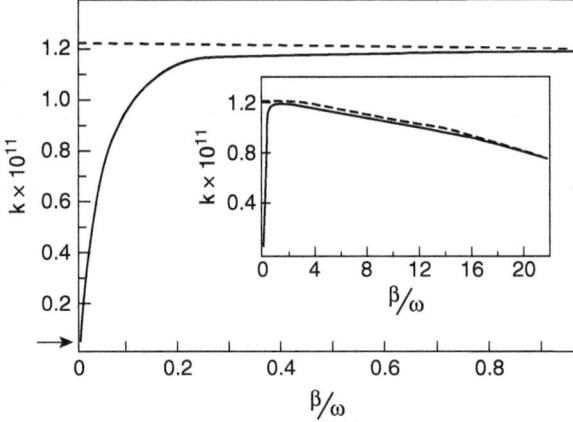

Figure 8.9 The rate constant of excited-state population decays from a Gaussian sink plotted against friction parameter. The SE result is given by dashed line. The arrow at the lower left corner indicates the limiting zero-friction value. Reprinted from *Chem. Phys. Lett.* **99**, B. Bagchi, S. Singer, and D. W. Oxtoby, Non-monotonic dependence of electronic relaxation rate on solvent viscosity, p. 285 (1983). Copyright (1983) with permission from Elsevier.

limit, the solute particle oscillates several times in the well and passes through the origin with maximum speed (as in the case of harmonic oscillators), thus spending less time at the origin than predicted by SE. The period of oscillation coincides with the frequency of the harmonic well.

The dependence of the population decay rate on the friction parameter β, for a Gaussian sink centered at the origin, is shown in Fig. 8.9. The decay rate reaches a maximum around $\beta \approx \omega$ as the friction parameter β is increased. In the case when β approaches zero, the rate attains a limit value shown by an arrow in the Fig. 8.9. Trajectory calculations of the Fokker–Planck equation show a nonmonotonic dependence of the rate on viscosity. This dependence is similar to the well-known nonmonotonic viscosity dependence (Kramers' turnover) of rate in the high-activation-barrier limit. The only difference is that in Kramers' solution the rate becomes zero when $\beta \to 0$, whereas in the case of barrierless reactions the rate attains a limiting value.

The reason for the nonmonotonic dependence of rate on β can be easily understood in the following way. The rate of population decay at low viscosities is governed by two competing factors. One is the amount of time the solute particle takes to reach the origin from its initial position and the other is the amount of time the particle spends near the origin (where $k_{nr}S(X)$ is large, note the Gaussian behavior of $S(X)$ centered at origin). As $\beta \to 0$ (i.e., viscosity of the solvent is very low) the time to reach the origin as well as the amount of time the solute particle spends near the origin decreases. As a result of this competition, the rate can attain a maximum value at nonzero β. This feature is not predicted from the Smoluchowski equation because it neglects the oscillatory behavior of the solute particle in the potential well. The major experimental difficulty to find this crossover is that the solute molecules have very low solubility in solvents with low viscosities.

The above approach is fully classical. However, in the low friction limit the quantum nature of the vibrational motion will begin to have an effect. We may especially need to consider the effects of frequency-dependent friction on the vibrational population relaxation.

8.4 MEMORY EFFECTS IN BARRIERLESS REACTIONS

A limitation of all the preceding descriptions of barrierless reactions is the use of an overdamped, Markovian description of the dynamics. Thus, the response of the liquid to the motion of the reactant was assumed to be purely viscous. As discussed in Chapter 7, in the case of high barrier reactions, the above assumption leads to a completely wrong viscosity dependence of the rate. A similar situation can arise in the case of those barrierless reactions that are very fast (with time constants in the subpicosecond range) and thus probe only the short-time response of the solvent. This calls for a generalization of the above theories (discussed in **Section 8.2**) to include the bimodal response of the solvent.

A non-Markovian description of friction for the reactive motion can remove the preceding limitations. As discussed in Chapter 7, the usual approach is to start with the generalized Langevin equation (GLE). The GLE for a solute in a one-dimensional harmonic potential under the influence of a stochastic force, $R(t)$, is given as

$$\frac{dv(t)}{dt} = -\omega^2 X(t) - \int_0^t \zeta(t-\tau)v(\tau)d\tau + \frac{1}{m}R(t), \qquad (8.34)$$

where ω is the frequency of the excited potential surface, $\zeta(t-\tau)$ is the memory kernel of friction, T is the temperature, and k_B is the Boltzmann constant. The random force $(R(t))$ and the memory kernel $(\zeta(t-\tau))$ are related with the following fluctuation–dissipation relation:

$$\langle R(t)R(\tau)\rangle = k_B T m \zeta(|t-\tau|). \qquad (8.35)$$

For a harmonic surface, the generalized Langevin equation can be transformed into a generalized Smoluchowski equation with a time-dependent diffusion coefficient $D(t)$ defined by

$$D(t) = \frac{-k_B T}{m\omega^2}\frac{d\ln\xi(t)}{dt}, \qquad (8.36)$$

where $\xi(t) = L^{-1}\hat{\xi}(z)$ is given as

$$\hat{\xi}(z) = \frac{1}{z + \omega^2\hat{\vartheta}(z)}, \qquad (8.37)$$

and

$$\hat{\vartheta}(z) = \frac{1}{z + \hat{\zeta}(z)}. \tag{8.38}$$

Thus, time-dependent diffusion coefficient, $D(t)$, can be obtained from the time-dependent friction, $\xi(t)$. However, to find $\hat{\vartheta}(z)$ we need to know the frequency-dependent friction, $\hat{\zeta}(z)$. Time-dependent diffusion coefficient can have a value at short times that is quite different from that at long times. The short-time value of $D(t)$ is determined by the short-time or the high-frequency frictional response of the solvent that is determined by the static correlations and hence by the local structure surrounding the reactant. This is to be contrasted with the long-time diffusion, which is often determined by the dynamical correlations. The friction at long time is proportional to the slow, zero-frequency viscous response, which might have no influence on ultrafast chemical reactions. The generalized Smoluchowski equation contains some (but not all) of the effects of underdamped motion in a simplified description.

The generalized Smoluchowski equation does not sustain oscillatory motion on the potential energy surface which is present in the low-friction (viscosity) or underdamped limit. To circumvent this difficulty, different relaxation times are used where the fast-time scale (low friction) relaxes faster until the minimum of the excited potential energy surface (S_1) and from there on a very slow relaxation (high friction) takes over.

The zero-frequency friction, $\zeta(z=0)$, can be obtained by the time integration over $\zeta(t)$. The important point here is that while the value of $\zeta(z=0)$ may be controlled by the slow time constant, τ_2, the rate of an ultrafast reaction can be coupled to the faster relaxation time, τ_1. One indeed needs a non-Markovian theory to capture this aspect of the dynamics. Clearly, this itself can give rise to a fractional viscosity dependence of the reaction rate. A simplified description of a relaxation reaction in terms of a time-dependent diffusion coefficient in a harmonic surface can be described by a slightly modified Smoluchowski equation for the probability distribution $P(X, t)$, with a sink term as [11],

$$\frac{\partial P(X,t)}{\partial t} = \frac{\partial}{\partial X}\left\{D(t)\frac{\partial P(X,t)}{\partial X} + \frac{D(t)}{k_B T}m\omega^2 XP(X,t)\right\} - S(X)P(X,t). \tag{8.39}$$

Here $S(X)$ is the sink function given by $S(X) = k_s\delta(X - X_s)$, k_s is the sink transfer rate constant.

A comparison of the time-dependent survival probability for the Markovian and non-Markovian analysis is shown in Fig. 8.10. As the figure indicates the population decay in the non-Markovian case is faster than the Markovian case suggesting that the excited-state population has mostly decayed even before the steady state is reached. The reason for the faster decay in the non-Markovian case is related to the transient diffusion that arises from the short-time ultrafast relaxation component.

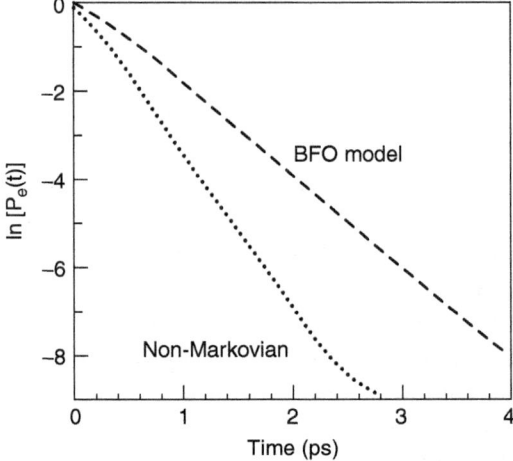

Figure 8.10 The decay of the excited-state population for the Bagchi-Fleming-Oxtoby (BFO) pinhole sink model (dotted line) and the current non-Markovian analysis (closed circles). Reprinted with permission from *J. Phys. Chem. A*, **103**, 9061 (1999). Copyright (1999) American Chemical Society.

8.5 UNUSUAL FEATURES OF BARRIERLESS CHEMICAL REACTIONS

8.5.1 Excitation Wavelength Dependence

This is truly a hallmark of barrierless reactions. The dependence of the population relaxation on the wavelength of the excitation light enters through the initial distribution position X_0 on the excited-state potential surface (S_1). When the sink transfer is rapid, the nuclear relaxation in the S_1 surface governs the overall rate and hence the initial excitation position greatly influences k_I. Figure 8.11 indicates k_I is inversely related to X_0, suggesting longer-wavelength excitation pulses will have higher isomerization rate. However, when the population transfer at the sink position is slow ($k_s \leq 1 \times 10^{11} \text{s}^{-1}$), the relaxation process becomes completely independent of the excitation wavelength.

8.5.2 Negative Activation Energy

One interesting feature of barrierless reactions is the possibility of small but measurable negative activation energy. In fact, experiments have reported a crossover from positive to negative activation energy. The Arrhenius plot calculations for a k_s value of $1 \times 10^{12} \text{s}^{-1}$ depicted in Fig. 8.12(a) show a crossover from positive to negative activation energy for decreasing values of τ_2. This crossover can be explained on the basis of a shift in the rate-governing process from relaxation in the excited surface (S_1) to the sink transfer. Due to the same reason a similar crossover behavior is also observed (Fig. 8.12(b)) on increasing the wavelength of the excitation light (*i.e.* changing the initial excitation position).

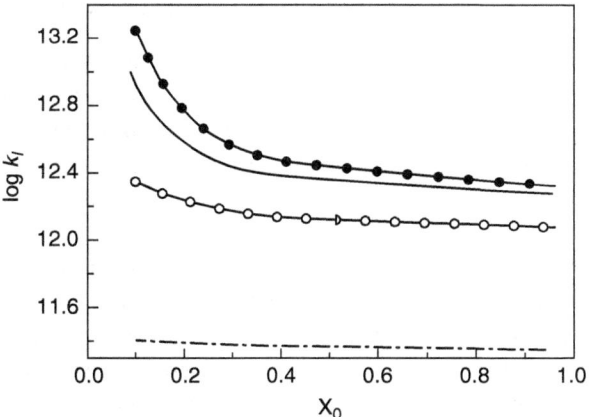

Figure 8.11 Semilog plot showing the dependence of the average rate constant (k_I) with the initial excitation position (X_0) for different sink transfer-rate-constant values $k_s = 1 \times 10^{14} s^{-1}$ (solid line with close circle), $k_s = 1 \times 10^{13} s^{-1}$ (solid line), $k_s = 1 \times 10^{12} s^{-1}$ (solid line with open circle), and $k_s = 1 \times 10^{11} s^{-1}$ (dash-dot line). Reprinted with permission from *J. Phys. Chem. A*, **103**, 9061 (1999). Copyright (1999) American Chemical Society.

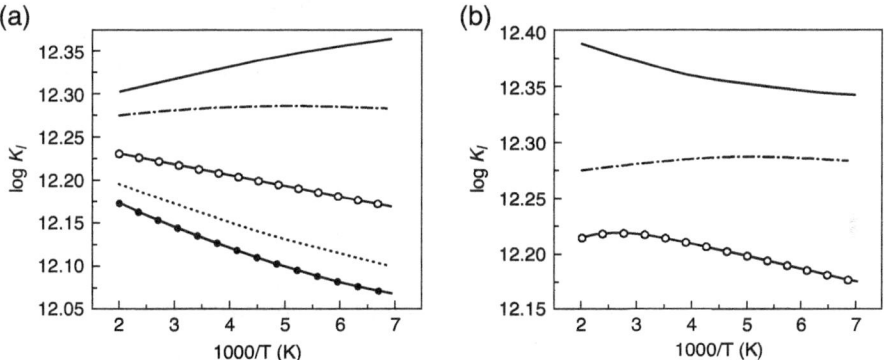

Figure 8.12 (a) Arrhenius plot for $k_s = 2.5 \times 10^{12} s^{-1}$ and $X_0 = 0.5$. The curves are calculated for τ_2 values of 0.1 (solid line), 0.2 (dash-dot line), 0.3 (solid line with open circle), 0.4 (dotted line), and 0.5 (solid line with closed circle). (b) Arrhenius plot obtained for τ_2 and k_s values of 0.2 ps and $2.5 \times 10^{12} s^{-1}$, respectively. Various curves obtained for the initial excitation position, X_0, values of 0.3 (solid line), 0.5 (dash-dot line), and 0.9 (solid line with open circle). Reprinted with permission from *J. Phys. Chem. A*, **103**, 9061 (1999). Copyright (1999) American Chemical Society.

8.6 MULTIDIMENSIONAL REACTION POTENTIAL ENERGY SURFACE

Owing to the absence of a significant activation barrier impeding the reactive motion, the rate can easily be influenced if the reaction coordinate (such as the twisting around a molecular axis, as in isomerization) is coupled to other coordinates, such as low-frequency vibrations. In such cases, the dependence of the decay rate on various properties can be different from what would be expected for one-dimensional reaction. The observed fractional viscosity dependence of the reaction rate has led to the suggestion that a one-dimensional reaction free-energy surface is inadequate in the case of barrierless reactions. A need for the multidimensional free-energy surface was further demonstrated by an apparent difference in the reaction rate when measured by three different experimental techniques, namely, ground-state recovery, excited-state absorption, and fluorescence quenching. An explanation of this interesting observation is presented by Åberg and coworkers [9] by using a two-dimensional reaction energy surface.

In barrierless reactions, the addition of extra dimension (coordinate) opens a new channel for the reaction. When the frictional resistance along the other coordinate is large, then the diffusive flux passes through this new channel and gives rise to the fractional viscosity dependence. The addition of an extra dimension may also play an important role in determining the excitation and probe wavelength dependences. For example, in the case of TPM dyes where the reactive motion is a low-frequency large amplitude motion (rotation of phenyl groups), then the Franck-Condon active mode may be some high frequency vibrational mode of the excited state. In such cases, the friction dependence of the nonreactive mode may not be of importance for the reaction. In the case of isomerization of trans-stilbene the second mode exhibits friction dependence. The theoretical approach developed in Section 8.2 for one dimension can be easily extended to the multidimensional potential energy surface.

8.7 ANALYSIS OF EXPERIMENTAL RESULTS

8.7.1 Photoisomerization and Ground-State Potential Energy Surface

In the experimental investigations on TPM dyes (crystal violet and ethyl violet) [3, 4] and 1,1'-diphenyl-4,4'-cyanine (1144C) [12], temperature-dependent crossover is observed in the Arrhenius plot when the viscosity is varied. As illustrated in the previous section, such turnover can arise due to the change in the rate-determining step from relaxation in the excited surface (S_1) to the sink transfer. Near this viscosity value, wavelength-mediated crossover is also expected. More recent data on 1144C photoisomerization at 7 cP for 450, 660, and 770 nm excitation show a similar trend.

The observation of viscosity-mediated temperature crossover and the absence of wavelength-mediated crossover in most of the experimental studies on barrierless isomerization reaction suggests: (i) a narrow S_0 surface potential curvature and/or (ii) the internuclear separation between the reactant and the product ground-state minima to be far away. For the latter to be true, the wave packet is expected to move a relatively long distance on the excited-state potential energy surface before it encounters the sink to decay. Thus, the motion of the wave packet on the S_1 surface becomes rate controlling. Under this condition, the isomerization kinetics is expected to be strongly coupled to the solvent viscosity and the turnover in temperature dependence is not expected. In contrast, the experimental studies on TPM dyes show viscosity-mediated crossover in temperature dependence.

For the former case, however, when the ground-state potential energy surface is narrow, the change in excitation wavelength does not appreciably change the excitation position X_0 in the S_1 surface. Thus, even when a viscosity-dependent temperature crossover is observed, the Arrhenius plot does not show a change in temperature dependence for varying wavelengths. On the contrary, for a reactant molecule with a broad potential curvature, different excitation pulses can create the initial distribution at different positions in the excited S_1 surface. This indicates the distance traveled by the wave packet can be controlled by the wavelength of the excitation pulse. Thus, by properly tuning the wavelength for a fixed viscosity, the crossover in temperature dependence is expected in the Arrhenius plot. The wavelength-dependent turnover in activation energy provides valuable information concerning the structure of the ground-state reactant potential energy surface.

8.7.2 Decay Dynamics of Rhodopsin and Isorhodopsin

Time-resolved spectral dynamics studies show that 11-*cis* to *all-trans* isomerization of rhodopsin completes within 200 fs. No red-shifted stimulated emission was observed in the time-resolved measurements, which suggest a rapid torsional motion of the excited species during its motion along the Franck–Condon region (low τ_1 and τ_2 values) [1, 13]. Excited-state relaxation is shown to be of the order of 100 fs. This fast relaxation in the S_1 surface is further confirmed by resonance Raman intensity analysis as well. Theoretical calculations show a fast sink transfer ($k_s = 1 \times 10^{14}$ s^{-1}) and short Brownian motion ($X_0 = 0.25 - 0.32$) along the excited-state surface to account for the observed 145 – 205 fs delay dynamics ($k_I \approx 6.8 - 4.9 \times 10^{12}$ s^{-1}) of rhodopsin. These values have several implications. The low value of X_0 suggests that the isomerization should be independent of the viscous drag of the solvent particles. The short nuclear motion is justified by the resonance Raman studies indicating the overall structural change associated with the isomerization is relatively small. A physical reason for this being that the -*cis* rhodopsin already exists in its twisted form along the reaction coordinate because of the steric interaction between the C-13 methyl and C-10 hydrogen groups. In an earlier analysis on rhodopsin isomerization, excited state vibrational coherence was invoked to account for the very low delay time observed. Theoretical analysis clearly shows a short-distance nuclear motion is also an important factor for this ultrafast femtosecond isomerization.

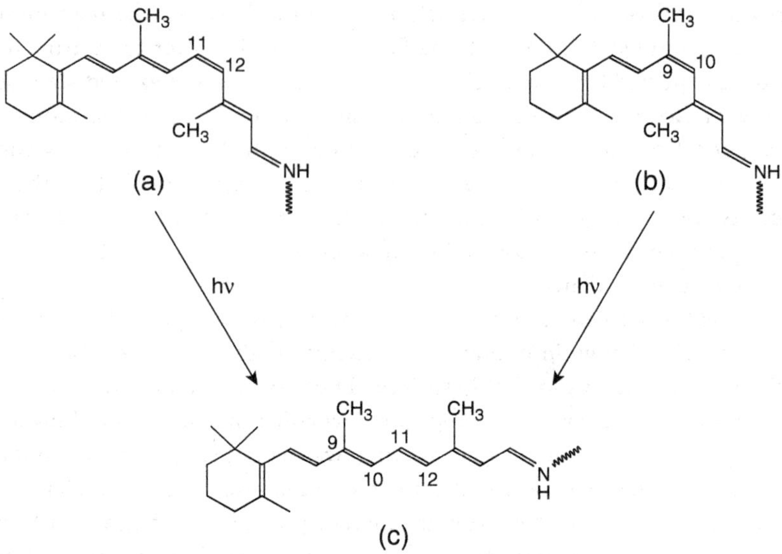

Figure 8.13 Schematic representations of the light-induced isomerization reactions of (a) cis- rhodopsin and (b) cis- isorhodopsin leading to the same photoproduct *all-trans* bathorhodopsin.

An interesting comparison can be made between the isomerization dynamics of rhodopsin and isorhodopsin (-cis rhodopsin). In contrast to rhodopsin, the isomerization in isorhodopsin occurs about the $C_9 - C_{10}$ double bond. Nevertheless, photoproduct of **9-cis** isomer is the same *all-trans* bathorhodopsin as that of rhodopsin. A schematic sketch of -cis rhodopsin and -cis rhodopsin and *all-trans* photoproduct (bathorhodopsin) is shown in Fig. 8.13.

Spectral measurements on isorhodopsin show a delay time of 600 fs, which is nearly three times longer than rhodopsin. Theoretical calculations for the isorhodopsin isomerization for $X_0 = 0.5$, $\tau_2 = 200$ fs suggest a time delay of 575 – 660 fs $\left(k_I \approx 1.7 - 1.5 \times 10^{12}\,\text{s}^{-1}\right)$ depending on the sink transfer rate constant $\left(k_s \approx 2 - 1.5 \times 10^{12}\,\text{s}^{-1}\right)$.

These calculations predict comparatively long-distant nuclear motion, i.e., $X_0 = 0.5$ for isorhodopsin as against 0.25 – 0.3 for rhodopsin. The high X_0 value can also be attributed to the lack of any strong intramolecular nonbonded interactions that distort the polyene skeleton in the vicinity of the *cis* bond. Prediction of relatively long-distant motion in the S_1 surface for isorhodopsin is in good agreement with the photochemical quantum yield analysis where the -*cis* isomerization shows a wavelength and temperature dependence while rhodopsin isomerization does not. This indicates the excited-state relaxation is the rate-limiting process in the isomerization dynamics of isorhodopsin. The wavelength dependence of the reaction also seems to indicate this fact.

In addition to the high X_0 value (compared to the rhodopsin isomerization), the slow relaxation dynamics of isorhodopsin is attributed to the long-time, slow component decay time constant, τ_2, from our theory. The high S_1 value indicates the excited-state electronic relaxation of isorhodopsin (-*cis* isomer) is controlled by the slow excited-state dynamics. Recent experimental results show the excited state absorption persists even after 150 fs for the -*cis* isomer. Note here, the S_1 surface absorption for rhodopsin is complete before 100 fs. In addition, the spectral breathing of the stimulated emission of isorhodopsin also indicates the wave packet motion in the S_1 surface is slower than the rhodopsin case. Thus, the relatively long-lived excited-state absorption explains the high τ_2 value.

8.7.3 Conflicting Crystal Violet Isomerization Mechanism

Another interesting problem is the isomerization dynamics of crystal violet [14]. Absorption studies on crystal violet suggest two ground-state conformers to be present under thermal equilibrium, one isomer to have D_3 symmetry structure with all three phenyl rings tilted in one direction and proposed the other to have C_2 symmetry (one phenyl ring is tilted in the opposite direction). Schematic representations of these conformers are shown in Fig. 8.14. The proposed C_2 symmetry structure has been under constant study and debate for the past five decades. In fact, the existence of two ground-state conformers itself has been questioned. Recently, it is also confirmed the presence of two ground-state isomers using femtosecond pump-probe measurements and predicted the other conformer to have a C_3 symmetry structure using molecular-orbital calculations.

The viscosity-mediated temperature crossover reported in higher alcohols can be confirmed from Fig. 8.12(a) as arising due to the competition between S_1 state relaxation and sink transfer for rate governing.

Temperature-variation studies on crystal violet isomerization in lower alcohol solvents indicate the overall rate constant is independent of temperature. Even though only a small temperature range between 283 and 313K has been studied, their results suggest from Fig. 8.12(b) that the sink transfer rate is comparable to or slightly less than the excited-state relaxation time constant in the lower alcohol solvents.

Wavelength dependence on barrierless isomerization kinetics of crystal violet has provided several interesting insights. Experimental studies reveal in this particular case no strong evidence of wavelength dependence on the isomerization kinetics. This is owing to the fast relaxation on the excited-state surface. This suggests the sink transfer rate to be the rate-governing step in the crystal violet isomerization.

8.8 SUMMARY

Barrierless chemical and biological reactions are surprisingly common in nature, ranging from electronic relaxation in dye molecules to ligand binding to enzymes. But those are not the only reasons that these reactions attracted so much attention from the scientific community. As these reactions proceed at an astonishingly fast rate, accurate measurement of rates of these reactions posed a great challenge to experimentalists. Such measurements had to await the advancement of ultrafast laser technology. When measurements indeed became available, a

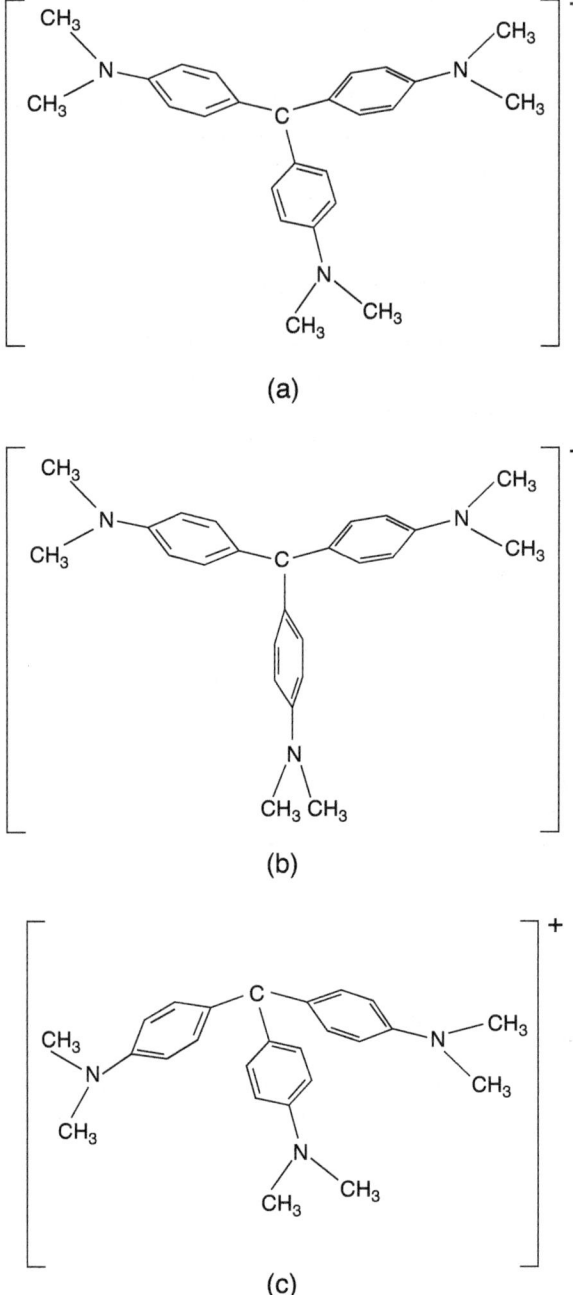

Figure 8.14 The proposed conformations of crystal violet in solution, (a) equilibrium ground-state conformer with D_3 symmetry (propeller structure) with all the phenyl rings tilted about the same molecular plane, (b) distorted propeller structure with C_2 symmetry as the other ground-state isomer, and (c) symmetry-breaking C_3 pyramidal structure as a solvation isomer to the propeller D_3 structure (a).

host of interesting results were obtained, such as nonexponential decay kinetics, fractional viscosity dependence, the excitation wavelength dependence of decay, and the weak negative departure coefficient of rate. A new class of models was proposed to explain these results. We termed them as the Standard Model of barrierless reactions. Theoretical studies showed that the observed results arise essentially from nonsteady-state population dynamics on the reactant surface. Although theoretical formulation proved rather easy, solution of the proposed equations was not simple because in most cases solutions had to be obtained numerically. Nevertheless, agreement between theory and experiment can be considered satisfactory.

In the next chapter we shall consider yet another class of barrierless reactions that are not driven by a reaction potential energy surface but need to overcome an entropic barrier.

REFERENCES

1. R. W. Schoenlein, L. A. Peteanu, R. A. Mathies, and C. V. Shank, *Science*, 254, 412 (1991).
2. T. Forster and G. Hoffmann, *Z. Phys.* 75, 63 (1971).
3. D. A. Cremers and M. W. Windsor, *Chem. Phys. Lett.* 71, 27 (1980).
4. D. Ben-Amotz and C. B. Harris, *J. Chem. Phys.* 86, 4856 (1987).
5. G. Oster and N. Nishijima, *J. Am. Chem. Soc.* 78, 1581 (1956).
6. B. Bagchi, G. R. Fleming, and D. W. Oxtoby, *J. Chem. Phys.* 78, 7375 (1983); B. Bagchi and G. R. Fleming, *J. Phys. Chem.* 94, 9 (1990).
7. H. Sumi and R. A. Marcus, *J. Chem. Phys.* 84, 4894 (1986).
8. N. Agmon and J. J. Hopfield, *J. Chem. Phys.* 78, 6947 (1983).
9. B. Bagchi, S. Singer, and D. W. Oxtoby, *Chem. Phys. Lett.* 99, 225 (1983).
10. B. Bagchi, *J. Chem. Phys.* 87, 6257 (1987).
11. R. A. Denny and B. Bagchi, *J. Phys. Chem. A* 103, 9061 (1999).
12. J.-L. Alvarez, A. Yartsev, U. Åberg, E. Åkesson, and V. Sundström, *J. Phys. Chem. B* 102, 7651 (1998).
13. L. A. Peteanu, R. W. Schoenlein, Q. Wang, R. A. Mathies, and C. V. Shank, *Proc. Natl. Acad. Sci. U.S.A.* 90, 11762 (1993).
14. G. N. Lewis, T. T. Mayel, and D. J. Lipkin, *J. Am. Chem. Soc.* 64, 1774 (1942).

9

Dynamical Disorder, Geometric Bottlenecks, and Diffusion-Controlled Bimolecular Reactions

9.1 INTRODUCTION

In the previous chapter we discussed in detail the dynamics of barrierless or activationless reactions in solutions that are often involved in isomerization and electron transfer. The isomerization reactions that are often studied involve large-amplitude motion of the bulky groups through space, as discussed extensively in the last chapter. Electron-transfer reactions involve the motion of the collective solvent polarization coordinate. While each solvent molecule might move a little, the total polarization might undergo substantial fluctuation during an electron-transfer reaction. These reactions are driven by a reaction potential energy surface that guides the reactant to the reaction window.

There are, however, many physicochemical and biological processes where such a guiding reaction potential energy surface is either absent or rather weak. The crucial rate-determining step can be a passage through a small constriction or bottleneck. That is, the free-energy barrier is *entropic* in origin. Sometimes this "gate" or bottleneck fluctuates with time and a successful crossing of the gate depends critically on this fluctuation. Such models have come to be known as "Dynamic Disorder Models" [1]. One important difference of this class of models from the one studied in the last chapter is that here one may use (in some, but not all, cases) a Hamiltonian description of the dynamics where the system is represented as a point particle moving in a potential given by the Hamiltonian such that *the energy and momentum conservation laws are obeyed.* They constitute an exciting class of models as one can understand through them the conditions for ergodicity of a system, and also approach to chaos.

In the following we shall develop the ideas described above with concrete examples. Many of the examples are taken from recent literature.

9.2 PASSAGE THROUGH GEOMETRIC BOTTLENECKS

In many complex systems, particularly in systems of biological interest, it has been observed that the rate-determining step is a passage through a narrow configurational space. Such processes are often very slow, even though no activation energy is involved in the rate-determining step. Examples are abundant in nature, for example, passage of molecules through zeolite cages, ligand binding to myoglobin, to name a few. Such processes have been modeled as barrierless processes. Sometimes the channels are periodic, as in the case of zeolite. In a series of papers, Zwanzig studied several elegant models. A few of them are discussed below [2, 4].

9.2.1 Diffusion in a Two-Dimensional Periodic Channel

The effective long-time diffusion coefficient (D^*) of a particle diffusing in a periodic force field, as shown in Fig. 9.1, can be much smaller than its local diffusion coefficient D because the particle has to go through a periodic array of potential barriers created by the narrow necks [2].

For diffusion of a particle in the preceding two-dimensional channel, unbounded in the x-direction but bounded by periodic walls in the y-direction, the corresponding diffusion equation is

$$\frac{\partial C(x, y, t)}{\partial t} = D\nabla^2 C(x, y, t), \qquad (9.1)$$

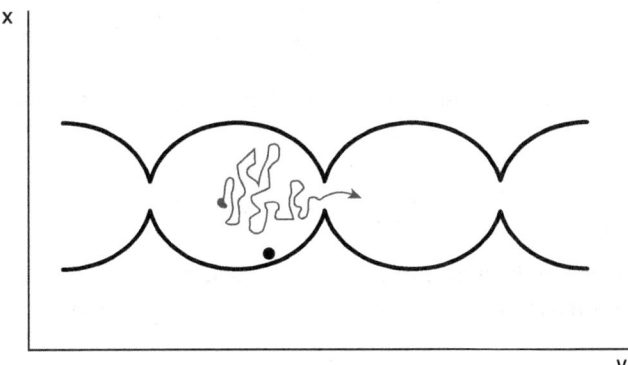

Figure 9.1 Two-dimensional periodic channel separated by narrow necks or constrictions. These necks act as entropic bottlenecks because of their low configurational weight.

where $C(x, y, t)$ is the concentration at position (x, y) at time t and the effects of the walls are accounted for by the usual no-flux boundary condition on the walls of the periodic cannel:

$$\hat{n} \cdot \nabla C = 0, \qquad (9.2)$$

where $\hat{n}$ is a unit vector normal to the wall. The region of diffusion is bounded by $-\infty < x < +\infty$, $y_1(x) < y < y_2(x)$. These functions are periodic with periodicity 2π.

One now exploits the useful analogy of the diffusion equation with the time-dependent Schrödinger equation. One follows a procedure analogous to the problem of finding an "effective mass" for a quantum-mechanical particle moving in a periodic medium, where the effective diffusion coefficient, D^*, can be calculated by solving the eigenvalue equation for the diffusion operator, satisfying the required boundary conditions on the walls,

$$\nabla^2 \psi = -\lambda \psi, \quad \hat{n} \cdot \nabla \psi = 0. \qquad (9.3)$$

As the system is periodic, the eigenfunctions satisfy the Bloch condition,

$$\psi(x + 2\pi, y) = \exp(2\pi i Q) \psi(x, y), \qquad (9.4)$$

where Q is an arbitrary wave vector. There is a set of eigenvalues $\{\lambda_m(Q)\}$ corresponding to each Q and for any given Q the smallest eigenvalue is $\lambda_0(Q)$. The effective diffusion coefficient is given by the limit

$$\frac{D^*}{D} = \lim_{Q \to 0} \frac{\lambda_0(Q)}{Q^2}. \qquad (9.5)$$

After some rigorous mathematical analysis using complex variable theory Zwanzig obtained the following rather unusual expression for the effective diffusion coefficient:

$$\frac{D^*}{D} = \left(1 + \frac{1}{2}a^2 \frac{\sinh 2V}{2V}\right)^{-1}. \qquad (9.6)$$

This result is for a symmetrical channel, where a and V are determined by

$$Y_{\min} = 2(V - a \sinh V), \qquad (9.7)$$

$$Y_{\max} = 2(V + a \sinh V). \qquad (9.8)$$

In the limit, when the maximum width approaches infinity (keeping the minimum width constant), the effective diffusion coefficient vanishes as

$$D^*/D \approx 4/Y_{\max}. \qquad (9.9)$$

Thus, the entropic barrier essentially kills the long-time diffusion coefficient in the limit of very wide "traps," that is, here the width in the y-direction acts as a barrier or trap. In general, the effective diffusion coefficient is always smaller than the local diffusion coefficient.

9.2.2 Diffusion in a Random Lorentz Gas

A random Lorentz gas is a model system that consists of a single point particle moving in a triangular array of immobile disk scatters. The point particle moves with a constant velocity between elastic collisions with the disks, so that the energy of the particle (which is purely kinetic) is conserved. The speed of the particle and the radius of the disks are taken as 1, while the lattice spacing is $2 + W$, where W is the separation parameter. Each collision changes the direction of motion of the point particle, that is, each collision changes the velocity while the speed remains constant, so that the momentum of the particle remains conserved. Each collision leads to a mixing of trajectories and many collisions should make the system forget its memory, that is, make it ergodic. At close packing, when $W = 0$, the moving particle is trapped in a single triangular region formed between three disks. See Fig. 9.2 for an illustration of the model.

Bunimovich and Sinai [3] obtained rigorous results for the high-density regime of this regular Lorentz model. One important corollary of their work is that the motion of the particle in this high-density regime is highly ergodic and the velocity auto-time-correlation function decays exponentially with the number of collisions.

In an interesting study, Machta and Zwanzig [4] studied the diffusion coefficient of the point-particle gas by using a simple and appealing approach. They derived an expression for the diffusion coefficient in terms of the separation

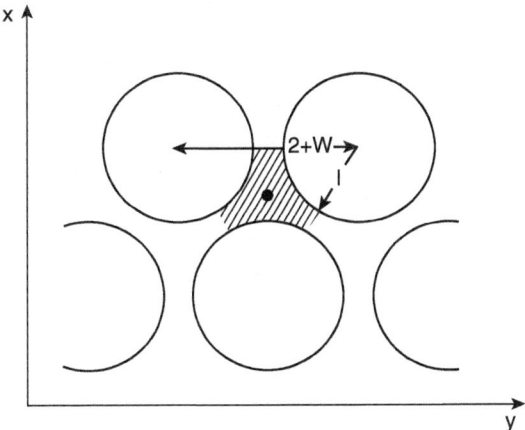

Figure 9.2 The geometry of the scatterers in the periodic Lorentz gas. The region marked with cross lines indicates a single triangular trapping region. Reprinted with permission from R. Zwanzig and J. Machta, *Phys. Rev. Lett.* **50**, 1959 (1983). Copyright (1983) American Physical Society.

parameter W, assuming that the exact motion of the particle in this high-density regime can be replaced by a random walk between triangular trapping regions (see shaded area in Fig. 9.2). The final expression for the diffusion coefficient is

$$D_{rw}(W) = \frac{W}{\pi}(2+W)^2 \left[\sqrt{3}(2+W)^2 - 2\pi\right]^{-1}. \quad (9.10)$$

Equation (9.10) has been derived based on the assumption that jumps between trapping zones are uncorrelated with one another. This feature arises in this model due to strong mixing properties of scattering from the surfaces of the negative curvature, thus the moving particle collides several times in each trap before moving to the next one. Therefore, based on these assumptions the periodic Lorentz gas model can be mapped onto a stochastic model.

The agreement between the above analytical expression for the diffusion coefficient and the values generated through computer simulations is very good for small values of W. As the value of W increases, which means the spacing between the hard discs increases, the agreement also decreases, indicating a breakdown of the ergodic behavior.

A subsequent extension of the above model to the diffusion of a point particle with constant energy in a continuous triangular lattice potential was carried out that demonstrated the role of focusing and defocusing of the particle trajectory by the potential.

9.3 DYNAMICAL DISORDER

A process in which the rate-determining step is a passage through a fluctuating bottleneck represents a model of dynamical disorder. The initial idea about the necessity of such a model or picture came from the kinetic studies of ligand binding to myoglobin where the movement of ligands inside the protein is dependent on the energy barriers that the ligand faces during its sojourn from solvent outside to interior of the protein [5]. The rate is also found to be dependent on solvent viscosity. The well-known Kramers' theory for the diffusion over a high potential barrier provides a simple dependence of rate on solvent viscosity and predicts that rate constants are inversely proportional to the viscosity and the decay kinetics is exponential. However, it is not clear how the motion of a ligand over a barrier inside the protein can be directly affected by the solvent outside the protein. Furthermore, experimental evidence suggests an approximately power-law temporal decay of binding kinetics at low temperatures [5]. This nonexponential (power-law) decay is due to the distribution of barrier heights as the different conformational substrates offer different (entropic) barriers to rebinding.

First let us discuss rate processes with *static* disorder. The following discussion follows the treatment of Zwanzig rather closely [1]. The rate-law equation for the ligand concentration can be written as

$$\frac{dC}{dt} = -k(X)C, \quad (9.11)$$

where $k(X)$ is the rate constant, which depends on the barrier height X:

$$k(X) = k_0 \exp(-X/k_B T). \tag{9.12}$$

If the probability of finding the barrier height between X and $X + dX$ is $\rho(X)dX$, then the time dependence of the average ligand concentration is

$$\langle C(t) \rangle = C(0) \int dX \rho(X) \exp(-k(X)t). \tag{9.13}$$

The time dependence of the concentration is no longer exponential in time. This is rather trivial and such a picture has been used over many years to explain experimentally observed nonexponential decay in glassy system.

However, the situation becomes considerably more complicated when the barrier X is a random function of time. As such a situation occurs in many systems, we separate it out from static disorder and call it a dynamic disorder. The solution of the time-dependent rate equation is

$$C(t) = C(0) \exp\left[-\int_0^t dt' k(X(t'))\right]. \tag{9.14}$$

If X fluctuates very fast, then one may replace $k(X(t'))$ by $\langle k \rangle$ which leads to the exponential decay, and this is the reason for the exponential binding kinetics at higher temperatures.

Let us now assume that dynamics of $X(t)$ can be determined by solving the following Langevin equation:

$$\frac{dX(t)}{dt} = -\lambda X(t) + R(t), \tag{9.15}$$

where λ is the decay rate and $R(t)$ is a Gaussian white noise, $\langle R(t_1) R(t_2) \rangle = 2\lambda \theta \delta(t_1 - t_2)$. The solution of Eq. (9.15) is given by

$$X(t) = X(0)e^{-\lambda t} + \int_0^t dt' e^{-\lambda(t-t')} R(t'). \tag{9.16}$$

Now the solution of Eq. (9.14) with X given by Eq. (9.16) is nontrivial due to the fact that the expression of $X(t)$ involves a random force term and one has to take the average over noise. However, to overcome this difficulty an alternative approach has been proposed, which is described below. The following approach is fairly simple and intuitively appealing.

If $P(C, X, t)$ denotes the joint probability distribution function of variables C and X, then the equation of motion for the noise averaged probability density function can be written as

$$\frac{\partial P}{\partial t} = -\frac{\partial}{\partial C}(-k(X)CP) - \frac{\partial}{\partial X}(-\lambda XP) + \frac{\partial}{\partial X}\left(\lambda\theta\frac{\partial P}{\partial X}\right). \quad (9.17)$$

Equation (9.17) is a Fokker–Planck equation described earlier in Chapter 2. The average of time-dependent concentration is

$$\langle C(t)\rangle = \int dX \int dC\, CP(C, X, t). \quad (9.18)$$

In this approach the highly nontrivial work of performing noise average is replaced by the task of solving a partial differential equation.

This approach is similar to the stochastic Liouville equation approach pioneered by Kubo to treat the effects of environmental fluctuations on magnetic line shape (discussed in detail in **Chapter 13**), and has been extended later to treat the line shape in other systems also. In many of these applications, one is able to derive a reduced equation of motion for $\langle C(t)\rangle$.

9.4 DIFFUSION OVER A RUGGED ENERGY LANDSCAPE

In many natural processes, a molecule or a reacting system has to move over a potential energy surface that has multiple maxima and minima. Well-known examples include protein folding, enzyme kinetics, and diffusion in supercooled liquids [6–9]. In the first case, the system may diffuse under an overall funnel-shaped potential which is populated at small length scales by multiple maxima and minima, as shown in Fig. 9.3 given below [6]. In the case of enzyme kinetics, enzyme conformational motion on a rugged energy landscape can explain results observed in single-molecule spectroscopic studies [8, 9]. In the case of diffusion in a supercooled liquid, the motion of the whole system might be considered as diffusion on a flat potential with many maxima and minima. A one-dimensional caricature of funnel-shaped potential is shown in the inset of Fig. 9.3.

Recently, one more example has gained importance. This is the diffusion of a nonspecifically bound protein along a DNA double helix in search of the specific binding site [8]. Here the apparently random sequence of base pairs in a DNA poses a rugged potential energy surface on which the protein executes a diffusive motion.

In all the above examples, the depth of the minima and the height of the maxima involved are rather small, of the order of one or two $k_B T$. However, even such small minima or maxima may play important roles in determining the magnitude of the diffusion coefficient. It is crucially important to understand that diffusion here is a macroscopic long-time process involving mean square displacement over distances much larger than the scale over which the rugged

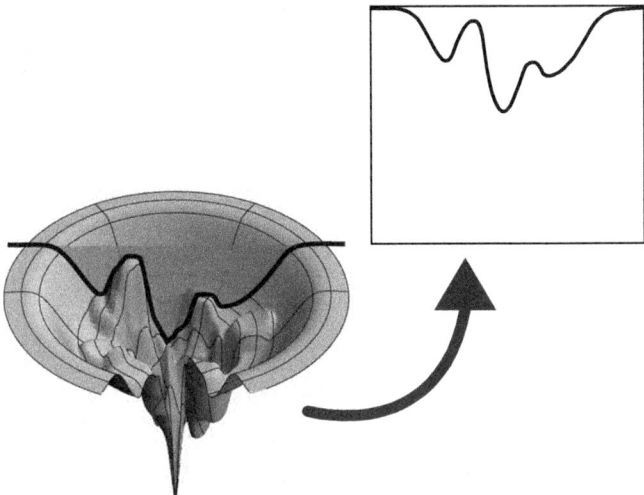

Figure 9.3 A rugged funnel-shaped landscape for protein folding. In the *inset* one-dimensional illustration of funnel shaped potential is shown [http://www.dillgroup.ucsf.edu/].

potential varies. This also limits the validity of the present discussion to specific cases, like protein diffusion on a long stretch of a DNA molecule.

Fortunately, a closed-form expression for the dependence of the diffusion coefficient on ruggedness is available [7]. Let us consider the situation depicted in Fig. 9.3, except we place the fluctuations in the potential minima and maxima regularly in a one-dimensional lattice, as shown in Fig. 9.4. We assume that the total potential energy acting on the diffusing particle is given by

$$V(X) = X^2 + \varepsilon \left(\cos k_1 X + \sin k_2 X \right), \qquad (9.19)$$

where ε is a measure of the "roughness" of the potential, as it determines the heights of the maxima and the depths of the minima.

The Brownian motion of the system on potential $V(X)$ is described by the Smoluchowski equation

$$\frac{\partial P(X,t)}{\partial t} = \frac{\partial}{\partial X} D e^{-\beta V(X)} \frac{\partial}{\partial X} e^{\beta V(X)} P(X,t), \qquad (9.20)$$

where D is the "bare" diffusion coefficient and $\beta = 1/k_B T$. The ruggedness renormalizes this "bare" diffusion coefficient. This equation describes the time (t) dependence of the probability distribution $P(X, t)$. Actually, the rough potential energy $V(X)$ can generally be assumed to be of the following separable form:

$$V(X) = V_0(X) + V_1(X), \qquad (9.21)$$

where $V_0(X)$ is a smooth background on which a random, oscillating perturbation $V_1(X)$ is superimposed. The expression of the renormalized diffusion coefficient

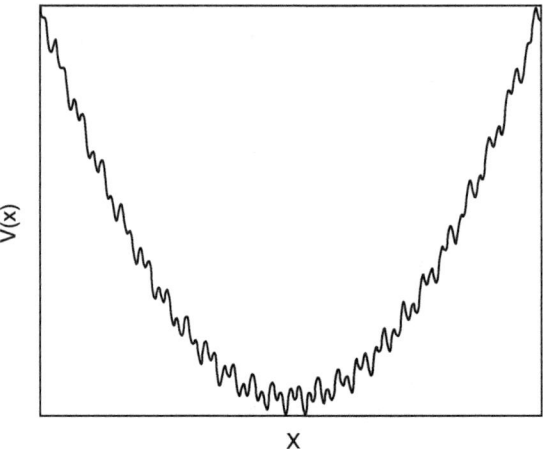

Figure 9.4 A rugged potential given by Eq. (9.19) is shown.

derived below also holds when $V_0(X) = 0$, that is, the motion occurs on a rugged but flat energy surface.

The mean time required for a system starting out at X_0 to reach X for the first time is known as the *mean first passage time* (MFPT) and is denoted by τ. We assume that there is a reflecting boundary condition at $X = a$. For convenience we consider only $a < X_0 < X$. The MFPT is found by solving the differential equation

$$D e^{\beta V} \frac{\partial}{\partial X} e^{-\beta V} \frac{\partial}{\partial X} \tau(X) = -1, \qquad (9.22)$$

with an absorbing boundary condition at $X = X_0$. The solution is given by

$$\tau = \int_{X_0}^{X} dy\, e^{\beta V(y)} \left(\frac{1}{D}\right) \int_{a}^{y} dz\, e^{-\beta V(z)}. \qquad (9.23)$$

It is worth noting here that the integrations in this equation have the effect of *spatial averages* (a trick used repeatedly by Zwanzig in many applications of statistical mechanics!). That is, the integral over a small distance ΔX may be approximated by

$$\int dz\, e^{-\beta V(z)} \simeq \int dz\, e^{-\beta V_0(z)} \left\langle e^{-\beta V_1(z)} \right\rangle, \qquad (9.24)$$

where the bracket, $\langle \ \rangle$, denotes the spatial average over the roughness to smooth the potential. If the amplitude of the fluctuations in V_1 varies with z, then the averaged quantity still remains a function of the coordinate. We denote

$$\langle e^{-\beta V_1(z)} \rangle = e^{\psi^-}(z) \text{ and } \langle e^{+\beta V_1(z)} \rangle = e^{\psi^+}(z).$$

With these approximations, the modified MFPT is given by

$$\tau = \int_{X_0}^{X} dy \, e^{\beta V_0(y) + \psi^+(y)} \left(\frac{1}{D}\right) \int_{a}^{y} dz \, e^{-\beta V_0(z) + \psi^-(z)}. \qquad (9.25)$$

On further simplification the above equation can be written as

$$\tau = \int_{X_0}^{X} dy \, e^{\beta V_0(y) - \psi^-(y)} e^{\psi^+(y)} \left(\frac{1}{D}\right) e^{\psi^-(y)} \int_{a}^{y} dz \, e^{-\beta V_0(z) + \psi^-(z)}$$

$$= \int_{X_0}^{X} dy \, e^{\beta V^*(y)} \left(\frac{1}{D^*}\right) \int_{a}^{y} dz \, e^{-\beta V^*(z)}. \qquad (9.26)$$

Thus, we approximate a rough potential $V(X) = V_0(X) + V_1(X)$ with an effective potential $V^*(X) = V_0(X) - \psi^-(X)/\beta$, and the diffusion coefficient is approximated by an effective diffusion coefficient $1/D^*(X) = e^{\psi^+(X)} \times (1/D) e^{\psi^-(X)}$.

The corresponding effective Smoluchowski equation, valid only for distances much larger than the characteristic length scale of the fluctuations in V, is expected to be

$$\frac{\partial P(X,t)}{\partial t} = -\frac{\partial J}{\partial X}, \qquad (9.27)$$

with

$$J = -D^*(X) \, e^{-\beta V^*(X)} \frac{\partial}{\partial X} e^{\beta V^*(X)} P(X,t). \qquad (9.28)$$

However, as Zwanzig has noted, this is only a conjecture; there is no direct derivation of the effective Smoluchowski equation for a rough potential. However, the MFPT predicted by this effective Smoluchowski equation agrees with the MFPT obtained by the spatial averaging process.

If the amplitude of the roughness in V_1 is independent of X and has a Gaussian distribution, with a probability proportional to $\exp\left(-V_1^2/2\varepsilon^2\right)$, in which ε is the root-mean-squared roughness, $\varepsilon^2 = \langle V_1^2 \rangle$, then we can easily find that

$$e^{\psi^+} = e^{\psi^-} = e^{\beta^2 \varepsilon^2/2}, \tag{9.29}$$

and the effective diffusion coefficient is

$$D^* = D \exp\left[-(\varepsilon/k_B T)^2\right]. \tag{9.30}$$

This strong dependence on ruggedness (ε) was not anticipated.

Note the super-Arrhenius dependence of the diffusion constant on ε that gives a measure of the ruggedness of the landscape. That is, the effective long-time diffusion constant decreases rapidly with increase in ε.

Equation (9.30) embodies a general result that is expected to be valid for small ε for many different types of systems. Note also the inverse quadratic temperature dependence of D^*. However, this simple expression is valid only for small ε, which, as was already noted, should be of the order of $k_B T$ or less.

9.5 DIFFUSION-CONTROLLED BIMOLECULAR REACTIONS

Here we shall discuss a class of common chemical reactions that are both barrierless and diffusion controlled yet they display certain unique features that are absent in all of the models discussed previously. These are bimolecular diffusion-controlled reactions we commonly encounter in organic and inorganic chemistry [10–12]. The simplest example of bimolecular reactions in solution is an $S_N 2$ reaction. Another currently popular example is fluorescence quenching, which is characterized by the following reaction:

$$A^* + B \rightarrow A + B$$

where A^* is an excited state of A, and B is a quencher. Another example is diffusive migration of electronic excitation between molecules that are randomly distributed in space [13].

When such a bimolecular reaction takes place in solution, the reactant and the product diffuse through solution until they encounter each other and react. There is often no *molecular* potential energy surface of the reaction.

The first theoretical study of this problem was carried out by Smoluchowski [10] at the beginning of the last century, who treated the kinetics of colloid coagulation as a diffusion-controlled process. The diffusion equation was then solved with an absorbing boundary condition where the probability is zero at the surface of a sphere of radius R. This sphere represents the distance of closest approach of two spheres and is called the reaction radius. If $c(r,t)$ is the time- (t) and separation- (r)

dependent concentration of the reactant pair, then the equation of motion has the form of a *dynamical disorder model*:

$$\frac{\partial c(r,t)}{\partial t} = D_M \nabla^2 c(r,t) - k(r)c(r,t), \tag{9.31}$$

where D_M is the mutual diffusion coefficient and is equal to sum of the diffusion coefficients of reacting species, $D_M = D_{A^*} + D_B$.

The boundary value calculation provides the following well-known expression for the concentration:

$$c(r,t) = c_0 \left[1 - \frac{R}{r} \operatorname{erfc} \frac{r-R}{\sqrt{4D_M t}} \right], \tag{9.32}$$

where complementary error function is defined as usual by

$$\operatorname{erfc} x = \frac{2}{\sqrt{\pi}} \int_x^\infty dy\, e^{-y^2}. \tag{9.33}$$

Here c_0 is the uniform initial concentration of the diffusing species. The rate is given by the flux across the spherical boundary at $r = R$. This rate is time dependent and is given by

$$k(t) = 4\pi D_M R c_0 \left(1 + \frac{R}{\sqrt{\pi D_M t}} \right). \tag{9.34}$$

The long-time limit of the above expression gives the well-known Smoluchowski rate

$$k = 4\pi D_M R c_0. \tag{9.35}$$

Although this expression is one of the most used expressions in chemical kinetics, it is limited by several factors.

(i) First, it assumes that the reaction occurs with unit probability as the two spheres reach a separation R. In practice, reactions occur only with a finite rate and there is always a certain probability that reactants escape from the reaction sphere unreacted. This will be the case particularly if the reaction requires an orientational match. Collins and Kimball [14] described the finite rate of reaction at contact distance by using a partially reflecting or radiation boundary condition at the contact distance.

(ii) The second limitation is that reactants often interact through a long-range interaction, which is neglected in deriving Eq. (9.35). If the reactant and the product are charged, then their mutual diffusion is influenced by the Coulombic interaction energy term. The most well-known study in this area was carried out by Onsager in 1938 [15] who presented a theoretical study of *"the probability that a pair of ions of given initial separation will recombine with each other,"* by using the laws of Brownian motion. Onsager used the Smoluchowski diffusion equation to describe the Brownian motion of ions under the influence of the external field and their own Coulombic attraction. Onsager obtained an analytic solution only for the steady-state escape probability. Later, Hong and Noolandi [16] gave an analytical solution of the time-dependent Onsager problem.

Let us now discuss the more general type of reactions in solution that occur under the influence of some potential, $V(r)$ [17]. For simplicity, let us assume that $V(r)$ depends only on the relative separation of the reacting pair (r). The reaction occurs with certain probability when the reacting species diffuse together and interact. The equation of motion for the probability of finding the reacting species at a relative separation r is given by the following Smoluchowski equation:

$$\frac{\partial p(r,t)}{\partial t} = D_M \nabla \cdot e^{-\beta V(r)} \nabla e^{\beta V(r)} p(r,t). \quad (9.36)$$

For the spatial dimension d, the above equation becomes

$$\frac{\partial p(r,t)}{\partial t} = \frac{D_M}{r^{d-1}} \frac{d}{dr} r^{d-1} e^{-\beta V(r)} \frac{d}{dr} e^{\beta V(r)} p(r,t). \quad (9.37)$$

The initial condition depends on the experimental condition. One natural choice for the initial condition is the equilibrium condition and is given by

$$p(r,0) = N e^{-\beta V(r)} \quad (N = \text{Normalization constant}). \quad (9.38)$$

The time-dependent rate constant is the normal inward component of total flux at contact distance, $r = R$. Mathematically,

$$k(t) = \int d\sigma \hat{n} \cdot D_M \left[e^{-\beta V(r)} \nabla e^{\beta V(r)} p(r,t) \right]_{r=R}. \quad (9.39)$$

The final expression for the time-dependent rate constant is again obtained from the flux and is given by

$$k(t) = \frac{2\pi^{d/2} D_M R^{d-1} e^{-\beta V(R)}}{\Gamma(d/2)} \left[\frac{d}{dr} e^{\beta V(r)} p(r,t) \right]_{r=R}. \quad (9.40)$$

Thus, in order to calculate the final time-dependent rate one has to solve the Eq. (9.37) to find the expression for $p(r,t)$, to be used in Eq. (9.40).

If the reaction probability is unity when the reacting species come into contact, then this is described by an absorbing (Smoluchowski) boundary at contact, $p(R, t) = 0$. However, if there is only a finite probability of reaction at contact, then the radiation boundary condition given as

$$\int d\sigma \hat{n} \cdot D_M \left[e^{-\beta V(r)} \nabla e^{\beta V(r)} p(r, t) \right]_{r=R} = k_0 p(R, t), \quad (9.41)$$

which is clearly more appropriate in most cases. Here k_0 is the intrinsic rate of reaction. The absorbing (Smoluchowski) boundary condition is the special case of radiation boundary condition in $k_0 \to \infty$ limit.

Let $k_S(t)$ be the time-dependent rate constant calculated from the Smoluchowski boundary condition, $k_{CK}(t)$ be the rate constant calculated from the radiation boundary condition, and $\hat{k}_S(z)$ and $\hat{k}_{CK}(z)$ be the Laplace transforms of the respective rates; then, for spherically symmetric molecules, an elegant relation of $\hat{k}_{CK}(z)$ can be derived in terms of $\hat{k}_S(z)$ as

$$\hat{k}_{CK}(z) = \frac{k_0 e^{-\beta V(R)} \hat{k}_S(z)}{k_0 e^{-\beta V(R)} + z \hat{k}_S(z)}. \quad (9.42)$$

Equation (9.42) provides a useful expression for the time-dependent rate for radiation boundary condition in terms of the time-dependent rate constant for absorbing or Smoluchowski boundary condition for a bimolecular reaction under the potential $V(r)$. Experiments and simulations indeed find a time-dependent rate, which can be at least partly fitted by Eq. (9.42). Szabo has presented several generalizations of Eq. (9.42), which can be obtained from Ref. 17.

9.6 SUMMARY

The concept of a chemical reaction as a passage through a narrow opening (or, hole) came originally from ligand (or, substrate) binding to proteins. Development in this area was driven by several simple but elegant models by Zwanzig who showed how escape of a particle from a small hole can be described as a reaction with an entropic barrier. He also developed the concept of dynamical disorder to include the effects of fluctuations in the size of the hole or constriction. This perspective leads to unification of many different chemical and biological problems, as discussed above.

In addition, the problems considered under dynamical disorder or entropic bottleneck can be regarded as a subfield of the general area of diffusion-controlled reaction. These are distinct from activated barrier crossing processes as the rate is often limited by the narrowness of the configuration space that can give rise to reaction. Dynamical disorder gives us a method to deal with such problems which are common in chemistry and biology.

REFERENCES

1. R. Zwanzig, *Acc. Chem. Res.* 23, 148 (1990).
2. R. Zwanzig, *Physica A* 117, 277 (1983).
3. L. A. Bunimovich and Y. G. Sinai, *Commun. Math. Phys.* 78, 479 (1980).
4. R. Zwanzig and J. Machta, *Phys. Rev. Lett.* 50, 1959 (1983).
5. R. H. Austin, K. W. Beeson, L. Eisenstein, H. Frauenfelder, and I. C. Gunsalas, *Biochemistry* 14, 5355 (1975).
6. J. N. Onuchic, and P. G. Wolynes, *Curr. Opin. Struct. Bio.* 14, 70 (2004).
7. R. Zwanzig, *PNAS* 85, 2029 (1988).
8. P. C. Blainey, G. Luo, S. C. Kou, W. F. Mangel, G. L. Verdine, B. Bagchi, and X. S. Xie, *Nature Struct. Mol. Bio.* 16, 1224–29 (2009).
9. W. Min, X. S. Xie, and B. Bagchi, *J. Chem. Phys.* 131, 065104 (2009).
10. M. Z. Smoluchowski, *Z. Phys. Chem.* 92, 129 (1917).
11. D. F. Calef and J. M. Deutch, *Ann. Rev. Phys. Chem.* 34, 493 (1983).
12. S. A. Rice, in *Diffusion-Limited Reactions*, edited by C. H. Bamford, C. F. H. Tipper, and R. G. Compton, Comprehensive Chemical Kinetics Vol. 25 (Elsevier, Amsterdam, 1985).
13. S. W. Haan and R. Zwanzig, *J. Chem. Phys.* 68, 1879 (1978).
14. F. C. Collins and G. E. Kimball, *J. Colloid. Sci.* 4, 425 (1949).
15. L. Onsager, *Phys. Rev.* 54, 554 (1938).
16. K. M. Hong and J. Noolandi, *J. Chem. Phys.* 68, 5163 (1978).
17. A. Szabo, *J. Phys. Chem.* 93, 6929 (1989).

10

Electron-Transfer Reactions

10.1 INTRODUCTION

Electron-transfer processes are involved in many important chemical, physical, and biological reaction systems. Because of the ubiquitous presence of these processes in diverse systems, the understanding of electron-transfer processes is an important task. Electron-transfer reactions (ETR) between donor–acceptor pairs in condensed phases are often strongly coupled to polar solvation dynamics and vibrational relaxation processes. The coupling to solvent relaxation is easy to understand, especially for electron transfer in a polar liquid as the electric field of the charge itself is strongly coupled to solvent polarization. Interestingly, the realization that the vibrational modes and their dynamics can play important roles in the electron-transfer reactions came only recently with the development of femtosecond (fs) laser spectroscopic techniques. In fact, vibrational energy relaxation plays an important role in many photoinduced-electron-transfer reactions than the solvent polarization relaxation. Another relevant recent discovery is that ion solvation dynamics in many common dipolar liquids contains an ultrafast component with a time constant of the order of 100 fs or even less. Moreover, the amplitude of this ultrafast component is found to be significant. This raises the interesting question regarding the role of this component in the electron-transfer reaction, especially in the presence of the participating vibrational modes. Although our understanding of ETR has achieved dramatic progress recently, it still remains imperfect in many cases. Nevertheless, it is now clear that both vibrational energy and solvent polarization relaxations significantly influence the ET processes.

10.2 CLASSIFICATION OF ELECTRON-TRANSFER REACTIONS

10.2.1 Classification Based on Ligand Participation

Depending on the involvement of the ligand molecules in the electron-transfer process, electron-transfer reactions are broadly divided into two categories.

(a) **Outer-Sphere-Electron Transfer:** In outer-sphere-electron-transfer reactions no bonds are broken or formed, i.e., there is no change in the coordination sphere of either donor or acceptor molecule. An example is the following reaction:

$$Fe(CN)_6^{4-} + IrCl_6^{2-} \rightarrow Fe(CN)_6^{3-} + IrCl_6^{3-}.$$

(b) **Inner-Sphere-Electron Transfer:** An inner-sphere mechanism is one in which the donor and acceptor molecules share a bridging ligand in their inner or primary coordination spheres and the electrons are transferred across that bridging group:

$$[CoCl(NH_3)_5]^{2+} + [Cr(H_2O)_6]^{2+} \rightarrow [Co(NH_3)_5(H_2O)]^{2+} + [CrCl(H_2O)_5]^{2+}.$$

In the above example, $-$ Cl$-$ is the bridging group.

10.2.2 Classification Based on Interactions between Reactant and Product Potential Energy Surfaces

On the basis of the interactions between the donor and acceptor potential energy surfaces (PES's), electron-transfer reactions are characterized into two categories, (i) the diabatic and (ii) the adiabatic reactions, as shown in Fig. 10.1. The idea is that the total Hamiltonian of the system can be partitioned as

$$H = H_0 + V, \tag{10.1}$$

where H_0 is the unperturbed Hamiltonian. Reactants and products are the eigenfunctions of H_0. V is the perturbation that gives rise to electron transfer. Often the diabatic surfaces are assumed to be the Born–Oppenheimer approximate wave functions of the unperturbed Hamiltonian, H_0. The motion on a diabatic surface does not change the electronic state of the system. In the diabatic picture ET is induced by the perturbation term V, which couples reactant and product electronic states $|1\rangle$ and $|2\rangle$. The strength of the coupling is characterized by the matrix element $H_{12} = \langle 1| V |2\rangle$. This approximation is valid in the weak-coupling limit, and the ETR is also called nonadiabatic. On the other hand, when coupling is large, this is called adiabatic ETR. Adiabatic PES's are Born–Oppenheimer approximate solutions to the full Hamiltonian, H, of the system. In contrast to the

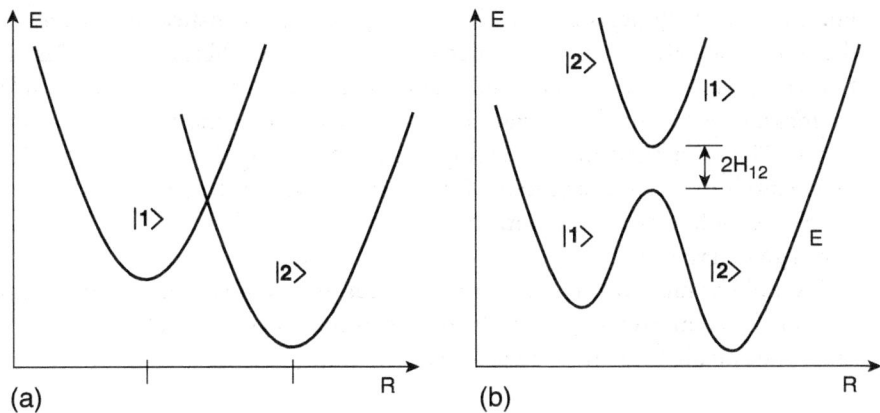

Figure 10.1 Schematic representation of diabatic (a) and adiabatic (b) potential energy surfaces.

diabatic PES, here the electronic state changes smoothly from that of the reactants to the products upon the proper change of nuclear coordinates. The adiabatic states can be constructed as a linear combination of diabatic states:

$$|\phi\rangle = c_1 |1\rangle + c_2 |2\rangle. \quad (10.2)$$

If the energies of $|1\rangle$ and $|2\rangle$ are E_1 and E_2, respectively, the energies of the new (adiabatic) surfaces are given by the following expression:

$$E_\pm \approx \frac{E_1 + E_2}{2} \pm \frac{\sqrt{\Delta_{12}^2 + 4H_{12}^2}}{2}, \quad (10.3)$$

where $\Delta_{12} = |E_1 - E_2|$. The electronic coupling matrix depends on distance of separation and is usually assumed to decrease exponentially with distance:

$$H_{12}^2 = J_0^2 \exp[-\beta(r - r_0)], \quad (10.4)$$

where β is the attenuation coefficient, and r_0 is the equilibrium distance between donor and the acceptor. J_0 is the magnitude of H_{12} at $r = r_0$.

Many of the common electron-transfer reactions fall into the weakly adiabatic limit, with the coupling ranging between 0.05 and 0.5 eV.

10.3 MARCUS THEORY

Electron-transfer reactions are inevitably described and understood in terms of the celebrated Marcus theory. Marcus first conceptualized the ET reaction as a process where the reactant crosses over the transition state to reach the product driven by the collective polarization fluctuations of the surrounding solvent molecules [1–5]. Although Marcus termed them as "nonequilibrium fluctuations," they are still described by equilibrium statistical mechanics and, therefore, they are just thermal fluctuations off the equilibrium state. Based on this view, Marcus developed a one-dimensional theory of ET with *the solvent polarization energy as the reaction coordinate* that defines the free-energy surface of ET. The celebrated Marcus theory has the following three basic ingredients: (A) identification and description of the reaction coordinate, (B) derivation of the expression of free-energy barrier, and (C) calculation of the rate starting from the Fermi golden rule.

It should be mentioned here that the Marcus expression of the electron-transfer rate was also derived independently by Levich and Dogonadze. However, here we follow essentially the derivation of Marcus.

10.3.1 Reaction Coordinate (RC)

The crucial point to realize is that the orientation of solvent dipoles around the reactant and the product are different. If an electron transfers from the reactant to the product while the surrounding solvent molecules are in the configuration in equilibrium to the reactant state (as in Franck–Condon principle where electron transfer is so fast that nuclear motions do not change during the excitation), then the energy is *not* conserved (see Fig. 10.2). This cannot be allowed because the reaction is not driven by an external excitation but by thermal fluctuations. Marcus, on the other hand, treats ET as a motion along solvent polarization as experienced by the reaction system. To a large extent, the merit and also the reason for the success of Marcus theory lies in the identification of the reaction coordinate. Note that this reaction coordinate is different from the reaction coordinates we usually encounter in chemical kinetics.

To motivate the definition of the reaction coordinate, let us consider the following example of electron transfer reaction:

$$Co^{2+} + Co^{3+} \rightarrow Co^{3+} + Co^{2+}. \tag{R-1}$$

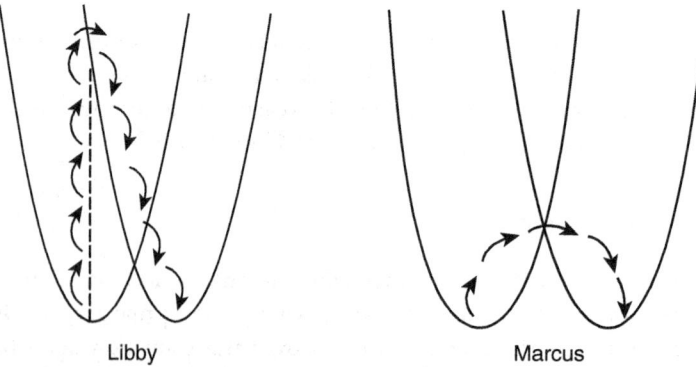

Libby Marcus

Figure 10.2 Schematic diagram of Libby's and Marcus's approach for ETR.

Unlike in the usual reactions, there is no appreciable structural (e.g., bond length, bond angle) change in ETR's in solution. So what is really changing? It is the charge state of the solute pair that is changing and, because the charge state is strongly coupled to the polarization of the solvent, the solvent polarization is also undergoing a change. Thus, we have several candidates for the choice of reaction coordinate. One is of course the distance R between the two reactant pair. The second is the solvent polarization. Owing to Coulombic interaction between the reactant pair, the reaction energy surface is dependent on the separation, R. So, let us consider electron transfer at constant separation R. Now, owing to the conservation of energy as mentioned above, electron transfer can occur only when the energy of the reactant and product are the same. The quantity that changes when the electron is transferred from one state to the other is the collective solvent polarization, which determines the solvation energy of the reactant and product. The stabilization due to solvation is a maximum for both reactant and product, but a minimum when the electron is in transit. Thus, when the electron is in the state intermediate between the reactant and the product (for example, half way towards electron transfer), that state can serve as the transition state of the reaction.

To describe this solvation energy difference, one must define a collective physical quantity as reaction coordinate, as in a phase transition. This collective physical quantity (X) is the interaction energy of the donor–acceptor pair with solvent just before and just after the electron transfer. That is, the reaction coordinate is the solvation energy difference between reactant and product at a given solvent polarization. This physical picture is quantified below.

Because in the dielectric continuum model the configuration of the solvent around solute pair is described by the orientational polarization function $\mathbf{P}(\mathbf{r})$, the expression of interaction energy (X) can be written as

$$X = \int d\mathbf{r} \left[\mathbf{D}_P(\mathbf{r}) - \mathbf{D}_R(\mathbf{r})\right] \cdot \mathbf{P}(\mathbf{r}), \tag{10.5}$$

where $\mathbf{D}_R(\mathbf{r})$ is the displacement vector due to the reactant and $\mathbf{D}_P(\mathbf{r})$ is the displacement vector due to the product. When the configuration of the solvent fluctuates, the polarization function $\mathbf{P}(\mathbf{r})$ changes with time so X also changes with time.

An alternative, simpler definition is to consider that the solvent polarization gives rise to a potential that is different at the donor and acceptor sites. Then the same reaction coordinate is given by

$$X = e(V_D - V_A), \tag{10.6}$$

where V_D and V_A are the electrostatic potentials at donor and acceptor sites due to permanent dipoles of solvent molecules.

The nature of the reaction coordinate can be understood easily by considering several different cases. If ETR is a symmetric self-exchange reaction, such as in reaction **R-1**, then the difference in displacement vector occurs due to the different spatial location of the charges before and after the reaction. Let us consider another common reaction:

$$A + B \rightarrow A^+ + B^-. \quad \text{(R-2)}$$

In this case, the initial state is not charged, but the final state is charge separated. Therefore, the reaction coordinate is essentially the energy of interaction between the electric field of the (A^+/B^-) pair with the polarization field of the solvent. That is, the reaction coordinate is essentially the solvation energy of the charge-transfer state. Therefore, the dynamics along the reaction coordinate could be largely controlled by the same dynamics that is probed by solvation dynamic experiments.

10.3.2 Free-Energy Surfaces: Force Constant of Polarization Fluctuation

Having identified the reaction coordinate, we now proceed to calculate the free-energy surface of the reaction and this is rather involved. Here we follow the succinct derivation given by Tachiya [5]. It is assumed that the free-energy surfaces of the reactant and the product are harmonic functions of the reaction coordinate (X) and have the same curvature, given by a force constant, k. Thus, the surfaces could be described as

$$G_1(X) = \frac{1}{2}kX^2, \quad (10.7)$$

$$G_2(X) = \frac{1}{2}k(X - X_0)^2. \quad (10.8)$$

One can find the force constant, k, and the equilibrium position, X_0, of free-energy surfaces in the following way. Let us consider a simple electron-transfer reaction as shown in reaction (R-2). In the dielectric continuum model, the configuration of the solvent around a donor and acceptor pair is described by the orientational polarization function $\mathbf{P}(\mathbf{r})$. The difference in the interaction energy of the donor–acceptor pair with the surrounding solvent molecules between initial and final states is given by

$$X = -\int_{\substack{|\mathbf{r}-\mathbf{r}_A|>a \\ |\mathbf{r}-\mathbf{r}_B|>b}} \left[\frac{e(\mathbf{r}-\mathbf{r}_A)}{|\mathbf{r}-\mathbf{r}_A|^3} - \frac{e(\mathbf{r}-\mathbf{r}_B)}{|\mathbf{r}-\mathbf{r}_B|^3} \right] \cdot \mathbf{P}(\mathbf{r})d\mathbf{r}, \quad (10.9)$$

where $\mathbf{r}_A$ and $\mathbf{r}_B$ stand for the positions of donor and acceptor molecules. We have assumed donor and acceptor as spherical molecules whose radii are a and b, respectively.

For the initial state, the free energy of a polarization field described by $\mathbf{P}(\mathbf{r})$ is given by

$$G_1 = 2\pi (\varepsilon_{op}^{-1} - \varepsilon_s^{-1})^{-1} \int\limits_{\substack{|\mathbf{r}-\mathbf{r}_A|>a \\ |\mathbf{r}-\mathbf{r}_B|>b}} [\mathbf{P}(\mathbf{r})]^2 d\mathbf{r}, \quad (10.10)$$

where ε_{op} and ε_s stand for optical and static dielectric constants, respectively. The screening effect of electronic polarization is neglected.

Because different polarization functions $\mathbf{P}(\mathbf{r})$ may produce the same interaction energy difference (X), we need to find the polarization function $\mathbf{P}(\mathbf{r})$ that minimizes the free energy G_1, along reaction coordinate (RC) given by Eq. (10.9). This can be determined by minimization of the following functional

$$F = 2\pi (\varepsilon_{op}^{-1} - \varepsilon_s^{-1})^{-1} \int\limits_{\substack{|\mathbf{r}-\mathbf{r}_A|>a \\ |\mathbf{r}-\mathbf{r}_B|>b}} [\mathbf{P}(\mathbf{r})]^2 d\mathbf{r}$$

$$+ \alpha \left[-X - \int\limits_{\substack{|\mathbf{r}-\mathbf{r}_A|>a \\ |\mathbf{r}-\mathbf{r}_B|>b}} \left[\frac{e(\mathbf{r} - \mathbf{r}_A)}{|\mathbf{r} - \mathbf{r}_A|^3} - \frac{e(\mathbf{r} - \mathbf{r}_B)}{|\mathbf{r} - \mathbf{r}_B|^3} \right] \cdot \mathbf{P}(\mathbf{r}) d\mathbf{r} \right]. \quad (10.11)$$

Minimizing F with respect to the variational parameter $\mathbf{P}(\mathbf{r})$ leads to the following Euler's equation:

$$\mathbf{P}(\mathbf{r}) = \frac{\alpha(\varepsilon_{op}^{-1} - \varepsilon_s^{-1})}{4\pi} \left[\frac{e(\mathbf{r} - \mathbf{r}_A)}{|\mathbf{r} - \mathbf{r}_A|^3} - \frac{e(\mathbf{r} - \mathbf{r}_B)}{|\mathbf{r} - \mathbf{r}_B|^3} \right]. \quad (10.12)$$

Using Eqs. (10.9), (10.10), and (10.12) one can write

$$G_1 = \frac{2\pi (\varepsilon_{op}^{-1} - \varepsilon_s^{-1})^{-1}}{I} X^2, \quad (10.13)$$

where integral I is given by

$$I = \int\limits_{\substack{|\mathbf{r}-\mathbf{r}_A|>a \\ |\mathbf{r}-\mathbf{r}_B|>b}} \left[\frac{e(\mathbf{r} - \mathbf{r}_A)}{|\mathbf{r} - \mathbf{r}_A|^3} - \frac{e(\mathbf{r} - \mathbf{r}_B)}{|\mathbf{r} - \mathbf{r}_B|^3} \right]^2 d\mathbf{r}. \quad (10.14)$$

Equation (10.14) can be approximated as

$$I = 4\pi e^2 \left(\frac{1}{a} + \frac{1}{b} - \frac{2}{R} \right), \tag{10.15}$$

where R is the distance between the donor and acceptor molecules. Now putting this expression of I into Eq. (10.13), we get

$$G_1 = \frac{X^2}{4\lambda_X}, \tag{10.16}$$

where

$$\lambda_X = \frac{e^2}{2} \left(\frac{1}{\varepsilon_{op}} - \frac{1}{\varepsilon_s} \right) \left(\frac{1}{a} + \frac{1}{b} - \frac{2}{R} \right). \tag{10.17}$$

λ_X is called solvent (or Marcus) reorganization energy. From Eqs. (10.7) and (10.16) force constant k is

$$k = \frac{1}{2\lambda_X}. \tag{10.18}$$

For the final state, the free energy of the system is given as

$$G_2 = 2\pi (\varepsilon_{op}^{-1} - \varepsilon_s^{-1})^{-1} \int\limits_{\substack{|\mathbf{r}-\mathbf{r}_A|>a \\ |\mathbf{r}-\mathbf{r}_B|>b}} [\mathbf{P}(\mathbf{r})]^2 d\mathbf{r}$$

$$- \int\limits_{\substack{|\mathbf{r}-\mathbf{r}_A|>a \\ |\mathbf{r}-\mathbf{r}_B|>b}} \left[\frac{e(\mathbf{r}-\mathbf{r}_A)}{|\mathbf{r}-\mathbf{r}_A|^3} - \frac{e(\mathbf{r}-\mathbf{r}_B)}{|\mathbf{r}-\mathbf{r}_B|^3} \right] \cdot \mathbf{P}(\mathbf{r}) d\mathbf{r}, \tag{10.19}$$

where the second term stands for the interaction energy between the charges and the polarization field. Following the same procedure adopted above one can determine G_2 as

$$G_2 = \frac{(X - 2\lambda_X)^2}{4\lambda_X}. \tag{10.20}$$

According to Eqs. (10.16) and (10.20), the reorganization energy λ_X is related to both the frequency of the harmonic free-energy surface and the separation between the positions of the minima of the initial and final free-energy surfaces.

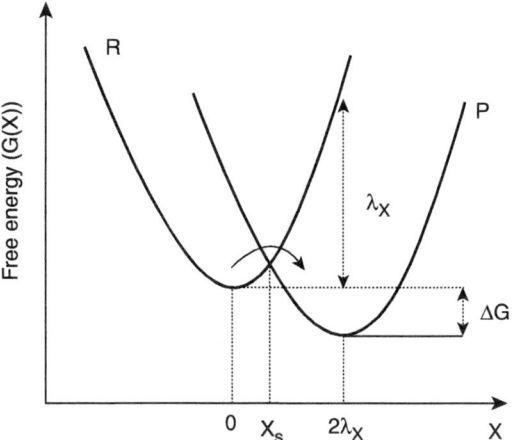

Figure 10.3 General schematic representation of the reactant (R) and product (P) free-energy surfaces as harmonic functions of the solvent reaction coordinate (X) for a one-dimensional Marcus model of ETR.

In general, one can also write the equation as

$$G_2 = \frac{(X - 2\lambda_X)^2}{4\lambda_X} + \Delta G, \quad (10.21)$$

where ΔG is the free-energy gap of the reaction. For simple reactions like $A + B \to A^+ + B^-$, ΔG is

$$\Delta G = IP - EA - \Delta g_s^{or} - \Delta g_s^{el} - \frac{e^2}{R}, \quad (10.22)$$

where Δg_s^{el} and Δg_s^{or} are the solvation energy due to electronic polarization of solvent and solvation energy due to the orientational polarization of the solvent, respectively. IP and EA are the ionization potential and electron affinity of the donor and acceptor molecules, respectively.

A simple schematic representation of the reactant and product surfaces, $G_1(X)$ and $G_2(X)$, respectively, is shown in Fig. 10.3. It is also explained in Fig. 10.3 that λ_X is the value of the free energy on the reactant surface at the position of the minimum of the product surface and ΔG is the free-energy difference between the minimum of the reactant surface and that of the product surface.

10.3.3 Derivation of ETR Rate

We now present a simple derivation of Marcus expression of electron transfer rate. Towards this goal, we consider the following simple situation

$$A + e \to B$$

as shown in Fig. 10.4.

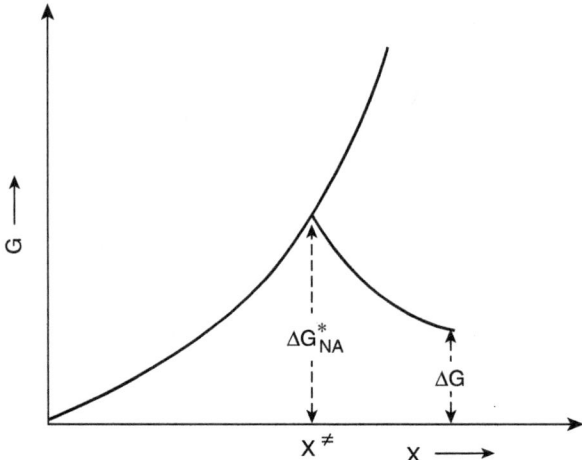

Figure 10.4 Activation energy for the diabatic case is the intersection point of reactant and product free-energy surfaces.

For this simple A to B reaction, transition-state theory (TST) predicts the reaction rate by considering primarily the following:

1. the activation energy (ΔG^*) to cross the barrier at the transition state,
2. the velocity (v) of the reactant A to cross the transition state, and
3. the probability (p) for reactant A to convert to product B.

In the case of ET, the probability "p" can be determined by the electronic matrix element, H_{12}. As shown in Fig. 10.4 the activation energy ΔG^*_{NA} is the intersection point of free-energy surfaces of reactant and product:

$$\Delta G^*_{NA} = \frac{X^{\neq 2}}{4\lambda_X}$$

$$= \frac{(X^{\neq} - 2\lambda_X)^2}{4\lambda_X} + \Delta G \quad (10.23)$$

$$= \frac{X^{\neq 2}}{4\lambda_X} - X^{\neq} + \lambda_X + \Delta G$$

or

$$X^{\neq} = \lambda_X + \Delta G. \quad (10.24)$$

Therefore, the activation energy is obtained as

$$\Delta G^*_{NA} = \frac{X^{\neq 2}}{4\lambda_X} = \frac{(\lambda_X + \Delta G)^2}{4\lambda_X}. \quad (10.25)$$

Note that ΔG_{NA}^* is independent of the coupling between the two diabatic surfaces. The transition-state rate for nonadiabatic reactions is given as

$$(\text{Rate})_{NA} = \frac{\text{Eqb. flux over barrier}}{\text{Initial population}} = \frac{\int_{-\infty}^{+\infty} k(X) P_{eq}(X) dX}{\int_{-\infty}^{\infty} dX P_{eq}(X)}, \quad (10.26)$$

where $k(X)$ is the intrinsic rate given by the Fermi golden rule as

$$k(X) = \frac{2\pi}{\hbar} H_{12}^2 \delta\left(G_1(X) - G_2(X)\right). \quad (10.27)$$

One can then calculate the rate constant, which is given by the following expression:

$$(\text{Rate})_{NA} = \frac{2\pi H_{12}^2}{\hbar (4\pi \lambda_X k_B T)^{1/2}} \exp\left(-\frac{\Delta G_{NA}^*}{k_B T}\right). \quad (10.28)$$

This is the famous Marcus expression for the ET rate in the nonadiabatic case, and in the adiabatic limit, the expression can be given by the well-known transition-state theory as

$$(\text{Rate})_A = \frac{\omega_0}{2\pi} \exp\left(-\frac{\Delta G_A^*}{k_B T}\right), \quad (10.29)$$

where ω_0 is the frequency in the reactant well. k_B is the Boltzmann constant and T is the temperature (in Kelvin). ΔG_{NA}^* and ΔG_A^* are the activation energies in nonadiabatic and adiabatic limits, respectively.

Based on the deviation of λ_X value from the free-energy gap ΔG, electron-transfer reactions are categorized into the following three major groups:

(a) $-\Delta G < \lambda_X$, the reaction site is in the normal (N) region.
(b) $-\Delta G = \lambda_X$, the reaction occurs at the minimum of the reactant surface and is the fastest as the reaction has no barrier.
(c) $-\Delta G > \lambda_X$, the reactive site is in the inverted region (I).

These three cases are schematically shown in Fig. 10.5. It is easy to see from Fig. 10.5 that while there is a barrier for the reaction to occur in the normal and inverted regions, there is no barrier for the activationless case.

The rate of ET exhibits a bell-shape energy-gap dependence as shown in Fig. 10.6. This is referred to as the Marcus parabolic dependence of the rate. The bell shape is generally not symmetric due to the participation of the vibrational modes. In some cases, the influence of the vibrational modes is so strong to lead a non-Marcus energy-gap dependence when the rate does not resemble a bell shape.

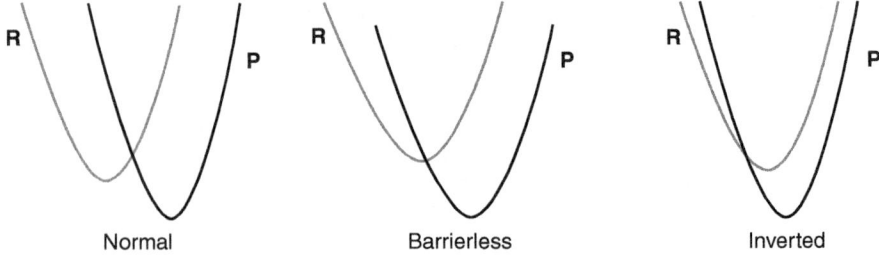

Figure 10.5 Schematic representation of normal, barrierless, and inverted region cases, respectively, describing the bell-shaped free-energy-gap dependence of ET rate.

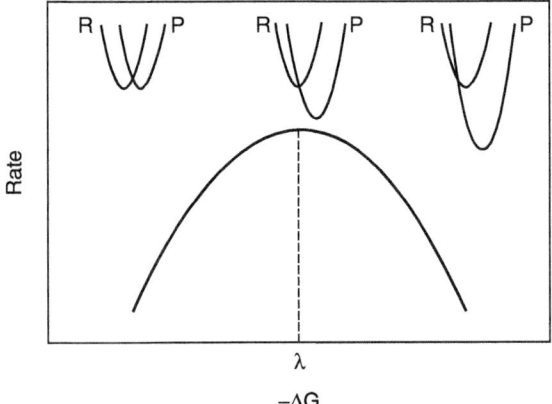

Figure 10.6 Bell-shaped energy-gap dependence of ET rate as predicted by Marcus's theory.

Note that Eqs. (10.28) and (10.29) do not contain any *dynamical* influence of solvent.

10.3.4 Experimental Verification of the Marcus Theory

In the famous Rehm and Weller experiments [6], nonparabolic dependence with the rates higher in the inverted region than in the normal region (as shown in Fig. 10.7) was observed.

Although it remained a paradox for a long time, this type of a "breakdown" of the parabolic-energy-gap dependence has now been satisfactorily explained. It is important to note that the bell-shaped dependence of the rate constant on ΔG is derived on the assumption that the donor–acceptor distance is fixed. In Rehm–Weller experiment, there were a number of acceptors distributed around a donor and they (donor and acceptors) move randomly. Therefore, the distance between the donor and the acceptor is not fixed but variable. The explanation is based on Marcus theory itself [7]. In the simplest theory of diffusion-mediated reactions, it is assumed that the ETR occurs only at one specified distance (encounter distance). For a relatively high value of ΔG, the above assumption seems good

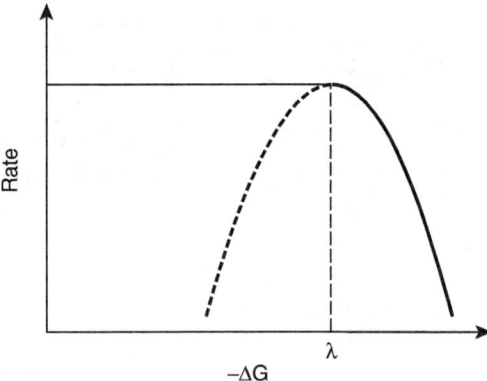

Figure 10.7 The solid line represents the Rehm–Weller behavior and the broken line the Marcus inverted region of the free-energy-gap dependence of ETR rate.

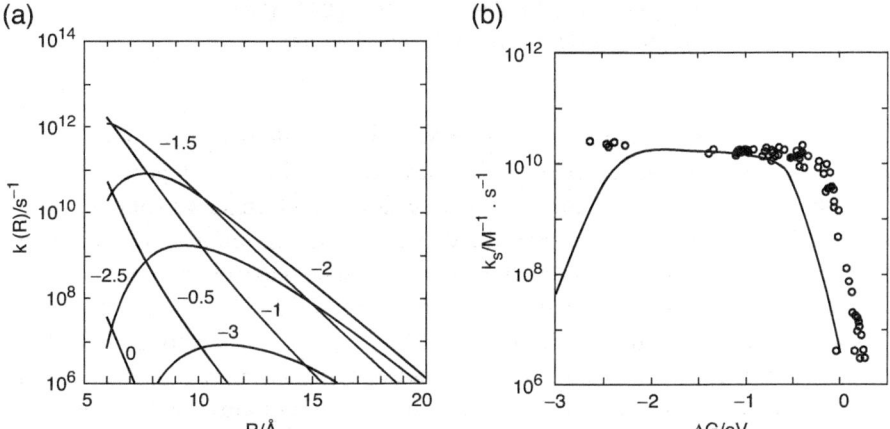

Figure 10.8 (a) Electron-transfer rate versus donor–acceptor distance for different values of free energy. Adapted with permission from *J. Phys. Chem.* **96**, 8441 (1992). Copyright (1992) American Chemical Society. (b) Theoretical prediction is shown by full Rehm–Weller experimental result by dotted line. Adapted with permission from *J. Phys. Chem.* **96**, 8441 (1992). Copyright (1992) American Chemical Society.

but for low values of $\Delta G (\Delta G < -1 \text{ eV})$ the rate constant has maxima at different distances depending on ΔG as shown in the Fig. 10.8(a). It is more reasonable to assume that electron transfer occurs at this maximum. In the case where electron transfer occurs at this maximum, both the reaction rate and the intrinsic rate are greater than in the case where it is assumed to occur at the encounter distance. Therefore, the rate constant should also be greater.

This is the reason that the inverted region appears at lower value of ΔG than expected value from Marcus theory as shown in Fig. 10.8(b).

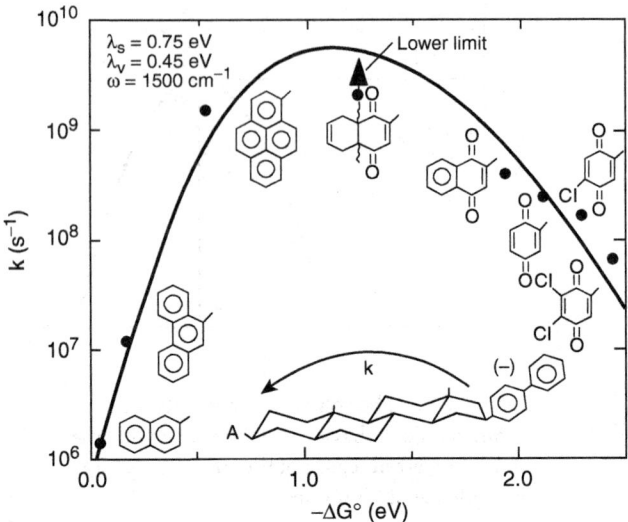

Figure 10.9 Experimental verification of parabolic-energy-gap dependence of the electron transfer rate. Reprinted with permission from *J. Am. Chem. Soc.* **106**, 3047 (1984). Copyright (1984) American Chemical Society.

The first experimental verification of parabolic dependence of rate on ΔG was obtained by Miller et al. [8]. They studied the rate dependence on the exothermicity of the intramolecular long-distance electron-transfer reactions in radical anions without changing the distance between donor and acceptor pairs. They found precisely the predicted parabolic dependence of rate on ΔG as shown in Fig. 10.9. Because the maximum of the curve is defined by λ_X, use of a less polar solvent should lead to a displacement of the curve at less negative ΔG. This was also confirmed by the results obtained in isooctane as solvent in place of MTHF (2-methyltetrahydrofuran). The maximum rate corresponds to a transfer in which the Franck–Condon factors are maximized. The fall-off at large exothermicity is caused by an increasing mismatch of the overlap of the vibrational wave functions.

10.4 DYNAMICAL SOLVENT EFFECTS ON ETRs (ONE-DIMENSIONAL DESCRIPTIONS)

Because the ETR rate is coupled to solvent polarization, it is expected that solvent polarization relaxation can influence the ET rate. An important work to analyze the dynamic solvent effects was carried out by Zusman [9] and was based on the one-dimensional Kramers' approach. According to the Zusman model, in the event of very slow (overdamped limit) solvent relaxation, the electron-transfer rate in the adiabatic limit is given by the following equation:

$$k_{ET} = \frac{1}{\tau_L} \left(\frac{\lambda_X}{16\pi k_B T} \right)^{1/2} \exp\left(-\frac{(\lambda_X + \Delta G)^2}{4\lambda_X k_B T} \right), \qquad (10.30)$$

where τ_L is the solvent longitudinal relaxation time, λ_X is the reorganization energy, ΔG is the free-energy gap of the reaction, k_B is the Boltzmann constant, and T is the temperature (in Kelvin). τ_L is equal to $\frac{\varepsilon_\infty}{\varepsilon_0}\tau_D$, where ε_0 and ε_∞ are the static and the infinite-frequency dielectric constants of the solvent, respectively, and τ_D is the dielectric relaxation time. The above expression implies that typically the electron-transfer rate cannot exceed $\left(\frac{1}{\tau_L}\right)$, because $\left(\frac{\lambda_X}{16\pi k_B T}\right)^{1/2}$ is in most cases less than unity.

Hynes [10] described the dynamical influence of solvent relaxation on the electron-transfer rate in terms of the frequency-dependent dielectric friction on the motion along the reaction coordinate. Hynes demonstrated that the Markovian description along the reaction coordinate adopted in the Zusman model can be generalized to include the more realistic non-Markovian dynamics. A notable feature of Hynes' work is the derivation of a general relation between frequency-dependent friction and the solvation time-correlation function — the latter being determined by the time-dependent Stokes shift. *The extent of solvent influence was found to vary with the degree of reaction adiabaticity.* Also, in some cases, the short-time solvent dynamics was shown to be more important than the longer time solvent relaxation.

Rips and Jortner [11], in a detailed study, modeled the outer-sphere electron-transfer reaction as a two-surface problem. In their treatment, a real-time path integral formalism was used to derive the influence functionals of the medium in the Gaussian approximation and for the electron-transfer rate. The work of Rips and Jortner allows the high barrier and the low barrier reactions to be considered *within the same formalism.* For a high-barrier electron transfer, the rate is given by

$$k_{ET}^{-1} = \frac{\sqrt{4\pi k_B T \lambda_X}}{k_0}\left[1 + \frac{2k_0 \tau_L}{\lambda_X}\right]\exp\left[\frac{\Delta G^*}{k_B T}\right], \qquad (10.31)$$

where k_0 refers to the intrinsic rate of surface crossing. This expression also interpolates between the adiabatic and the nonadiabatic limits. When $k_0 \gg \lambda_X/\tau_L$, we recover the expression for the adiabatic limit (Eq. (10.30)) where the rate, k_{ET}, is inversely proportional to τ_L. The opposite limit (i.e., $k_0 \ll \lambda_X/\tau_L$), gives the rate for the nonadiabatic case.

For symmetric exchange reactions ($\Delta G = 0$), a more general expression has also been derived (by generalization of Zusman model) for the average rate $\langle k_I \rangle$ of ET. This expression, valid for any barrier height, is given by the following simple form [15]:

$$\frac{1}{k_I} = \frac{\sqrt{4\pi k_B T \lambda_X}\exp[\Delta G^*/k_B T]}{2k_0}$$

$$+ \frac{1}{D_X}\int_{X_0}^{X_s} dX \exp[\beta G_1(X)] \int_{-\infty}^{X} dX' \exp[-\beta G_1(X')], \qquad (10.32)$$

where X_0 is the point on the reaction coordinate of the initial population (characterized here by a delta function) on the reactant surface, X_s is the point on the reaction coordinate corresponding to the intersection of the reactant and product surfaces (or sink), ΔG^* is the activation energy, D_X is the solvation energy diffusion coefficient, and $G_1(X)$ is the reactant free energy. A particular feature of the electron-exchange reactions is that they are nonadiabatic processes with weak electronic coupling and with a high activation barrier (on the order of 0.2 eV). In the limit of high barrier, the above rate expression reduces exactly to the rate expression of Rips and Jortner (Eq. (10.31)). As usual, $\beta = 1/k_B T$.

Experimental studies show that the slow component of electron-transfer dynamics is affected by τ_L, the longitudinal polarization relaxation time of the solvent. However, theory predicted a strong dependence of ET rate on solvent relaxation than was ever observed. This strong dependence on solvent relaxation was predicted not only for the reactions in the normal region but even for the barrierless ET. There are two reasons for this. First, solvent relaxation was assumed to be overdamped and the dielectric relaxation of the medium was assumed to be Debye-like. Under these approximations, solvation dynamics of a charge is a single exponential, with the time constant equal to the longitudinal relaxation time (τ_L). Second, only the motion along the classical polarization coordinate was assumed to be relevant in electron transfer. It is now known that both of these assumptions may have only limited validity.

More recent developments point to a rather different picture of dynamic solvent effects on photoinduced electron transfer. In contrast to the predictions of the Zusman and related models, recent experimental studies on systems like Nile Blue in N-N-dimethylaniline (DMA) reveal electron transfer rates to be much faster than τ_L. This is a barrierless case with a broad reactive region as in many low-barrier reactions. Another case already mentioned is ET in betaine-30. It is now clear from these experiments that the sink broadening effects from nonsolvent modes can be as important as the polar solvent modes. Therefore, a one-dimensional reaction description with a point reactive site is insufficient. We next describe the role of vibrational modes in ETR.

10.5 ROLE OF VIBRATIONAL MODES IN WEAKENING SOLVENT DEPENDENCE

As already mentioned, despite the initial claims, clear unambiguous evidence of dynamic solvent effects on photoinduced electron transfer has surprisingly been few, despite the large number of ET reactions studied in a polar medium. Recent experiments on electron-transfer reactions in femtosecond time scales seem to demonstrate that the electron transfer is often decoupled from solvent polarization. Two major works, that of Sumi and Marcus and of Jortner and Bixon, contributed notably to understand the role of vibrational modes in ET and are discussed below.

10.5.1 Role of Classical Intramolecular Vibrational Modes: Sumi–Marcus Theory

In order to treat the zero or low-barrier electron-transfer reactions, Sumi and Marcus [12] proposed a quasi-two-dimensional model where the electron transfer

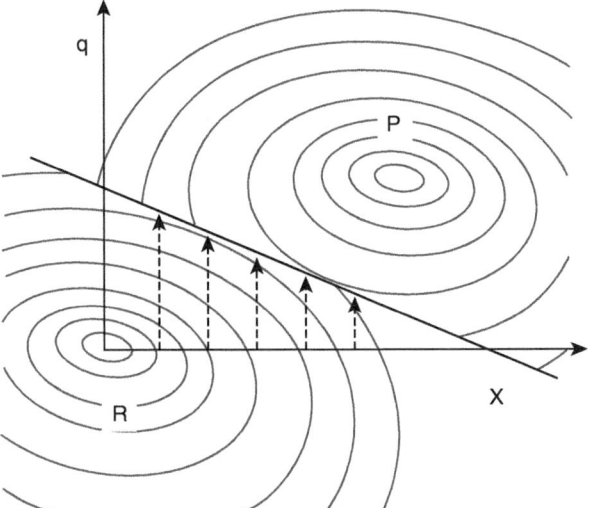

Figure 10.10 A two-dimension model of electron-transfer reactions taking into consideration low-frequency solvent modes (q) along with solvent polarization coordinate (X).

is coupled to a low-frequency solvent mode (X) and to a fast relaxing "average" low-frequency vibrational mode (q). The q-mode can also be interpreted as a rapidly relaxing solvent mode, which is different from the usual bulk solvent polarization mode. The presence of the two classical modes broadens the reactive region as the reaction can now proceed along the curve intersecting the two free-energy surfaces. The projection of this intersection curve onto the X–q plane results in a continuous line, as shown in Fig. 10.10. ETR can occur along this line. Now, an extended sink near the potential minimum reduces the solvent influence on the electron transfer because different relaxation trajectories can now reach the reactive region. This may result in pronounced nonexponential kinetics in low-barrier cases. The above features are mainly responsible in effecting a change from the solvent control to the vibration control of the ET dynamics.

In the Sumi–Marcus model, the influence of dynamics along the vibrational mode (q) was included implicitly by assuming the relaxation along this mode to be infinitely fast. The reaction was then described by assuming a one-dimensional Brownian diffusive motion along the solvent coordinate (X) while the reaction occurs at each X with a rate constant $k(X)$ during the diffusion. It is the solvent coordinate–dependent rate, $k(X)$, that contains the effects from the intramolecular mode through its dependence on the effective free energy of activation, $\Delta G^*(X)$, of the reaction along the solvent coordinate. The effects from the internal modes are, therefore, indirectly reflected in the solvent coordinate–dependent rate, $k(X)$, which is given by

$$k(X) = v_q \exp\left(-\frac{\Delta G^*(X)}{k_B T}\right), \tag{10.33}$$

where ν_q is a frequency factor and $\Delta G^*(X)$ is the X-dependent free energy of activation. $\Delta G^*(X)$ is given by the expression

$$\Delta G^*(X) = \frac{\lambda_X}{2\lambda_q}(X - X_c)^2, \qquad (10.34)$$

where X_c is the value of X corresponding to $q = 0$ on the sink line projected on the two-dimensional plane spanned by the solvent and vibrational coordinates, X and q, respectively. $\lambda_i (i = X, q)$ are the reorganization energies corresponding to the two reaction coordinates.

The extent of influence of the internal modes on the dynamics is roughly determined by the relative inner shell and outer shell reorganization energies, λ_q and λ_X. When $\lambda_X/\lambda_q \ll 1$, the reaction window is broad in X and there is a weak X-dependence of reaction. When $\lambda_X/\lambda_q \gg 1$, the sink reaction window is narrow in X and the dynamics becomes solvent-controlled, as predicted by the Zusman model.

In order to explicitly treat dynamic solvent effects, Sumi et al. used a theoretical description of the *barrierless* electron-transfer reactions based on a modified Smoluchowski equation for the time evolution of the probability distribution $(P_1(X, t))$:

$$\frac{\partial P_1(X,t)}{\partial t} = D_X\left(\frac{\partial^2 P_1(X,t)}{\partial X^2}\right) + \frac{D_X}{k_B T}\frac{\partial}{\partial X}\left[P_1(X,t)\frac{dG_1(X)}{dX}\right] - k(X)P_1(X,t), \qquad (10.35)$$

where $G_1(X)$ is the reactant surface and D_X is the diffusion constant. The first and the second terms on the right account for the motion of the system on surface 1 in a potential $G_1(X)$ and the third term accounts for the decay resulting from the transfer of electrons to the product state. Note that a one-surface description is used here. In many low-barrier reactions, it is the full two-surface situation with significant back transfer of electrons that is generally encountered. The one-surface description can be relevant only in certain limiting conditions when the general two-surface problem reduces to a one-surface problem, for example, an asymmetric situation where the forward reaction from surface G_1 to surface G_2 is a barrierless process, while the back transfer of electron from the surface G_2 to surface G_1 is a high-barrier one.

An effective one-surface situation where the reaction occurs with unit probability once the reactants arrive at the origin (or minimum of the reactant surface) can be modeled by placing a pinhole sink at the origin of the surface. In such a case, the following expression for the survival probability on the surface 1 is obtained:

$$P_1(t) = \int_{-\infty}^{0} dX \left[P_0(X) + P_0(-X)\right] \text{erfF}(X, t), \qquad (10.36)$$

where

$$F(X, t) = \frac{1}{\sqrt{2\lambda_X}} \left[2k_B T \left[1 - \exp\left(-\frac{2t}{\tau_s}\right) \right] \right]^{-1/2} \exp\left(-\frac{t}{\tau_s}\right), \quad (10.37)$$

where τ_s is the solvent relaxation time. The error function, erf a, is defined as usual by

$$\text{erf } a = \frac{2}{\sqrt{\pi}} \int_0^a dy \exp(-y^2). \quad (10.38)$$

$P_0(X)$ is the initial probability distribution in the reactant well. If $P_0(X)$ is given by the equilibrium Boltzmann distribution, then $P_1(t)$ reduces to the following, much simpler and elegant form, first derived by Szabo:

$$P_1(t) = \frac{2}{\pi} \sin^{-1} \left[\exp\left(-\frac{t}{\tau_s}\right) \right]. \quad (10.39)$$

This equation predicts that the decay of the reactant population is, in general, nonexponential in nature. The long-time decay, however, is still given by a single exponential with a rate constant equal to τ_s.

Recently, good agreement between experimental results and the Sumi–Marcus model was observed in the temperature dependence of intermolecular electron transfer between oxazine1 (OX1) and electron-donating solvent, aniline (AN). But the predictions of the Sumi–Marcus model in the inverted region were found to be slower by more than 10^6 times in such systems as 4-(9-anthryl)-N,N,-dimethylaniline (ADMA), Bis-{N,N'-dimethylaminophenyl}-sulfone (DMAPS), and betaine-30. Here ETR rates were found to greatly exceed the solvent relaxation rate. This discrepancy has been attributed to the classical approximation of the intramolecular modes. This limitation was soon rectified by using a quantum-mechanical approach in the work of Jortner and Bixon.

10.5.2 Role of High-Frequency Vibration Modes

In marked departure from the Sumi–Marcus theory [12], Jortner and Bixon [13] argued that the relevant vibrational modes in an electron-transfer reaction, particularly for reactions in the Marcus-inverted regime, are the high-frequency vibrational modes, as only they can act in the electron transfer via the Franck–Condon overlaps. The quantum "jumps" of a high-frequency vibrational mode can open several new point reaction sinks for electron transfer, as shown in Fig. 10.11, in contrast to the broadening effect by a low-frequency vibrational mode.

The work of Jortner and Bixon, which also included dynamic solvent effects, was based on a quantum-mechanical treatment of the high-frequency vibrational coordinate in place of the classical one of Sumi–Marcus. The predicted rates were closer to the experimental results in fast-relaxing polar aprotic solvents than

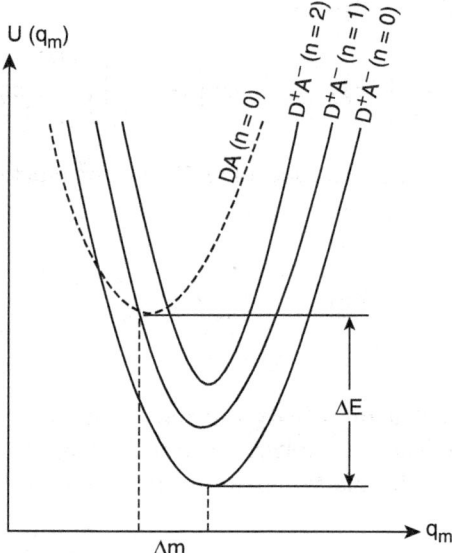

Figure 10.11 Schematic one-dimensional free-energy-surface representation for the electronic ground state ($n = 0$) for reactant and the manifold vibronic excited states for product. Reprinted with permission from *J. Chem. Phys.* **88**, 167 (1988). Copyright (1988) American Institute of Physics.

those predicted by the Sumi–Marcus model. This increase in the rate observed was because of the additional sinks that arise from the intersection of the ground reactant surface with the multiple vibronic product surfaces. The microscopic rate from the ground level of the reactant state to the nth vibronic level of the product state was assumed to be given by the relation

$$k^{0 \to n} = \frac{2\pi H_{12}^2}{\hbar (4\pi \lambda_X k_B T)^{1/2}} \exp\left(-\frac{\Delta G_{0 \to n}}{k_B T}\right), \quad (10.40)$$

where H_{12} is the effective electronic coupling and $\Delta G_{0 \to n}$ is the effective activation energy for the $0 \to n$ channel given by the following expression:

$$\Delta G_{0 \to n} = \lambda_X + \Delta G - 2X\lambda_X + n h \nu_q. \quad (10.41)$$

Here λ_X is the solvent reorganization energy, ΔG is the free-energy gap, and ν_q is the frequency of the quantum mode. The net rate is the sum of rates $k^{0 \to n}$ corresponding to the n vibronic levels of the product $D^+ A^-$ state. The merit of this model is that it correctly explains the great enhancement of the rate over the Sumi–Marcus prediction and, in fact, gives rise to values close to those observed experimentally with realistic values of the parameters. However, as already mentioned, in the original model, the reaction channels were still assumed to be point sinks with no width. The diffusive motion between the widely spaced point sinks is still solvent-controlled and as a result, this model failed to describe the crossover to the solvent-independent behavior observed in experiments on systems such as betaine-30 in highly viscous and slowly relaxing solvents.

Later, Bixon and Jortner analyzed the effects of medium-induced dynamics with model calculations to determine energy- (E) dependent rates, $k(E)$, for multimode harmonic systems with displaced potential surfaces using average Franck–Condon densities that were evaluated using quantum and classical formalisms. For activationless electron transfer, $k(E)$ is proportional to $\sqrt{(E+n\varepsilon)}$, where $n\varepsilon$ is the total zero-point energy for a case that involves n high-frequency modes, indicating a weak energy dependence. This, in turn, implies weak solvent dependence as the reaction can proceed without any solvent relaxation. Thus, the model of Jortner and Bixon seems to describe the activationless and inverted-region electron transfer, as experimental rates exhibit weak variation between limits of slow and fast medium-induced dynamics.

There are, however, limitations to this approach as well. As already mentioned, the reaction sites remain point sinks. Thus, one cannot really rule out solvent independence. Moreover, the theoretical treatment of Jortner and Bixon did not consider the dynamics of motion on the potential surfaces.

10.5.3 Hybrid Model of ETR: Crossover from Solvent to Vibrational Control

In order to reconcile the observed large differences between the theoretical predictions and experimental results, Walker et al. [14] proposed a combined Sumi–Marcus [12] and Jortner–Bixon [13] model. The authors pointed out that *any model of electron-transfer reaction in the inverted regime should minimally include a low-frequency solvent mode (classical), an intramolecular low-frequency mode (classical), and a high-frequency intramolecular mode (quantum mechanical)*. The role of the low-frequency vibrational mode is to broaden the point sinks formed by the high-frequency modes. If the dynamics along the low-frequency mode is assumed to be infinitely fast, the microscopic rate $k^{0 \to n}$ can be given by the following form:

$$k^{0 \to n} = \frac{2\pi H_{12}^2}{\hbar(4\pi \lambda_Q k_B T)^{1/2}} \exp\left(-\frac{(\Delta G_{0 \to n} + \lambda_Q)^2}{4\lambda_Q k_B T}\right), \quad (10.42)$$

where λ_Q is the reorganization energy of the low-frequency, classical vibrational mode, which is not solvent dependent. Note that λ_Q is now used in place of λ_X in Eq. (10.25) to account for the low-frequency vibrational contribution. $\Delta G_{0 \to n}$ is given by Eq. (10.41). In the "hybrid" model, therefore, λ_Q, λ_X, and ΔG are the three relevant parameters. However, the values of these parameters are often not known *a priori*. It is customary to obtain these relevant parameters by fitting the experimentally obtained absorption intensity profiles to line-shape models. Also, the exact identification of the nature of the vibrational modes that are involved is often difficult.

The "hybrid model" now predicted rates close to the experimental results of betaine-30 and tert-butylbetaine in a wide range of solvent environments. However, they did not consider the dynamics of the low-frequency vibrational mode explicitly. In addition, in the original form of this model, only the Markovian response of the solvent mode was included. This is equivalent to assuming that the solvation dynamics is single exponential. Clearly, generalization of this model is required in view of the observation that in most polar solvents, solvation dynamics is nonexponential and seems to contain two vastly different time scales. While the origin of these different time scales is well understood in ultrafast liquids (like water and acetonitrile), the same is certainly not clear for more complex liquids, like formamides and higher alcohols. However, it appears to be certain that even in these commonly believed slow liquids there exists an ultrafast component of time constants two to three orders smaller in magnitude than the time constant commonly observed in dielectric relaxation measurements. Theoretical study of ET rate in the presence of such diverse time scales poses an interesting challenge because the conventional approaches to solve the reaction–diffusion equations become inefficient, especially so in the presence of multiple delocalized sinks.

10.6 THEORETICAL FORMULATION OF MULTIDIMENSIONAL ELECTRON TRANSFER

In this section, a theoretical technique is discussed to solve the equation of motion for any reaction occurring on a multidimensional potential energy surface where the reaction occurs via a delocalized sink [15]. The merit of this scheme is that one can describe the situation where the motion along the reaction coordinate is non-Markovian and the reaction window is broad. The importance of this theory is clearly evident as almost all polar liquids seem to have a significant ultrafast component. Thus, one must include the non-Markovian relaxation behavior. The second aspect is equally important as the treatment of the reaction in the presence of a delocalized sink is notoriously difficult as most of the known techniques fail here.

The electron-transfer reaction is in general a multidimensional process involving both the polar solvent modes and the intramolecular vibrational modes of the molecule. In order to keep the problem tractable, all the potentials are usually assumed to be harmonic. Let us assume an n-dimensional electron-transfer process with a single classical solvent coordinate (X) and $(n-1)$ nonsolvent vibrational coordinates $(Q_1, Q_2, \ldots \ldots Q_{n-1})$. For simplicity, normalized coordinates are often used to describe ET. Each coordinate is normalized for the bottom of the reactant energy surface to be at 0, and that of the product at 1.

The n-dimensional potential energy surfaces (PES), V_1 and V_2, for the reactant and product states can, therefore, be described by the following equations:

Reactant:

$$V_1(X, Q_1, Q_2, \ldots, Q_{n-1}) = \frac{1}{2} 2\lambda_X X^2 + \sum_{i=1}^{n-1} \frac{1}{2} 2\lambda_{Q_i} Q_i^2. \quad (10.43)$$

Product:

$$V_2(X, Q_1, Q_2, \ldots, Q_{n-1}) = \frac{1}{2} 2\lambda_X (X-1)^2 + \sum_{i=1}^{n-1} \frac{1}{2} 2\lambda_{Q_i} (Q_i - 1)^2 + \Delta G, \quad (10.44)$$

where $\lambda_X, \lambda_{Q_i} (i = 1, \ldots, n-1)$ are the corresponding reorganization energies of these modes, and ΔG is the free-energy gap of the reaction, i.e., the vertical energy gap between the two surface minimums. Because we are primarily interested in the photoelectron transfer reactions, the reactant surface 1 is often referred to as the locally excited (LE) state and the product surface P as the charge transfer (CT) state.

As mentioned previously, the reaction occurs along the sink curve obtained by the intersection of the two potential energy surfaces. For an n-dimensional potential energy surface, the sink surface is always an $(n-1)$-dimensional one. For the sake of simplicity, we will use $\mathbf{Q}$ to represent the set of the vibrational coordinates $Q_1, Q_2, \ldots, Q_{n-1}$ in the rest of the chapter.

The time evolution of the probability distribution of the system on the locally excited surface $P_1(X, \mathbf{Q}, t)$ is assumed to be given by the following equation:

$$\frac{\partial P_1(X, \mathbf{Q}, t)}{\partial t} = \left(L_X^{(1)} + L_\mathbf{Q}^{(1)} \right) P_1(X, \mathbf{Q}, t) - S(X, \mathbf{Q}) P_1(X, \mathbf{Q}, t)$$
$$+ S(X, \mathbf{Q}) P_2(X, \mathbf{Q}, t), \quad (10.45)$$

$$\frac{\partial P_2(X, \mathbf{Q}, t)}{\partial t} = \left(L_X^{(2)} + L_\mathbf{Q}^{(2)} \right) P_2(X, \mathbf{Q}, t) - S(X, \mathbf{Q}) P_2(X, \mathbf{Q}, t)$$
$$+ S(X, \mathbf{Q}) P_1(X, \mathbf{Q}, t), \quad (10.46)$$

where $L_\mathbf{Q}^{(k)} = \sum_{i=1}^{n-1} L_{Q_i}^{(k)}$. $k = 1, 2$ stand for the reactant and product potential energy surfaces, respectively. The first term simulates the diffusion in a potential well $V_{(k)}(X, \mathbf{Q})$. The second and the third terms take into account the actual transfer and the back transfer along the sink curve. $S(X, \mathbf{Q})$ is the position-dependent sink function that describes the path along which the electron transfer takes place between the LE and the CT surfaces.

The operator $L_\xi (\xi = X, Q_i)$ is the general Smoluchowski operator and is given in the following form:

$$L_\xi^{(k)} = D_\xi(t) \left(\frac{\partial^2}{\partial \xi^2} + \frac{1}{k_B T} \frac{\partial}{\partial \xi} \left[\frac{dV_{(k)}(\xi)}{d\xi} \right] \right), \quad (10.47)$$

where $D_\xi(t)$ is the time-dependent diffusion coefficient of motion along the reaction coordinate. $D_\xi(t)$ is given by the relation [10]

$$D_\xi(t) = -k_B T \frac{d(\ln \Delta_\xi(t))}{dt}, \tag{10.48}$$

where $\Delta_\xi(t)$ is the time-correlation function of the ξth reaction coordinate and is usually assumed to be of the form $\sum_j w_j \exp(-t/t_j)$, where $\sum_j w_j = 1$.

The diffusion coefficient (D_ξ) is time independent for a Markovian single-exponential decay (as in the case of a Debye solvent) and is time dependent when the relaxation is characterized by a non-Markovian multiexponential time decay (as in the case of a non-Debye solvent). The effective relaxation time τ_{eff} is given by

$$\tau_{eff} = \int_0^t dt \, \Delta_\xi(t). \tag{10.49}$$

The solvent reaction coordinate time-correlation function, $\Delta_X(t)$, is defined as

$$\Delta_X(t) = \frac{\langle X(0)X(t) \rangle}{\langle X^2(0) \rangle}, \tag{10.50}$$

where $\langle \ldots \rangle$ denotes the average over the solvent degrees of freedom in equilibrium with the reactant state. It is this quantity $\Delta_X(t)$ that reflects the dynamics of the solvent polarization fluctuations. For homogeneous redox reactions, $\Delta_X(t)$ is the solvation time-correlation function of an ion, while for the photoinduced electron transfer in bianthryl, $\Delta_X(t)$ is that of a dipole.

Green's function technique has been used to solve Eqs. (10.45) and (10.46) and discussed in the **Appendix**. The main advantage of this technique is that an almost analytic solution can be obtained in the Laplace frequency plane. The time-dependent solution can subsequently be obtained by a simple Laplace inversion.

Average rate

The average rate of the reaction (k_I) is defined as

$$k_I^{-1} = \tau_a = \int_0^\infty dt P_1(t) = P_1(z = 0). \tag{10.51}$$

In order to obtain the average rate one has to calculate the survival probability $P_1(t)$. $P_1(z)$ is first obtained by numerically solving the system of equations and then Laplace inverting $P_1(z)$ to obtain $P_1(t)$. However, when the solvation time-correlation function is biphasic with widely different time scales, this method is not robust because evaluation of $P_1(t)$ faces stability problems as this procedure is computationally intensive. Fortunately, there is a direct, almost analytical, method to obtain this average rate. This method uses the well-known Kramers' technique to first obtain the solution of the system of linear equations by the matrix equation (Eq. (A14)). In the zero-frequency limit, this solution gives the average rate in the following form:

$$k_I^{-1} = \lim_{z \to 0} \frac{\left(\det \mathbf{B} - \sum_s k_s \det \mathbf{B}^j\right)}{\sum_s k_s \det \mathbf{B}}, \qquad (10.52)$$

where "det" represents the determinant of a matrix and the matrix elements of $\mathbf{B}$ are given (Eqs. (A.15)–(A.18)). The elements of matrix $\mathbf{B}^j$ are obtained from matrix $\mathbf{B}$ by replacing the jth column by the column vector $\mathbf{G_0}$.

However, Eq. (10.52) becomes ill-defined in the z tending to zero limit as the Laplace transform of the Green's function diverges in that limit. Therefore, if a straightforward numerical evaluation of k_I^{-1} is attempted by using Eq. (10.51), then this divergence makes this method useless. This problem is especially severe when the solvation dynamics is biphasic with a large difference between the rates of the two relaxation time components of $\Delta(t)$. Note that such a large difference (even to the extent of a factor of 100 or so) has been observed in recent solvation experiments. This difficulty can be circumvented by using the following method.

Let us first define a quantity $\Delta G\left(X_s, \mathbf{Q}_s, z | X_{m'}, \mathbf{Q}_{m'}\right)$ in the following form:

$$\Delta G\left(X_s, \mathbf{Q}_s, z | X_{m'}, \mathbf{Q}_{m'}\right)$$

$$= \lim_{z \to 0} \int_0^\infty dt \, \exp(-zt) \left[G_1\left(X_m, \mathbf{Q}_m, t | X_{m'}, \mathbf{Q}_{m'}\right) - G_{eq}\left(X_m, \mathbf{Q}_n\right)\right], \qquad (10.53)$$

where $G_{eq}\left(X_m, \mathbf{Q}_n\right) = G_1\left(X_m, \mathbf{Q}_n, t = \infty\right)$ is the equilibrium distribution. Note that this ΔG (not to be confused with the free energy) is well defined in the $z \to 0$ limit. A simple substitution of $\Delta G\left(X_s, \mathbf{Q}_s, z | X_{m'}, \mathbf{Q}_{m'}\right)$ in place of $G_1\left(X_m, \mathbf{Q}_n, z | X_{m'}, \mathbf{Q}_{m'}\right)$ in Eqs. (A.15)–(A.18)) gives the average rate in the following form:

$$k_I^{-1} = \frac{\left(\det \mathbf{C} - \sum_s k_s \det \mathbf{C}^j\right)}{\sum_s k_s \det \mathbf{C}^{j'}}, \qquad (10.54)$$

where the matrix elements of $\mathbf{C}$ are of the form

$$[c_{mn}] = \delta_{mn} + k_n \Delta G\left(X_m, \mathbf{Q}_m, z | X_n, \mathbf{Q}_n\right). \qquad (10.55)$$

The elements of matrix $\mathbf{C}^j$ are obtained from matrix $\mathbf{C}$ by replacing the elements c_{ij} in the jth column with the elements $\Delta G\,(X_i, \mathbf{Q}_i, z|X_0, \mathbf{Q}_0)$. Similarly, matrix $\mathbf{C}^{j'}$ is obtained by replacing the elements of the jth column by the equilibrium values $G_{eq}(X_i, \mathbf{Q}_i)$.

The theoretical results based on the preceding multidimensional model indicate that the enhanced rate in the inverted region is not only due to the additional reaction windows provided by the high-frequency vibrational modes but also due to the involvement of the ultrafast component of the solvation dynamics. While the high-frequency modes enhance the rate by opening up several efficient decay channels near the barrierless region, the ultrafast component in the solvation dynamics allows rapid relaxation of the excited-state population toward the potential minima.

10.7 EFFECTS OF ULTRAFAST SOLVATION ON ELECTRON-TRANSFER REACTIONS

When electron transfer is modeled as a Marcus two-surface problem, the activation barrier can be rather sharp, even cusp-like for a weakly adiabatic reaction. In such a situation, the act of electron transfer across the barrier can couple only to the high-frequency polar response of the liquid. The recent discovery of the ultrafast component in solvation dynamics has shown that this high-frequency response can be considerably different from that at zero frequency.

How should we study the effects of ultrafast solvent modes on adiabatic electron-transfer reactions? One simple approach would be to study the effects of ultrafast modes on the friction that acts on the reaction system as it moves along the reaction coordinate. An electron-transfer reaction comprises two distinct steps: (a) the energy-diffusion step to reach the barrier region and (b) the crossing of the barrier. The friction itself can have rather different effects on these two steps. The high-frequency librational and intermolecular vibrational modes, probed during the barrier crossing, may offer only a small friction of the motion along the remainder of the reaction coordinate. Therefore, these underdamped modes may lead to significantly less damping of the rate of the barrier crossing than would appear from the value of the zero-frequency (i.e., macroscopic) friction.

10.7.1 Absence of Significant Dynamic Solvent Effects on ETR in Water, Acetonitrile, and Methanol

Zusman predicted that the rate of an adiabatic electron-transfer reaction is inversely proportional to the longitudinal relaxation time, τ_L (see **Section 10.4**). The physical reason for such dependence is that, in Marcus picture, ETR is driven by nonequilibrium solvent polarization. This interesting prediction, however, has not been observed for dipolar liquids such as water, acetonitrile, and methanol. It has been suggested that this absence of coupling of ETR to solvent polarization relaxation is due to the presence of an ultrafast component (the libration and intermolecular vibration modes of the solvent, along with inertial motion) in these liquids.

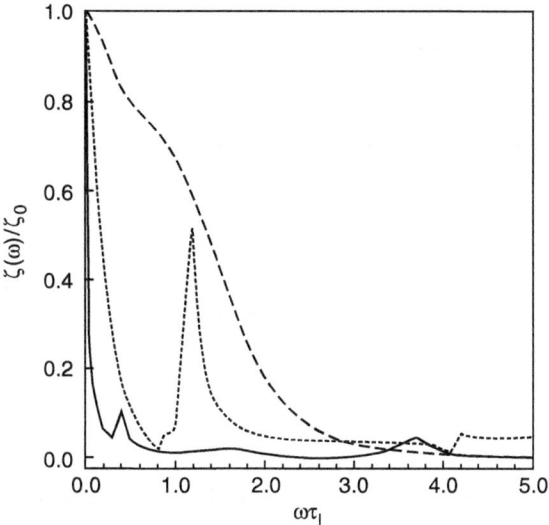

Figure 10.12 The real part of the frequency dependent friction scaled with the zero frequency friction, $\zeta(\omega)/\zeta_0$, acting along the reaction coordinate for an adiabatic electron transfer reaction, is plotted as a function of the Laplace-Fourier frequency, ω, for an outer sphere electron transfer reaction in methanol (solid line), water (short dashed line) & acetonitrile (long dashed line). The frequency is scaled by τ_I (time constant for the rotational inertial motion for the solvent molecules) which is equal to 0.06 ps, 0.1 ps and 0.45 ps for methanol, water and acetonitrile, respectively. Reprinted with permission from *J. Chem. Phys.* **102**, 6719 (1995). Copyright (1995) American Institute of Physics.

As we have already discussed, water, methanol, and acetonitrile are all characterized by the presence of a significant ultrafast component in the solvation dynamics and polarization relaxation. For water and acetonitrile, the ultrafast component is 60–70%, while for methanol, the component is about 25%. The frequency-dependent friction is found to be considerably smaller for the ultrafast librational mode than the zero-frequency friction.

Theoretical calculations show that the contribution from the ultrafast modes is sufficient to drive an adiabatic reaction. Because this coupling is to the ultrafast modes, little dependence on τ_L could be observed. The real part of the frequency-dependent friction acting along the reaction coordinate for an adiabatic electron-transfer reaction in methanol, water, and acetonitrile is shown in Fig. 10.12.

10.8 SUMMARY

Since the pioneering theory of Marcus in 1956, the study of electron-transfer reaction has steadily progressed. Many aspects of the theory came under scrutiny when accurate rate determination became possible in the early 1980s with the advent of laser spectroscopy. Once the validity of Marcus theory was established beyond doubt, attention shifted to the study of the dynamics of electron transfer. This was also required due to detailed dynamical studies that were made with ultrashort laser pulses. Thus, not just the overall rate but the time dependence of the population distribution could be measured. Also, a new class of electron transfer reactions, namely, the barrierless ETRs, came under focus. This necessitated the development of a new class of theories because the vibrational degrees of freedom of the reactant and product could now play an important role in promoting electron transfer, weakening the effect of solvation dynamics in the electron transfer dynamics.

APPENDIX

Solution of Eq. (10.45) and Eq. (10.46) by Green's Function Technique

By Laplace transformation, the dynamical equations for the two potential energy surfaces (Eqs. (10.45) and (10.46)) are converted into the following equations,

$$\left[z - (L_X^{(1)} + L_{\mathbf{Q}}^{(1)})\right] P_1(X, \mathbf{Q}, z) = P_1(X, \mathbf{Q}, t=0) - S(X, \mathbf{Q}) P_1(X, \mathbf{Q}, z)$$
$$+ S(X, \mathbf{Q}) P_2(X, \mathbf{Q}, z), \quad (A.1)$$

$$\left[z - (L_X^{(2)} + L_{\mathbf{Q}}^{(2)})\right] P_2(X, \mathbf{Q}, z) = P_2(X, \mathbf{Q}, t=0) - S(X, \mathbf{Q}) P_2(X, \mathbf{Q}, z)$$
$$+ S(X, \mathbf{Q}) P_1(X, \mathbf{Q}, z), \quad (A.2)$$

where $P_i(X, \mathbf{Q}, t=0)$ denote the initial probability distribution on the ith surface and z is the (Laplace) frequency conjugate to the time t. By definition, the Green's function follows these equations for the reactant and product surfaces:

$$\left[z - (L_X^{(1)} + L_{\mathbf{Q}}^{(1)})\right] G_1(X, \mathbf{Q}, z | \tilde{X}, \tilde{\mathbf{Q}}) = \delta(X - \tilde{X})\delta(\mathbf{Q} - \tilde{\mathbf{Q}}), \quad (A.3)$$

$$\left[z - (L_{X'}^{(2)} + L_{\mathbf{Q}'}^{(2)})\right] G_2(X', \mathbf{Q}', z | \tilde{X}', \tilde{\mathbf{Q}}') = \delta(X' - \tilde{X}')\delta(\mathbf{Q}' - \tilde{\mathbf{Q}}'), \quad (A.4)$$

where X' is equal to $X - 1$ and $\mathbf{Q}'_\xi$ is equal to $\mathbf{Q}_\xi - 1$ as the minimum of the product PES is at the value of 1 along each normalized reaction coordinate. $\delta(\mathbf{Q} - \tilde{\mathbf{Q}})$ denotes $\prod_{i=1}^{n} \delta(Q_i - \tilde{Q}_i)$. Now, for a one-dimensional harmonic potential surface j described by the reaction coordinate A_i, the Green's function is given by the following form [10]:

$$G_j(A_i, t | \tilde{A}_i) = \frac{1}{\sqrt{2\pi \sigma_i^2 \left[1 - \Delta_i(t)^2\right]}} \exp\left[-\frac{[A_i - \tilde{A}_i \Delta_i(t)]^2}{2\sigma_i^2 \left[1 - \Delta_i(t)^2\right]}\right], \quad (A.5)$$

where $\sigma_i^2 = k_B T/(2\lambda_i)$. Note that the Green's function for the surface V_2 is given by $X' = X - 1$ and $Q' = Q - 1$.

The composite Green's function for an N-dimensional surface (in the absence of any coupling between the reaction coordinates) is given by the product of the Green's function for the N individual coordinates

$$G_j(A_1, A_2, \ldots, A_N, t|\tilde{A}_1, \tilde{A}_2, \ldots \tilde{A}_N) = \prod_{i=1}^{N} G_j(A_i, t|\tilde{A}_i). \tag{A.6}$$

Note that the above Green's function is the solution of the dynamical equation in the absence of the sinks. The solution in the presence of the sinks is required. This is achieved by using the Green's function in a Laplace-transformed dynamical equation and subsequently obtaining the values for $P_1(X_s, \mathbf{Q}_s, z)$, where z is the Laplace frequency.

The solutions for $P_1(X, \mathbf{Q}, z)$ and $P_2(X, \mathbf{Q}, z)$ are given in terms of the two coupled equations of the following form:

$$P_1(X, \mathbf{Q}, z) = \int d\tilde{\mathbf{Q}} \int d\tilde{X} G_1\left(X, \mathbf{Q}, z|\tilde{X}, \tilde{\mathbf{Q}}\right) \left[P_1\left(\tilde{X}, \tilde{\mathbf{Q}}, 0\right)\right.$$
$$\left. - S\left(\tilde{X}, \tilde{\mathbf{Q}}\right) P_1\left(\tilde{X}, \tilde{\mathbf{Q}}, z\right) + S\left(\tilde{X}, \tilde{\mathbf{Q}}\right) P_2\left(\tilde{X}, \tilde{\mathbf{Q}}, z\right)\right], \tag{A.7}$$

$$P_2(X, \mathbf{Q}, z) = \int d\tilde{\mathbf{Q}}' \int d\tilde{X}' G_2\left(X, \mathbf{Q}, z|\tilde{X}', \tilde{\mathbf{Q}}'\right) \left[P_2\left(\tilde{X}', \tilde{\mathbf{Q}}', 0\right)\right.$$
$$\left. - S\left(\tilde{X}', \tilde{\mathbf{Q}}'\right) P_2\left(\tilde{X}', \tilde{\mathbf{Q}}', z\right) + S\left(\tilde{X}', \tilde{\mathbf{Q}}'\right) P_1\left(\tilde{X}', \tilde{\mathbf{Q}}', z\right)\right]. \tag{A.8}$$

For simplicity, the initial population excited on the reactant (LE) surface may be characterized as a delta function source at $(X_0, \mathbf{Q}_0)$. Mathematically, this is written as

$$P_i(X, \mathbf{Q}, t = 0) = \delta(X - X_0)\delta(\mathbf{Q} - \mathbf{Q}_0)\delta_{1i}. \tag{A.9}$$

The sink function, $S(X, \mathbf{Q})$ can be written as

$$S(X, \mathbf{Q}) = \int d\tilde{\mathbf{Q}} \int d\tilde{X} S(\tilde{X}, \tilde{\mathbf{Q}})\delta(X - \tilde{X})\delta(\mathbf{Q} - \tilde{\mathbf{Q}}). \tag{A.10}$$

This property is exploited to divide the continuous sink curve into a number of intervals. The sink function was then assumed to be of the following form in the discretized representation:

$$S(\tilde{X}, \tilde{\mathbf{Q}}) = \sum_s k_s \delta(X - X_s)\delta(\mathbf{Q} - \mathbf{Q}_s), \tag{A.11}$$

where k_s is the combined strength of each sink interval represented at the point $(X_s, \mathbf{Q}_s)$. The determination of k_s is described in the next subsection. Note that this discretization is perfectly general and valid for any delocalized sink.

$P_1(X, \mathbf{Q}, z)$ and $P_2(X, \mathbf{Q}, z)$ can then be expressed as

$$P_1(X, \mathbf{Q}, z) = G_1(X, \mathbf{Q}, z|X_0, \mathbf{Q}_0) - \sum_s k_s G_1(X, \mathbf{Q}, z|X_s, \mathbf{Q}_s) P_1(X_s, \mathbf{Q}_s, z)$$

$$+ \sum_s k_s G_1(X, \mathbf{Q}, z|X_s, \mathbf{Q}_s) P_2(X_s, \mathbf{Q}_s, z), \tag{A.12}$$

$$P_2(X, \mathbf{Q}, z) = G_2(X, \mathbf{Q}, z|X_0, \mathbf{Q}_0) - \sum_s k_s G_2(X, \mathbf{Q}, z|X_s, \mathbf{Q}_s) P_2(X_s, \mathbf{Q}_s, z)$$

$$+ \sum_s k_s G_2(X, \mathbf{Q}, z|X_s, \mathbf{Q}_s) P_1(X_s, \mathbf{Q}_s, z). \tag{A.13}$$

The sum is over the sink points where the populations are given by $P_i(X_s, \mathbf{Q}_s, z)$. By finding $P_1(X_s, \mathbf{Q}_s, z)$ and $P_2(X_s, \mathbf{Q}_s, z)$ using the preceding two equations, a set of linear equations are generated, which can be written in matrix form as

$$\mathbf{B} \cdot \mathbf{P} = \mathbf{G}_0. \tag{A.14}$$

The elements of $\mathbf{B}$, $\mathbf{P}$, and $\mathbf{G}_0$ are given as

$$[B_{mm'}] = \delta_{mm'} + k_{m'} G_1(X_m, \mathbf{Q}_m, z|X_{m'}, \mathbf{Q}_{m'}), \tag{A.15}$$

$$[B_{(m+k)m'}] = -k_{m'} G_2(X_k, \mathbf{Q}_k, z|X_{m'}, \mathbf{Q}_{m'}), \tag{A.16}$$

$$[B_{m(m'+k')}] = -k_{k'} G_1(X_m, \mathbf{Q}_m, z|X_{k'}, \mathbf{Q}_{k'}), \tag{A.17}$$

$$[B_{(m+k)(m'+k')}] = \delta_{(m+k)(m'+k')} - k_{s'} G_2(X_k, \mathbf{Q}_k, z|X_{k'}, \mathbf{Q}_{k'}), \tag{A.18}$$

$$[P_m] = P_1(X_m, \mathbf{Q}_m, z), \tag{A.19}$$

$$[P_{m+k}] = P_2(X_k, \mathbf{Q}_k, z), \tag{A.20}$$

$$[G_{0m}] = G_1(X_m, \mathbf{Q}_m, z|X_0, \mathbf{Q}_0), \tag{A.21}$$

$$[G_{0(m+k)}] = 0. \tag{A.22}$$

The population on the reactant surface as a function of the Laplace frequency can be written in the following form:

$$P_1(z) = \frac{1}{z}\left[1 - \sum_s k_s P_1(X_s, \mathbf{Q}_s, z) + \sum_s k_s P_2(X_s, \mathbf{Q}_s, z)\right]. \tag{A.23}$$

$P_1(t)$ is then obtained by the Laplace inversion of $P_1(z)$.

REFERENCES

1. R. A. Marcus, *J. Chem. Phys.* 24, 966 (1956); 24, 679 (1956).
2. R. A. Marcus and N. Sutin, *Biochim. Biophys. Acta* 811, 265 (1985).
3. (a) D. F. Calef and P. G. Wolynes, *J. Phys. Chem.* 87, 3387 (1983); (b) D. F. Calef and P. G. Wolynes, *J. Chem. Phys.* 78, 470 (1983).
4. W. J. Albery, *Electrode Kinetics*, Clarendon Press, Oxford (1975).
5. M. Tachiya, *J. Phys. Chem.* 97, 5911 (1993).
6. D. Rehm and A. Weller, *Isr. J. Chem.* 8, 256 (1970).
7. M. Tachiya and S. Murata, *J. Phys. Chem.* 96, 8441 (1992).
8. J. R. Miller, L. T. Calcaterra, and G. L. Closs, *J. Am. Chem. Soc.* 106, 3047 (1984).
9. L. D. Zusman, *Chem. Phys.* 49, 295 (1980).
10. J. T. Hynes, *J. Phys. Chem.* 90, 3701 (1986).
11. I. Rips and J. Jortner, *J. Chem. Phys.* 87, 2090, 6513 (1987).
12. H. Sumi and R. A. Marcus, *J. Chem. Phys.* 84, 4894 (1986).
13. J. Jortner and M. Bixon, *J. Chem. Phys.* 88, 167 (1988).
14. G. C. Walker, E. Akesson, A. E. Johnson, N. E. Levinger, and P. F. Barbara, *J. Phys. Chem.* 96, 3728 (1992).
15. B. Bagchi and N. Gayathri, *Adv. Chem. Phys.* 107, 1 (1999).
16. S. Roy and B. Bagchi, *J. Chem. Phys.* 102, 6719 (1995); *ibid.*, 102, 7937 (1995).

11

Förster (or, Fluorescence) Resonance Energy Transfer (FRET)

11.1 INTRODUCTION

Förster resonance energy transfer (FRET) is a process where radiationless transfer of excitation energy takes place from an optically excited state of a donor (D^*) molecule to the ground state of an acceptor (A) molecule. This is a long-range process where energy in favorable cases can be transferred between molecules 10 nm apart. The main utility of FRET is that the rate of such energy transfer depends strongly on the distance of separation between the donor and the acceptor.

FRET is sometimes referred to as Fluorescence Resonance Energy Transfer, although no fluorescence takes place during the transfer. However, the letter "F" in *FRET* has a different and indirect validity as the famous *Förster* expression of the rate of transfer, discussed below, is expressed in terms of the product of the fluorescence spectrum of the donor and the absorption spectrum of the acceptor molecules. Also, the event of transfer is often evidenced/detected by the appearance of fluorescence from the acceptor.

The most celebrated example of resonance energy transfer in nature is provided by photosynthesis, where the initially absorbed solar energy is transferred from carotenoids to chlorophylls, or between two chlorophylls.

This radiationless energy transfer occurs via long-range Columbic (for example, dipole–dipole) interaction. Because of this energy transfer, the donor molecule returns to the ground state and simultaneously the acceptor molecule gets excited. Förster energy transfer mostly happens when the energy gap between the ground and the excited states of the donor and acceptor molecules is nearly similar. See Fig. 11.1 for a schematic illustration of the process.

It was in 1948 that Th. Förster developed an elegant theory for this radiationless-energy-transfer process by using expressions derived from quantum mechanics and by employing several insightful approximations. Since then FRET has developed into a major experimental technique [1, 2], being used in studies in physics, chemistry, biology and materials science [4–9].

Förster expression for the rate of nonradiative energy transfer (k_{DA}) is commonly written as

$$k_{DA} = \frac{1}{\tau_D} \left(\frac{R_0}{R}\right)^6, \qquad (11.1)$$

where R is a distance between the centers of the D and A molecules, τ_D is the excited-state lifetime of the donor molecule in the absence of an acceptor, $(\tau_D^{-1} = k_{rad})$ and R_0 is known as Förster critical radius (defined as a distance at which the efficiency of energy transfer via a nonradiative process reduces by 50%).

Förster expressed R_0 in terms of spectral overlap between the emission spectrum of the donor (D) and the absorption spectrum of the acceptor (A) as

$$R_0^6 = \frac{9000 \ln(10) \kappa^2 Q_D c^4}{128 \pi^5 \eta^4 N_A} \int dv \frac{F_D(v) \varepsilon_A(v)}{v^4}. \qquad (11.2)$$

In this expression, $F_D(v)$ is the normalized fluorescence spectrum of the donor while $\varepsilon_A(v)$ is the extinction coefficient of the acceptor molecule. The integral $\int dv \frac{F_D(v)\varepsilon_A(v)}{v^4}$ is known as spectral overlap integral. Q_D is the quantum yield of the donor molecule, η is the refractive index of the solvent, N_A is Avogadro's number, and κ^2 is an orientation factor that takes into account the orientation of transition dipoles associated with D and A and is usually set to 2/3 which is a result of isotropic orientational averaging.

We shall present the full derivation of expressions (11.1) and (11.2) below.

The R^{-6} distance dependence of FRET makes it a highly sensitive technique to monitor the change in separation distance R between D and A. Owing to the sensitivity of FRET to separation R, it has often been termed as a "Spectroscopic Ruler" [9]. Because it provides a probe of structure and dynamics at nanoscopic lengths, applications of FRET technique are enormous. FRET is routinely used for understanding the conformational dynamics of macromolecules. With the advent of single-molecule spectroscopy, FRET has been used successfully to understand a range of biological processes, such as protein folding. This technique now is being intensely utilized in the area of materials science too, as discussed below.

In FRET experiments, fluorescent dyes are normally used as donor molecules. The acceptor molecule can also be a fluorescent dye. The efficiency of energy transfer via nonradiative process is expressed as

$$E_{ET}^{non\text{-}rad} = \frac{k_{DA}^{non\text{-}rad}}{k_{rad} + k_{DA}^{non\text{-}rad}} = \frac{R_0^6}{R_0^6 + R^6}, \qquad (11.3)$$

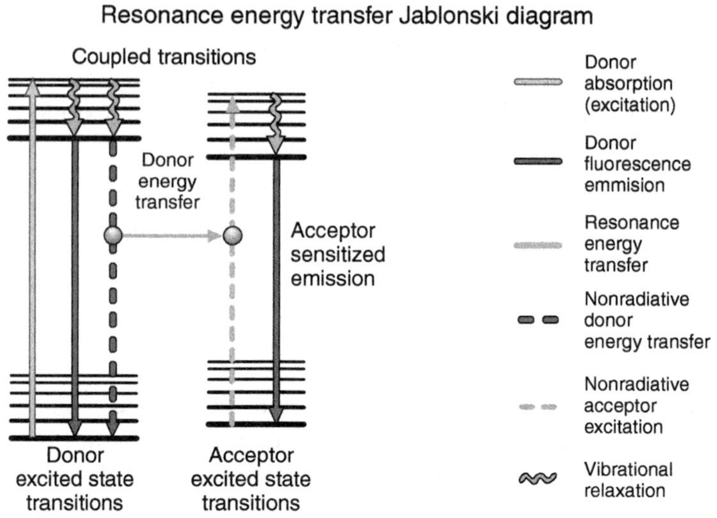

Figure 11.1 Jablonski diagram illustrating the coupled transitions between the donor and acceptor molecules involved in Förster resonance energy transfer. Absorption and emission transitions are represented by straight vertical arrows, while vibrational relaxation is indicated by wavy arrows. The coupled transitions are drawn with dashed lines. The phenomenon of FRET is illustrated by a gray arrow connecting the two states. Note that no photon is emitted in FRET, although the rate of FRET is conveniently calculated by using the emission spectrum of the donor and the absorption spectrum of the acceptor. (This figure has been modified. The original figure was taken from www.olympusmicro.com/primer/techniques/fluorescence/fret/fretintro.html.)

and in experiments where both donor and acceptor molecules are fluorescent, the efficiency in terms of fluorescence intensity $(I_{D/A})$ of D and A is written as

$$E_{ET}^{non\text{-}rad} = \frac{I_A}{I_A + I_D}. \qquad (11.4)$$

With the use of Eqs. (11.3) and (11.4), experimentalists determine the separation distance (R) between D and A.

The nonradiative mode of energy transfer is a near-field phenomenon (that is, distance between donor and acceptor is less than one wavelength of the light). In other words, this is a dominant mode of energy transfer only when the distance between donor and acceptor molecules is less than the wavelength of the "to be" emitted photon. Typically using conventional FRET technique, D-A separation distances up to 8–10 nm can be measured. The accuracy of these measurements depends on the value of R_0 of a given D-A pair. In general, the R_0 values of commonly used dyes, depending on their optical properties, lie in the range of 2–6 nm. Thus, FRET finds great use in the study of structure and dynamics.

Recently, FRET using metal (gold, silver) nanoparticles and also semiconductor quantum dots have been used to enhance the range of energy transfer.

The aim of this chapter is to give a brief overview of both FRET technique and Förster theory. The knowledge of the history of any scientific development enhances our understanding of the subject. In the following section, we shall briefly discuss the main developments that helped Förster in formulating the theory for nonradiative energy transfer. A rather elaborate discussion on the historical background of FRET can be found in Ref. 3.

11.2 A BRIEF HISTORICAL PERSPECTIVE

Nonradiative energy transfer was first observed in 1922 during an experiment in which a mixture of thallium and mercury atoms was excited at a wavelength that could only excite the mercury atoms. But surprisingly, an emission from thallium atoms was also observed and that too at distances larger than the collision radii of the two atoms. This fluorescence emission from thallium was termed as "sensitized fluorescence."

By the end of the 19th century, the concept of electromagnetic field and the work of Hertz in Göttingen (Germany) in the near-field were well established. These developments gave rise to the idea that energy could be transferred from excited donor atoms to acceptor atoms through space via electrodynamical interactions. Further experiments demonstrated an increase in the efficiency of nonradiative energy transfer as the energy levels of donor and acceptor atoms come in resonance with each other. These experiments highlighted the importance of resonance condition. By 1927, it was known that dipole–dipole interactions broaden the atomic spectra and could extend the radius of interatomic interactions beyond the radius expected from the collision theory. Some of these theories had already shown the R^{-6} dependence. In 1928, for the first time, Kallman and London used quantum-mechanical ideas to describe the process of nonradiative energy transfer between atoms in the vapor phase. They were able to explain the experimental results correctly by assuming that the dipole–dipole interactions between donor and acceptor atoms facilitate this mode of energy transfer under the "almost resonance" condition between the participating atoms.

At the same time, the nonradiative energy transfer was also observed between molecules in solution. The classical theory for this phenomenon in the condensed phase was first proposed by J. Perrin in 1927. He took into account the dipole–dipole interactions between the molecules, their relative orientation in the space, and the competition between the radiative and nonradiative modes of energy transfer. However, his theory predicted that energy transfer could take place even if the D, A molecules are separated by *100* nm, which was far greater than the experimental results. The failure of his approach was due to the assumption of "exact resonance" between the donor and the acceptor molecules. Later, in 1933 F. Perrin developed the corresponding quantum-mechanical theory of the same phenomenon but he also assumed the "exact resonance" condition and obtained similar results. In order to improve the agreement between the theory and the experiments, he considered the *collisions* of solvent molecules with D and A. These collisions broadened the energy distribution of D and A. As a result at any given time the probability of donor and acceptor molecules being in resonance with each other was no longer unity (i.e., *no* "exact resonance"). This assumption did bring down the value of distances over which energy transfer could take place, but the disagreement between theoretical and experimental results still persisted.

Finally, in 1948 Theodor Förster built upon the Perrins' idea and developed a quantitative quantum-mechanical treatment of nonradiative energy transfer [1]. He was already aware of the importance of the effect of spectral broadening on the rate of energy transfer. In his theory, Förster incorporated the broadening caused by the vibrational degrees of freedom associated with D and A, and their *interactions* with the solvent molecules. The results obtained by Förster were in good agreement with the experiments. Although F. Perrin also considered the broadening effect in his theory, the assumed distribution was much narrower than the actual energy distribution.

The main reason behind the widespread use of Förster theory and hence the FRET technique is that the rate of energy transfer is completely expressed in terms of quantities that can easily be determined through experiments. Förster was able to correlate the interaction between D and A with the emission and absorption spectra of the two, although in FRET no actual emission or absorption of photons take place, as mentioned previously.

11.3 DERIVATION OF FÖRSTER EXPRESSION

Before deriving the rate equation for FRET, we first discuss the conditions required for FRET to take place.

1. The separation distance between donor and acceptor molecules should lie within the range of 1–10 nm. For distances greater than 10 nm, the dipolar interactions between D and A are no longer strong enough to mediate the energy transfer (i.e., perturbation becomes negligible), while at distances less than 1 nm, the wavefunctions of D and A begin to overlap with each other. As a result, the exchange interactions dominate over dipole–dipole (Coulombic) interactions. The process of energy transfer from D to A mediated via the exchange interaction is termed as Dexter mechanism (see Fig. 11.2).

    The Dexter mechanism [4] is also a nonradiative process, but, unlike in FRET, here the energy transfer can take place between a spectroscopic-allowed transition in D to a forbidden transition in A. According to Dexter, the rate of energy transfer in this case depends exponentially on the separation distance between D and A and is written as

$$k_{DA} \propto J e^{-\left(\frac{2R}{L}\right)}, \tag{11.5}$$

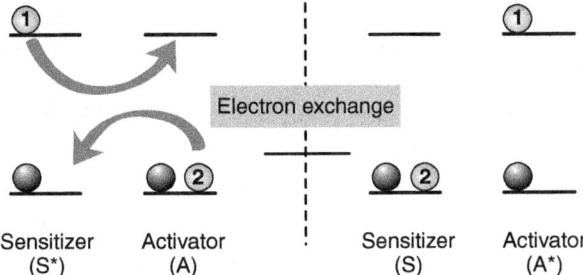

Figure 11.2 The Dexter mechanism of energy transfer that involves double electron transfer. The energy transfer takes place via the exchange of electrons between donor and acceptor molecules. This mechanism operates only at very small separation when direct overlap between orbitals of donor and acceptor molecules is possible.

where J is a spectral overlap integral, R is a separation distance between D and A, and L is a sum of van der Waals radii of D and A.
2. The quantum yield of the donor dye molecule should be high.
3. The emission spectrum of the donor molecule should have a considerable overlap with the absorption spectrum of the acceptor molecule (the resonance condition).
4. The orientation of the transition dipole moments of D and A should be favorable.

To derive his famous rate expression, Förster made the following assumptions:

1. Subsequent to the excitation of the donor molecule, the rate of vibrational relaxation is faster than the rate of FRET.
2. The process of energy transfer is incoherent, which means the coupling between D and A is weak and, as a result, the wave functions of D and A are independent of each other.
3. D and A molecules are assumed to be point dipoles. This approximation is known as the point dipole approximation. This assumption has been a subject of discussion over many years. However, this assumption is crucial to obtain the final simple and elegant expression.
4. Although not necessary, but for FRET in solution phase, an averaging over orientation of the dipolar angles is also performed.

Now, let us proceed and derive Förster rate equation. Consider a D-A system in which D is in its excited state while A is in the ground state (see Fig. 11.1). To describe the excitation energy transfer (EET) in D-A system, we introduce independent wave functions for both the molecules as $\Psi_{jl}(r_j; R_j^{nuc}) = \varphi_{jl}(r_j; R_j^{nuc})\chi_{jl}(R_j^{nuc})$ ($j = D$ or A; $l = g$ (ground, S_0) or e (excited, S_1); r_j and R_j^{nuc} are the electronic and nuclear coordinates, respectively). In writing the total wave function as a product of electronic and nuclear wave functions, we have used the Born–Oppenheimer approximation. We denote the vibronic energy levels associated with the ground and the excited states as u_j and v_j, respectively. The initial state in which the excitation is at D is represented by $|i\rangle = |\Psi_{Dev_D}\rangle |\Psi_{Agu_A}\rangle$ and characterized by energy $E_{Dev_D} + E_{Agu_A}$.

Similarly, the final state with energy $E_{Aev_A} + E_{Dgu_D}$ is given by $|f\rangle = |\Psi_{Aev_A}\rangle |\Psi_{Dgu_D}\rangle$. The $S_0 \to S_1$ transition energy is given by $E_{Dev_D} + E_{Agu_A} - E_{Aev_A} - E_{Dgu_D}$. As discussed previously, Förster assumed the D-A system to be in thermal equilibrium with its surroundings. Therefore, the transition rate from an initial state $|i\rangle$ to final state $|f\rangle$ can be calculated using the golden rule [5]:

$$k_{DA} = \frac{2\pi}{\hbar} \sum_u \sum_v f(E_{Dev_D}) f(E_{Agu_A}) |\langle \Psi_{Aev_A}, \Psi_{Dgu_D} |H_I| \Psi_{Dev_D}, \Psi_{Agu_A} \rangle|^2$$

$$\times \delta(E_{Dev_D} + E_{Agu_A} - E_{Aev_A} - E_{Dgu_D}). \tag{11.6}$$

Here the thermal distribution of the initial population of the vibrational states of D and A is denoted by $f(E_{Dev_D})$ and $f(E_{Agu_A})$, respectively. H_I accounts for the Coulombic interactions between D and A. Assuming that the electronic coupling matrix elements $(V_{DA} = \langle \varphi_{Aev_A}, \varphi_{Dgu_D} |H_I| \varphi_{Dev_D}, \varphi_{Agu_A} \rangle)$ are independent of nuclear coordinates (Condon approximation), we can write

$$\langle \Psi_{Aev_A}, \Psi_{Dgu_D} |H_I| \Psi_{Dev_D}, \Psi_{Agu_A} \rangle = V_{DA} \langle \chi_{Dgu_D} | \chi_{Dev_D} \rangle \langle \chi_{Aev_A} | \chi_{Agu_A} \rangle \tag{11.7}$$

Using the point dipole approximation, V_{DA} reduces to

$$V_{DA} = \frac{1}{\eta^2 R^3} \left(\vec{\mu}_{eg}^{(D)} \cdot \vec{\mu}_{eg}^{(A)} - \frac{3}{R^2} (\vec{\mu}_{eg}^{(D)} \cdot \vec{R})(\vec{\mu}_{eg}^{(A)} \cdot \vec{R}) \right). \tag{11.8}$$

The above expression of the coupling matrix can be simplified further into separation distance and orientation factors:

$$V_{DA} = \kappa \frac{\left|\vec{\mu}_{eg}^{(D)}\right| \left|\vec{\mu}_{eg}^{(A)}\right|}{\eta^2 R^3}, \tag{11.9}$$

where κ is a dimensionless geometric factor that is determined by the orientation of donor and acceptor transition dipoles $\vec{\mu}_{eg}^{(j)} (= \langle \varphi_{je} | \hat{\mu}_j | \varphi_{jg} \rangle)$ in space and is given by the unit vectors

$$\kappa = \left(\hat{\mu}_{eg}^{(D)} \cdot \hat{\mu}_{eg}^{(A)} - 3(\hat{\mu}_{eg}^{(D)} \cdot \hat{R})(\hat{\mu}_{eg}^{(A)} \cdot \hat{R}) \right).$$

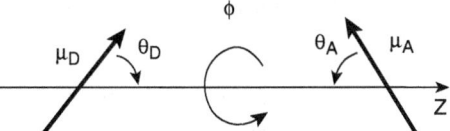

Figure 11.3 Illustration of the dependence of the orientation factor (κ^2) on the relative orientations of the donor emission dipole and the acceptor absorption dipole. The interaction energy between the two dipoles depends on the orientation of the transition dipole moment vector.

The polar angles θ and ϕ specify the orientation of donor and acceptor transition dipoles with respect to the separation vector $\hat{R}$. If we assume $\hat{R}$ is directed along the z-axis, then the orientation factor defined as κ^2 is given by

$$\kappa^2(\theta_A, \theta_D, \Phi_{AD}) = (\sin\theta_A \sin\theta_D \cos\phi_{AD} - 2\cos\theta_A \cos\theta_D)^2, \quad (11.10)$$

with $\phi_{AD} = (\phi_A - \phi_D)$. ϕ_{AD} is equivalent to a dihedral angle between the planes containing D and A molecules, while θ_D and θ_A are the angles that the dipoles of D and A form with $\hat{R}$, the axis joining the centers of D and A molecules (Fig. 11.3).

As discussed previously, the rate of non-radiative excitation energy transfer can be visualized as a process in which simultaneous optical emission at the donor and the optical absorption at the acceptor takes place. In order to rewrite the rate of energy transfer in terms of $S_1 \to S_0$ donor emission spectrum and $S_0 \to S_1$ acceptor absorption spectrum, the delta function of Eq. (11.6) is cleverly rewritten as a product of two delta functions describing the process of emission and absorption of energy, E, by the donor and acceptor, respectively, i.e.,

$$\delta(E_{Dev_D} + E_{Agu_A} - E_{Aev_A} - E_{Dgu_D}) = \int_{-\infty}^{\infty} dE\, \delta\left(E_{Dev_D} - E_{Dgu_D} - E\right)$$

$$\times \delta(E + E_{Agu_A} - E_{Aev_A}). \quad (11.11)$$

Using the above equations, Eq. (11.6) becomes

$$k_{DA} = 2\pi\kappa^2 \frac{\left|\vec{\mu}_{eg}^{(D)}\right|^2 \left|\vec{\mu}_{eg}^{(A)}\right|^2}{\eta^4 R^6} \int d\omega \sum_u \sum_v f(E_{Dev_D}) f(E_{Agu_A}) \left|\langle\chi_{Dev_D}|\chi_{Dgu_D}\rangle\right|^2$$

$$\times \left|\langle\chi_{Agu_A}|\chi_{Aev_A}\rangle\right|^2 \delta\left(E_{Dev_D} - E_{Dgu_D} - E\right) \delta(E + E_{Agu_A} - E_{Aev_A}). \quad (11.12)$$

We shall now separately derive the expressions for emission and absorption spectra and finally use these expressions to derive the famous Förster rate equation.

11.3.1 Expressions for Emission (or Fluorescence) Spectrum

The rate of transition from an excited state ($|\varphi_{ev}\rangle |\chi_{ev}\rangle$) to the ground state ($|\varphi_{gu}\rangle |\chi_{gu}\rangle$) of the molecule, accompanied with the emission of a photon of particular wave vector $\vec{k}$ and polarization direction ξ, is obtained by applying the golden rule,

$$R_{ems} = \frac{2\pi}{\hbar} \sum_{u,v} f(E_{ev}) \left| \langle \varphi_{ev} | \langle \chi_{ev} | H_I | \chi_{gu} \rangle | \varphi_{gu} \rangle \right|^2 \delta(E_{ev} - \hbar\omega - E_{gu}), \tag{11.13}$$

where, as before, $f(E_{ev})$ is a thermal distribution of the initial population of the vibrational states of the excited molecule. H_I is the *time-independent* interaction Hamiltonian which accounts for the matter–field interactions.

The *time-dependent* interaction Hamiltonian $H_I(t)$ for molecule–light interaction is given by

$$H_I(t) = \frac{-(-e)\eta}{mc} \vec{A} \cdot \vec{P}, \tag{11.14}$$

where m is a mass of an electron, $-e$ is the charge of the electron, and c is the velocity of light. $\vec{A}$ is a vector field potential and $\vec{P}$ is a momentum operator. The vector potential $\vec{A}$ is of the form

$$\vec{A} = \vec{A}_0 \cos(\vec{k} \cdot \vec{r} - \omega t) = \frac{\vec{A}_0}{2} \left(e^{i(\vec{k} \cdot \vec{r} - \omega t)} + e^{-i(\vec{k} \cdot \vec{r} - \omega t)} \right), \tag{11.15}$$

which is obtained as a solution to the wave equation, $\nabla^2 \vec{A} - \frac{1}{v_{em}^2} \frac{\partial^2 \vec{A}}{\partial t^2} = 0$, where the velocity of light in the medium, $v_{em} = c/\eta$.

One next substitutes the value of $\vec{A}$ in Eq. (11.14) to obtain the time-dependent interaction Hamiltonian as

$$H_I(t) = \frac{e\eta}{2mc} \left(e^{i(\vec{k} \cdot \vec{r} - \omega t)} + e^{-i(\vec{k} \cdot \vec{r} - \omega t)} \right) \vec{A}_0 \cdot \vec{P}. \tag{11.16}$$

The first term in the equation with $e^{-i\omega t}$ leads to stimulated absorption, while the second term with $e^{i\omega t}$ leads to stimulated emission. Because we are here interested only in the emission process, we ignore the first term of the time-dependent interaction Hamiltonian,

$$H_I(t) = \frac{e\eta}{2mc} e^{-i\vec{k}\cdot\vec{r}} A_0 \hat{\varepsilon}_\xi \cdot \vec{P} e^{i\omega t} = H_I e^{i\omega t}. \tag{11.17}$$

$\hat{\varepsilon}_\xi$ denotes the unit vector in the direction of $\vec{A}_0$. In order to arrive at an expression used by Förster, we need to rewrite the interaction matrix element in terms of the dipole operator instead of the momentum operator:

$$\langle \varphi_{gu} | H_I | \varphi_{ev} \rangle = \frac{i\omega \eta}{2c} A_0 \hat{\varepsilon}_\xi \cdot \vec{\mu}_{eg}. \tag{11.18}$$

Note that in Eq. (11.17) the factor $e^{-i\vec{k}\cdot\vec{r}}$ can be assumed to be equal to unity as the size of the molecule is much smaller than the wavelength of the light. Substituting Eq. (11.18) into the rate expression Eq. (11.13) and taking an orientational average of this rate, we get

$$\langle R_{ems} \rangle_{or} = \frac{\pi}{3\hbar} \frac{\omega^2 \eta^2}{2c^2} A_0^2 |\vec{\mu}_{eg}|^2 \sum_{u,v} f(E_{ev}) |\langle \chi_{ev} | \chi_{gu} \rangle|^2 \delta(E_{ev} - \hbar\omega - E_{gu}). \tag{11.19}$$

So far we have been concerned with an emission process in which a photon of definite wave vector $\vec{k}$ and polarization ξ is emitted. Now, let us consider the rate of photon emission $d\Gamma_{ems}$ with the magnitude of wave vector in the range between $\vec{k}$ and $\vec{k} + d\vec{k}$. Then,

$$d\Gamma_{ems} = \langle R_{ems} \rangle_{or} \rho(E) dE, \tag{11.20}$$

where $\rho(E)dE$ is the number of allowed states of the photon field in an energy interval E and $E + dE$ corresponding to a photon of wave vector in the range between $\vec{k}$ and $\vec{k} + d\vec{k}$. Expressing $\vec{k}$ in polar coordinates and taking into account the two polarization directions, we obtain the following expression for energy density:

$$\rho(E) dE = 2 \left(\frac{L}{2\pi} \right)^3 4\pi k^2 dk.$$

Note to determine $\rho(E)dE$, we have considered photon field in a cubic box with sides L (volume, $V = L^3$).

We next substitute this expression into Eq. (11.20) and integrate over an entire k space ($k = \omega\eta/c$) to obtain the total rate of decay in terms of ω as

$$\Gamma_{ems} = \int \frac{V}{3\hbar} \frac{\omega^4 \eta^5}{2\pi c^5} A_0^2 |\vec{\mu}_{eg}|^2 \sum_{u,v} f(E_{ev}) |\langle \chi_{ev} | \chi_{gu} \rangle|^2 \delta(E_{ev} - \hbar\omega - E_{gu}) d\omega. \tag{11.21}$$

Finally, we need to determine A_0 in order to find an expression for emission spectrum. From the theory of electromagnetic field, we can derive a relation between the time-averaged energy density $\langle u \rangle_t$ of the electromagnetic radiation and A_0 as

$$\langle u \rangle_t = \frac{N\hbar\omega}{V} = \frac{1}{8\pi}\frac{\omega^2\eta^4}{c^2}|\vec{A}_0|^2.$$

We rearrange this to obtain

$$|\vec{A}_0|^2 = 8\pi\frac{N\hbar\omega}{V}\frac{c^2}{\omega^2\eta^4}. \tag{11.22}$$

Here, V is the volume and N is the number of photons emitted. We next use Eq. (11.22) in Eq. (11.21) to obtain the emission rate in the following form:

$$\Gamma_{ems} = \int \frac{4}{3}N\frac{\omega^3\eta}{c^3}|\vec{\mu}_{eg}|^2 \sum_{u,v} f(E_{ev})|\langle \chi_{ev}|\chi_{gu}\rangle|^2 \delta(E_{ev} - \hbar\omega - E_{gu})d\omega. \tag{11.23}$$

This equation is almost correct but it predicts that, for $N = 0$, the total emission rate is zero, which implies that there can be no spontaneous emission. This is of course not correct. The rate equation for spontaneous emission can be rigorously derived using quantum electrodynamics. However, if we just replace N by $(N+1)$ in Eq. (11.23), the correct expression for the rate of emission is obtained. The rate of spontaneous emission is thus given by

$$\Gamma_{ems} = \frac{4}{3}\frac{\eta}{c^3}|\vec{\mu}_{eg}|^2 \int \omega^3 G(E)d\omega, \tag{11.24}$$

where $G(E) = \sum_{u,v} f(E_{ev})|\langle \chi_{ev}|\chi_{gu}\rangle|^2 \delta(E_{ev} - \hbar\omega - E_{gu})$.

We know that the radiative rate of an excited electronic state is given by $\Gamma_{ems} = \int d\omega I(\omega)$ where $I(\omega)$ is the emission rate per frequency interval, which can be identified with the emission spectrum of the considered molecule. Note that $I(\omega)$ is not normalized. In order to obtain Förster expression, we normalize the emission spectrum from donor as

$$F_D(\omega) = \frac{\omega^3 G_D(E)}{\int \omega^3 G_D(E)d\omega}.$$

To summarize, the expression for normalized emission spectrum is given by

$$F_D(\omega) = \frac{4\eta\omega^3}{3c^3}\tau_D \left|\mu_{eg}^{(D)}\right|^2 \sum_{u_D v_D} f(E_{Dev_D})|\langle \chi_{Dev_D}|\chi_{Dgu_D}\rangle|^2 \delta\left(E_{Dev_D} - E_{Dgu_D} - \hbar\omega\right), \tag{11.25}$$

where the radiative lifetime is given $\tau_D = 1/\Gamma_{ems}$.

11.3.2 Absorption Spectrum

As in the case of derivation of emission spectrum, we again begin with the golden rule where the rate of transition from a molecule's ground state to an excited state is given by

$$R_{abs} = \frac{2\pi}{\hbar} \sum_{u,v} f(E_{gu}) |\langle \varphi_{ev}| \langle \chi_{ev}| H_I |\chi_{gu}\rangle |\varphi_{gu}\rangle|^2 \delta(E_{gu} + \hbar\omega - E_{ev}), \quad (11.26)$$

and similarly the interaction Hamiltonian is given by

$$H_I(t) = \frac{e\eta}{2mc} e^{i\vec{k}\cdot\vec{r}} \vec{A}_0 \cdot \vec{P} e^{-i\omega t} = H_I e^{-i\omega t}. \quad (11.27)$$

The difference between interaction Hamiltonian for the absorption and emission spectra lies in exponential factors. Following the same derivation as in the case of emission spectrum we get

$$H_I = -\frac{i\omega\eta}{2c} A_0 \hat{\varepsilon}_\xi \cdot \vec{\mu}_{eg}. \quad (11.28)$$

The orientational average rate of transition is given by

$$\langle R_{abs}\rangle_{or} = \frac{\pi}{3\hbar} \frac{\omega^2 \eta^2}{2c^2} |\vec{A}_0|^2 |\vec{\mu}_{eg}|^2 \sum_{u,v} f(E_{gu}) |\langle \chi_{ev}| \chi_{gu}\rangle|^2 \delta(E_{gu} + \hbar\omega - E_{ev}). \quad (11.29)$$

The orientational averaging introduces a factor of 1/3. The rate of energy absorption is

$$\frac{dE_{abs}}{dt} = \hbar\omega \langle R_{abs}\rangle_{or}. \quad (11.30)$$

If the absorbing medium is placed transverse to the electromagnetic beam of intensity, $I_{e.m.}$ and σ is the absorption cross section of the medium, the rate at which it absorbs energy is

$$\frac{dE_{abs}}{dt} = \sigma I_{e.m.}. \quad (11.31)$$

$I_{e.m.}$ is the average energy flux density transported by the electromagnetic field. The energy flux density (energy per unit area per unit time) for the electromagnetic field is given by the Poynting vector:

$$\vec{S} = \frac{c}{4\pi}(\vec{E} \times \vec{H}). \tag{11.32}$$

Again, using the relations in electromagnetism, we write $I_{e.m.}$ in terms of $\vec{A}_0$ as

$$I_{e.m.} = \langle S \rangle_{time} = \frac{\omega^2 \eta^3}{8\pi c}|\vec{A}_0|^2. \tag{11.33}$$

Substitute this expression into Eq. (11.31). The rearrangement gives the absorption cross section of absorbing medium:

$$\sigma = \frac{4\pi^2 \omega}{3\eta c}|\vec{\mu}_{eg}|^2 \sum_{u,v} f(E_{gu})|\langle \chi_{ev}|\chi_{gu}\rangle|^2 \delta(E_{gu} + \hbar\omega - E_{ev}). \tag{11.34}$$

The decadic extinction coefficient is related to absorption cross section as

$$\sigma(\omega) = \ln 10 \left(\frac{1000\varepsilon(\omega)}{N_A}\right). \tag{11.35}$$

N_A is Avogadro's number. Substituting this into Eq. (11.34) and rearranging, we get

$$\varepsilon_A(\omega) = \frac{4\pi^2 N_A \omega}{3000(\ln 10)\eta c}|\vec{\mu}_{eg}|^2 \sum_{u,v} f(E_{gu})|\langle \chi_{ev}|\chi_{gu}\rangle|^2 \delta(E_{gu} + \hbar\omega - E_{ev}). \tag{11.36}$$

$\varepsilon_A(\omega)$ is expressed in units of $L mol^{-1} cm^{-1}$.

11.3.3 The Final Förster Expression

Let us now return to Eq. (11.12) and rewrite it in terms of absorption and emission spectra of acceptor and donor, respectively. This leads to the following expression for the rate of transfer:

$$k_{DA} = \frac{9000(\ln 10)}{8\pi N_A \tau_D} \frac{\kappa^2 c^4}{\eta^4 R^6} \int d\omega \frac{F_D(\omega)\varepsilon_A(\omega)}{\omega^4}. \tag{11.37}$$

However, this expression is not the familiar form of the Förster rate equation. In order to derive that, we express ε_A and F_D on frequency scale ν, and the rate equation is given by

$$k_{DA} = \frac{9000(\ln 10)}{128\pi^5 N_A \tau_D} \frac{\kappa^2 c^4}{\eta^4 R^6} \int d\nu \frac{F_D(\nu)\varepsilon(\nu)}{\nu^4}, \quad (11.38)$$

which is the famous Förster expression. In a simpler form the Eq. (11.38) is expressed as Eq. (11.1). τ_D is typically of the order of 1 ns for commonly used dyes in EET experiments. τ_D is the radiative rate of donor.

11.4 APPLICATIONS OF FÖRSTER THEORY TO CHEMISTRY, BIOLOGY, AND MATERIALS SCIENCE

The strong distance dependence of the FRET rate has been exploited in many different ways. Applications range from sensors to studies of fundamental biological processes.

11.4.1 FRET-Based Glucose Sensor

In Figure 11.4 we show the operation of a glucose sensor based on competitive binding of dextran (labeled with malachite green, MG) to a sugar-binding protein labeled with allophycocyanin (APC). The replacement of dextran from the binding site by a glucose molecule results in the change in the donor–acceptor (APC-MG) distribution, which affects the fluorescence decay of the sugar–protein system. Experimentally, this change in decay is used to calculate the donor–acceptor distribution in the system and hence the concentration of the glucose.

11.4.2 FRET and Macromolecular Dynamics

Figure 11.5 depicts an application of FRET to macromolecular dynamics. With a suitable D-A pair attached to a macromolecule, the rate of EET provides information about favorable configurations and conformations of the macromolecule. For a fixed distance between D and A, as in the case of rigid biopolymers, the EET experiments can provide information directly on separation distance R between D and A sites. But for flexible macromolecules in solution, the study of long-distance energy transfer between two segments in a polymer chain is a nontrivial problem because the Brownian motion of the individual monomeric units of the polymer, i.e., the fluctuations in separation distance R and the relative orientation of the monomers, is not random but is correlated even at long lengths owing to the connectivity among the monomers.

What perhaps limits the use of FRET is a lack of quantitative theories in these systems that could allow extraction of detailed microscopic- or mesoscopic-level correlations from FRET data. As discussed below, one can gain considerable insight about the fluctuations in structure and dynamics if one could find the distribution of excitation energy that is now available, at least in principle in all cases and in

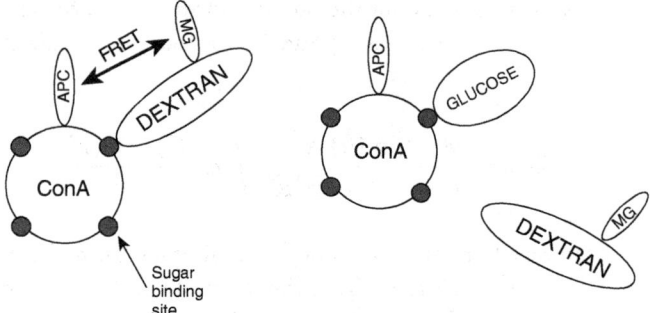

Figure 11.4 Illustration of a FRET-based glucose sensor. Glucose binding protein ConA is labeled with donor molecule allophycocyanin (APC) and dextran is labeled with acceptor malachite green (MG). Reprinted with permission from Olaf J. Rolinski, David J. S. Birch, Lydia J. McCartney, and John C. Pickup, Fluorescence nanotomography using resonance energy transfer: demonstration with a protein-sugar complex, *Phys. Med. Biol.* **46**, p. N221, 2001. Copyright (2001) Institute of Physics Publishing.

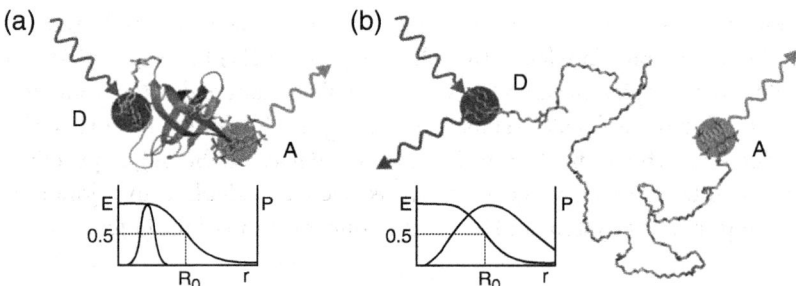

Figure 11.5 FRET study of protein unfolding with schematic structures of protein (Csp *Tm*) helix labeled with donor (D) and acceptor (A): (a) Folded Csp *Tm*; (b) Unfolded Csp *Tm*. In each case, functional form of FRET efficiency, *E* versus distance is shown along with the representation of the probability distribution of distances between donor and acceptor dyes. Reprinted with permission from Macmillan Publishers Ltd: *Nature*, **419**, p. 743 (2002). Copyright (2002).

practice in favorable ones, through single-molecule spectroscopy. The reason for the difficulty is the R^{-6} distance dependence of the rate, which although allowing straightforward determination of the Förster distance R_0, makes, at the same time, theoretical evaluation of dynamics of polymers difficult, as detailed below.

WILEMSKI–FIXMAN THEORY

Fortunately, for systems that are near equilibrium before and during energy transfer, the Wilemski–Fixman (WF) theory [6] along with the Förster expression can accurately describe the intrachain energy migration. WF theory gives nearly analytic solution for Rouse dynamics of a polymer chain for any arbitrary sink function, $S(R)$. Often the Rouse model (although rather unrealistic) is used to describe the polymer dynamics because theoretically this limit can easily be treated. An elegant alternative derivation of WF theory is provided by Portman and Wolynes using the variational approximation method [7]. They obtained upper and lower bounds for the survival probability $S_p(t)$ and showed that for a rapidly relaxing system, WF theory defines the upper bound on $S_p(t)$.

The many-body nature of polymer dynamics can be described by a joint, time-dependent probability distribution $P(\vec{r}_N, t)$, where $\vec{r}_N$ denotes the position of all the N polymer beads, at time t. The time dependence of the probability distribution $P(\vec{r}_N, t)$ is described by the following reaction–diffusion equation [6]:

$$\frac{\partial}{\partial t} P(\vec{r}_N, t) = L_B(\vec{r}_N) P(\vec{r}_N, t) - S(R) P(\vec{r}_N, t), \tag{11.39}$$

where L_B is the full $3N$-dimensional diffusion operator given by

$$L_B(\vec{r}_N) P(\vec{r}_N, t) = D \sum_{j=1}^{N} \frac{\partial}{\partial r_j} P_{eq}(\vec{r}_N) \frac{\partial}{\partial r_j} \frac{1}{P_{eq}(\vec{r}_N)} P(\vec{r}_N, t), \tag{11.40}$$

where the subscript "eq" denotes equilibrium, R is the distance between the two ends of the polymer chain, and D is the diffusion coefficient of a monomer. The last term on the right-hand side of Eq. (11.39) is the sink term. It can be of different forms. In our case, $S(R)$ describes Förster energy transfer, given by Eq. (11.1).

Wilemski–Fixman (WF) introduced a different sink function to describe the chemical reaction. The experimental observable is the survival probability $S_p(t)$ of the reactant and is given by

$$S_p(t) = \int P(\vec{r}_N, t) d\vec{r}_1 d\vec{r}_2 \ldots\ldots d\vec{r}_N. \tag{11.41}$$

The Wilemski–Fixman method of solution uses the Green's function $G(R_1, R_2, t)$ for the end-to-end separation distance R to define a sink–sink time-correlation function (SSTCF) $D(t)$ given by

$$D(t) = \int_0^\infty d^3R_1 \int_0^\infty d^3R_2 S(R_1) S(R_2) G(R_1, R_2, t) P_{eq}(R_2), \tag{11.42}$$

where $P_{eq}(R)$ is the equilibrium end-to-end distribution function. The Green's function $G(R_1, R_2, t)$ apart from the positions of reactants $(R_1$ and $R_2)$ also depends on $\rho(t)$, the normalized time-correlation function of end-to-end vector defined as $\langle \vec{R}(0) \cdot \vec{R}(t) \rangle / \langle R^2 \rangle$.

In order to obtain $S_p(t)$, WF made a closure approximation, according to which the Laplace transform of $S_p(t)$ is approximated as

$$\hat{S}_p(s) = \frac{1}{s} - \frac{k v_{eq}}{s^2(1 + k\hat{D}(s)/v_{eq})}, \quad (11.43)$$

where k is the momentum transfer variable and s is the Laplace transform variable. $\hat{D}(s)$ is a Laplace transform of the $D(t)$ defined above. v_{eq} (in Eq. (11.43)) is the rate when the distribution of polymer ends is at equilibrium.

The above expression (Eq. (11.43)) for the frequency-dependent survival probability is the main result of the Wilemski–Fixman theory. Let us analyze this nontrivial result. The first term on the right-hand side ensures that in the limit of time t going to zero, we have $S_p(t)$ equal to unity. The second term gives the decay in population due to the sink term. Owing to the complex time dependence of $D(t)$, a simple analytical solution of $S_p(t)$ is not available, although numerical evaluation shows a long-time exponential-like decay.

Equation (11.43) requires a nontrivial input in the form of $\hat{D}(s)$ which is the Laplace transform of $D(t)$. The final expression is a bit complicated. It is obtained after averaging over all the angles involved with vectors R_1 and R_2 over the three-dimensional Green's function of the Smoluchowski operator. The final expression for the sink–sink time-correlation function $D(t)$ is given by

$$D(t) = \left(\frac{3}{2\pi L^2}\right)^3 \left(\frac{1}{[1 - \rho^2(t)]^{3/2}}\right)$$

$$\times \int_0^\infty 4\pi R_1^2 S(R_1) dR_1 \int_0^\infty 4\pi R_2^2 S(R_2) dR_2 \exp\left(-\frac{3(R_1^2 + R_2^2)}{2L^2(1 - \rho^2(t))}\right)$$

$$\times \frac{\sinh\{[3\rho(t)R_1R_2]/(L^2(1 - \rho^2(t)))]\}}{[3\rho(t)R_1R_2]/(L^2[1 - \rho^2(t)])}. \quad (11.44)$$

Once the choice of the sink function is specified, it is straightforward to calculate numerically the survival probability by utilizing the above set of equations. WF's choice was the Heaviside sink function. For the Förster sink function, $S(R)$ is assumed to be given by

$$S(R) = \frac{k_F}{1 + (R/R_0)^6}, \quad (11.45)$$

where k_F is the rate of EET when the D-A separation is vanishingly small (i.e., $R/R_0 \to 0$). The above form of $S(R)$ is different (somewhat trivially) from the commonly used form of the Förster rate. The modification is necessary as otherwise the rate will show a divergence with $R \to 0$, which is allowed for a Rouse chain (note that in Rouse chain the beads can pass through one another without any hindrance).

In order to underline the reliability of FRET as a dynamic marker in a polymer folding process, Brownian dynamics simulations on model homo–polymer systems were carried out by Srinivas and Bagchi [8]. In these studies, survival probability, $S_p(t)$, defined as the probability of reaction between D-A pair (reaction imply transfer of energy between donor and acceptor) was used to study the folding dynamics. As is clear from the definition, $S_p(t)$ is a theoretical counterpart of fluorescence intensity of donor which by itself is a measure of "un-reacted" donor concentration. For FRET to be useful in the study of folding, it is essential to satisfy the following conditions:

(1) To choose a donor–acceptor system with a value of R_0 that allows study of both folded and unfolded states as FRET in principle, and is capable of providing information about structural changes only for certain values of R_0.
(2) The three time scales $\tau_{E,fold}$, $\tau_{E,unfold}$, and $\tau_{q,fold}$ should obey the $\tau_{E,fold} \ll \tau_{q,fold} \ll \tau_{E,unfold}$ condition where $\tau_{E,fold}$ and $\tau_{E,unfold}$ are time scales required by polymer in equilibrium state to fold and unfold, respectively, while $\tau_{q,fold}$ is the time required for polymer to fold subsequent to a quench in the temperature.

Figure 11.6(a) shows the decay of the survival probability $S_p(t)$ as obtained from BD simulations during the polymer-folding process. It can be seen that during the folding process, $S_p(t)$ follows the decay path of the unfolded state for $t < 100\,\tau$, after which it starts deviating and decay becomes more rapid. It is important to note that the major part of FRET occurs within a time of $200\,\tau$, which is the same as the average time taken by the polymer to collapse approximately to one-half of the mean square end-to-end distance of its initial unfolded configuration (compare Fig. 11.6a and Fig. 11.6b).

Experimentally, FRET was first used by Stryer and Haugland to study the FRET efficiency as a function of separation distance in stiff oligomers [9]. The sensitivity of FRET toward the changing distances inspired Stryer to use the term *spectroscopic ruler* to describe the technique of FRET. Recent developments in basic optical techniques used in FRET experiments and the availability of better dye molecules have made the FRET technique all together more exciting.

11.4.3 FRET and Single-Molecule Spectroscopy

The combination of FRET with the recently developed single-molecule spectroscopy (SMS) has elevated its applicability to new levels [10]. The SMS-FRET has been used to identify various subpopulations of the different species present in the system (Fig. 11.7). The ability to identify the different populations has proven to be useful in the study of dynamics of various biomolecular processes.

SMS-FRET has been used to examine the docking transition and the overall folding process in ribozyme molecule (*Tetrahymena thermophila*). For FRET measurements, the fluorescent dyes, Cy3 and Cy5, are attached to the ribozyme molecule. The docking transition studied here is a simple transition from

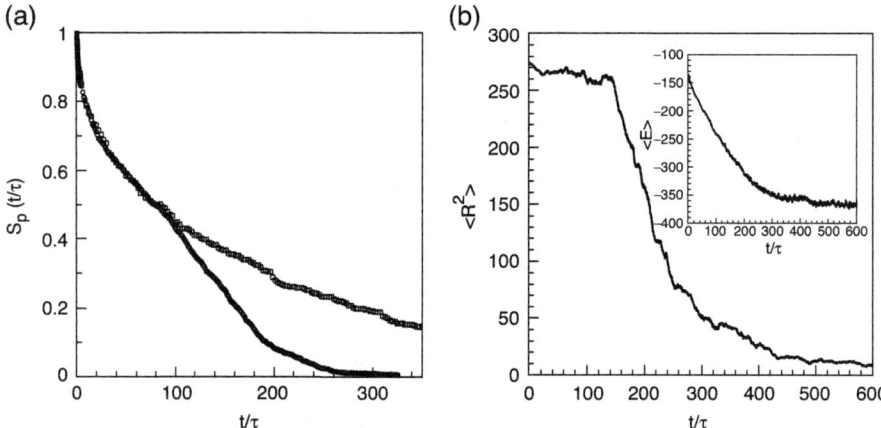

Figure 11.6 (a) Survival probability $S_P(t)$ for the polymer during the folding process (lower curve) compared to the same function for the equilibrium unfolded state (upper curve). (b) The mean square end-to-end distance $\langle R^2 \rangle$ and the average total energy (inset) as a function of time during folding process. Reprinted with permission from *J. Phys. Chem. B* **105**, 2475 (2001). Copyright (2001) American Chemical Society.

secondary to tertiary structure. The structures of the undocked and docked state of the ribozyme (see Fig. 11.8(a)) suggest that the separation distance between the D-A pair is larger in the undocked state compared to the docked state. FRET measurements show that the methylation of hydroxyl group of U-3 residue on S greatly destabilizes the docked state (lower panel of Fig. 11.8(b)). This confirms that the interaction between hydroxyl group of the U-3 residue on S (modified S, -3mS) with the ribozyme core favors the docking transition. The rate constants for docking and undocking transition in the case of all-ribose S were found to be $k_{dock} = 1.62 \pm 0.08$ s^{-1} and $k_{undock} = 0.224 \pm 0.015$ s^{-1} and for modified S, -3mS these were $k_{dock} = 1.25 \pm 0.06$ s^{-1} and $k_{undock} = 11.5 \pm 1.0$ s^{-1}. These values were obtained from the histograms of dwell times in corresponding docked and undocked states (Fig. 11.8(c)). The increase in the rate of undocking upon methylation with no significant decrease in the rate of docking suggests that the tertiary contact required for the docking transition forms subsequent to the transition state. If this contact would have formed in the transition state, the increase in the free-energy difference between the undocked state and the transition state of -3mS will result in an appreciable decrease in the rate of docking. But because the tertiary contact forms afterward, the absence of this contact in -3mS destabilizes the docked state and, therefore, only an increase in the rate of undocking is observed.

The preceding examples demonstrate that, with cleverly designed experiments, FRET can provide us with a wide variety of information on various aspects of molecular processes. The further advances in the FRET technique, especially the development of a new class of fluorescent dyes, such as green fluorescent proteins and its derivatives, have made it possible to observe the biomolecular processes *in vivo*, i.e., inside the living cells [11].

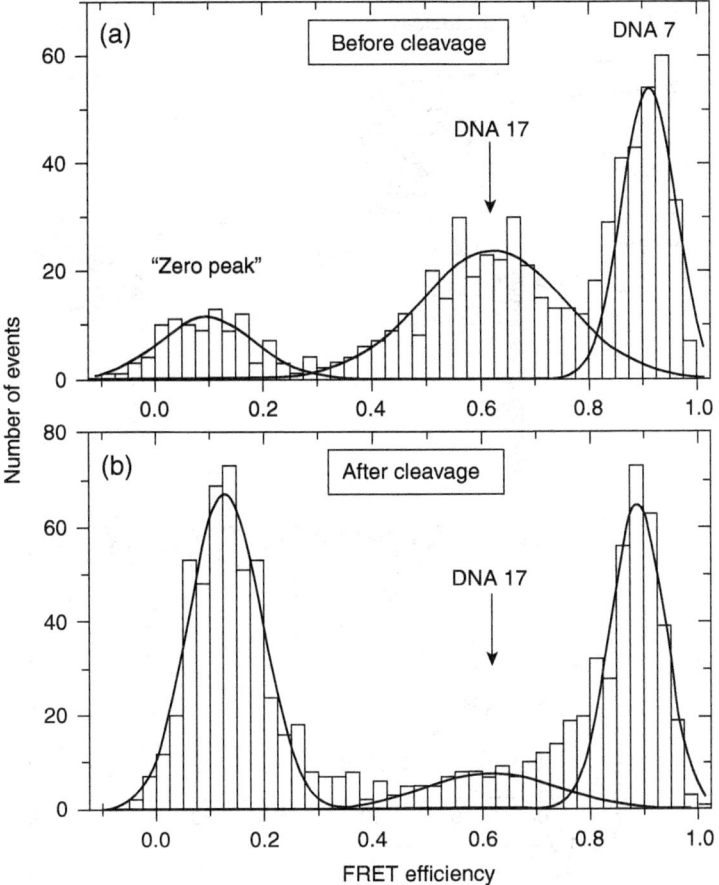

Figure 11.7 FRET histogram of a sample containing a 1:1 mixture of two different double-stranded DNA molecules with 7- and 17-base pair (bp) separations between sites labeled with donor and acceptor molecules. The two peaks at E = 0.6 and E = 0.9 corresponds to 17-bp and 7-bp separation, respectively. This demonstrates the ability of FRET to identify the subpopulations based on the conformational differences between the molecules. The peak at E = 0 arises from fluorescence of only donor-labeled molecules. The comparison of the (a) upper and (b) lower panels of the figure shows the cleavage of DNA 17 by a restriction endonuclease enzyme as suggested by the decrease in intensity of FRET peak corresponding to DNA 17 and simultaneous increase in the intensity of zero peak. Reprinted with permission from *Proc. Natl. Acad. Sci. U.S.A.* **96**, 3670, 1999. Copyright (2008) National Academy of Sciences, U.S.A.

The huge success and popularity of the FRET technique can be attributed to various factors such as development in optical techniques, availability of better fluorescent dyes, and, most importantly, to the success of the Förster theory. However, there are certain limitations of Förster theory that will be discussed in next section. These difficulties arise largely from the use of the point-dipole

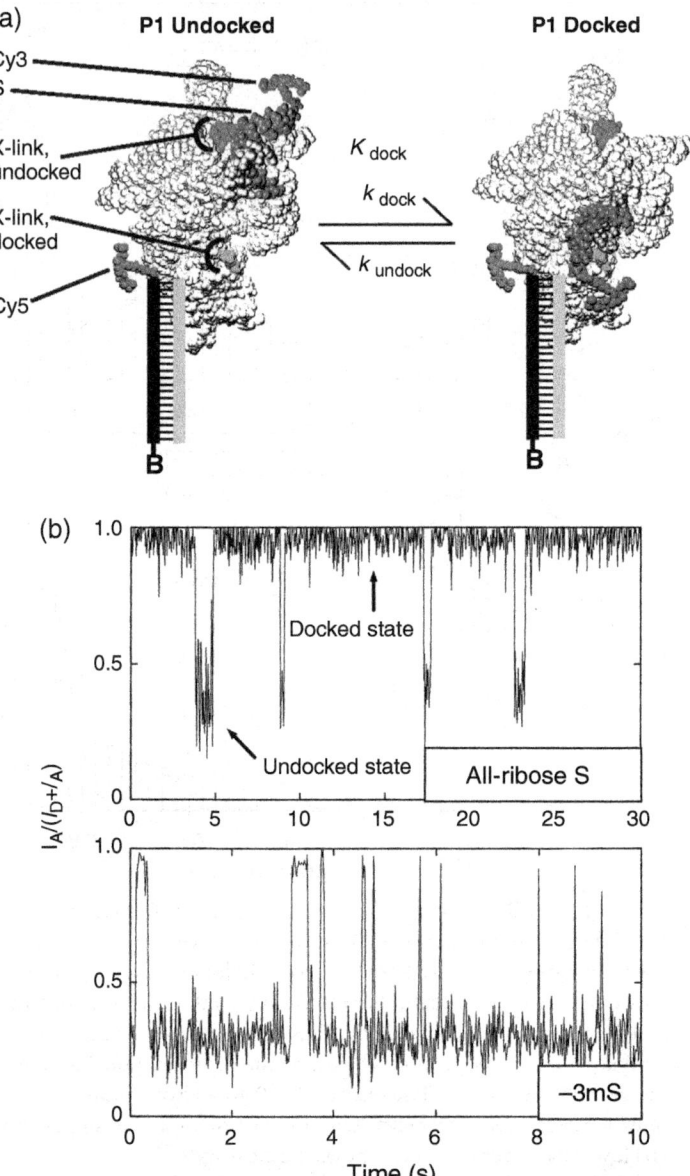

Figure 11.8 FRET study of the docking transition of P1 duplex in *Tetrahymena* ribozyme. (a) P1 duplex is formed when the susbstrate S binds to the 3' end of the ribozyme. The P1 duplex formed can reversibly dock into the ribozyme. X-link, undocked is the binding site of S before docking while X-link, docked is the site at which S binds during the docking transition. FRET between Cy3 on S and Cy5 on tether (dark black line) gives a signal for docking. (b) FRET signal from single ribozyme molecules showing P1 docking and undocking. FRET efficiency is defined as in Eq. (11.4). The FRET values corresponding to docked and undocked states are ∼ 0.9 and ∼ 0.3, respectively. The lower panel of the figure shows that the methylation of the hydroxyl group of residue U-3 on S greatly destabilizes the docked state.

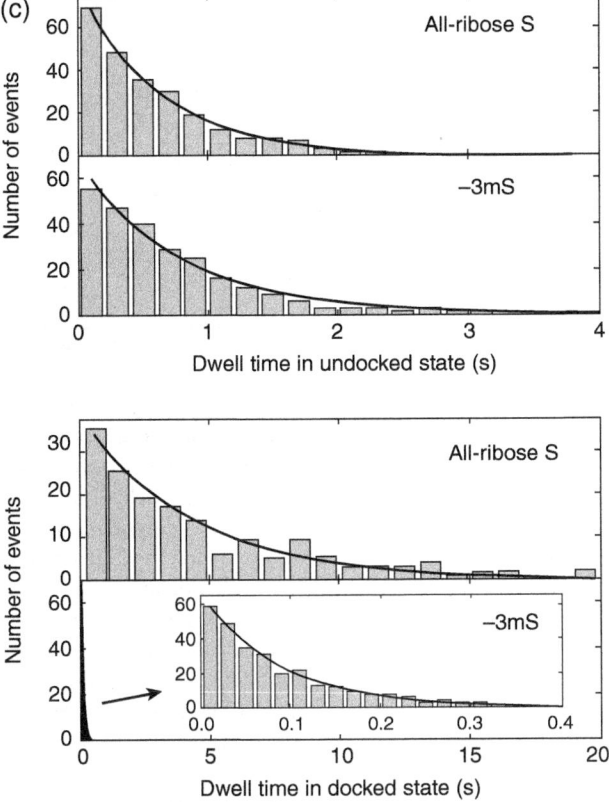

Figure 11.8 (Cont'd) (c) Histograms of the dwell times in the undocked (upper panel) and docked (lower panel) states obtained from FRET time trajectories. The solid lines are single exponential fits to the data giving rate constants for docking and undocking events. From *Science* **288**, 2048, 2000. Reprinted with permission from AAAS.

approximation. As mentioned already, the elegant Förster expression would also be absent in the absence of the point-dipole approximation.

11.4.4 Beyond Organic Dyes as Donor–Acceptor Pairs

Apart from the limitations of the Förster theory, the applicability of FRET is further restricted because of the small range of separation distances that can be measured via this technique. The recent use of noble metal nanoparticles as acceptors of excitation energy has almost doubled this separation range. This development comes in handy when one is interested in the study of conformations and dynamics of large and complex macromolecules. It has been known for a long time that the fluorescence from a dye molecule is strongly affected by the proximity of a metal surface. At intermediate separation distances (2–30 nm), the emission from dye molecule is strongly quenched due to surface energy transfer (SET), i.e., the

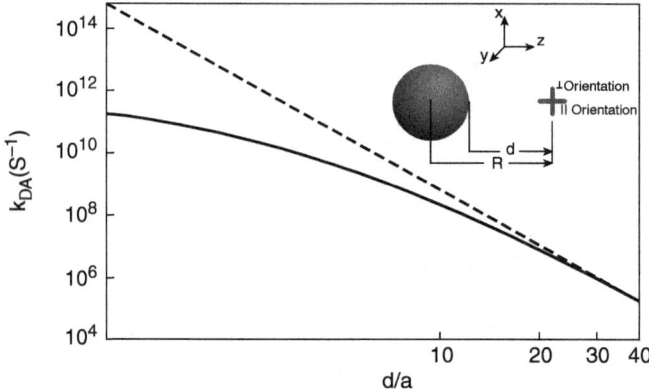

Figure 11.9 Distance dependence of the energy-transfer rate from a dye to a spherical nanoparticle of radius (a) 3 nm. The solid line depicts the rate dependence on distance measured from center of the dye molecule to the surface of nanoparticle (d/a), while the dotted line shows the rate dependence on center-to-center distance (R/a). For spherical nanoparticles, the EET rate follows Förster-type distance dependence at all separations. The inset shows the geometric arrangement of the spherical nanoparticle and the dye molecule in two different orientations, in parallel and perpendicular orientation with respect to center-to-center distance vector $\vec{R}$. $\vec{d}$ is the distance measured from the surface of the nanoparticle. Here, the rate has been calculated for D-A system in parallel orientation. Reprinted with permission from *J. Phys. Chem. B* **113**, 1817, (2009). Copyright (2009) American Chemical Society.

energy is nonradiatively transferred from a dye molecule to a metal surface. The rate of energy transfer via SET follows $1/d^4$ distance dependence, where d is the distance of the center of a dye molecule from a planar metallic surface. In case of energy transfer between a dye and a spherical nanometal particle, the rate follows $1/R^6$ distance dependence, where R is the distance between the centers of a dye molecule and a nanometal particle. However, the rate of energy transfer shows deviation from $1/d^6$ dependence. Therefore, whenever the difference between d and R is appreciable, it becomes important to make a distinction between the two while studying the distance dependence of rate of energy transfer (Fig. 11.9). In fact, this observation of different R and d dependencies can play an important role in studying conformation dynamics of large macromolecules.

The deviation from $1/R^6$ dependence becomes more prominent with the non-sphericity of the nanoparticles under consideration (Fig. 11.10). The applicability of Förster theory for D-A pairs involving metal nanoparticles and the effect of the nanoparticles' shape on the rate of energy transfer has been discussed in a recent review [12].

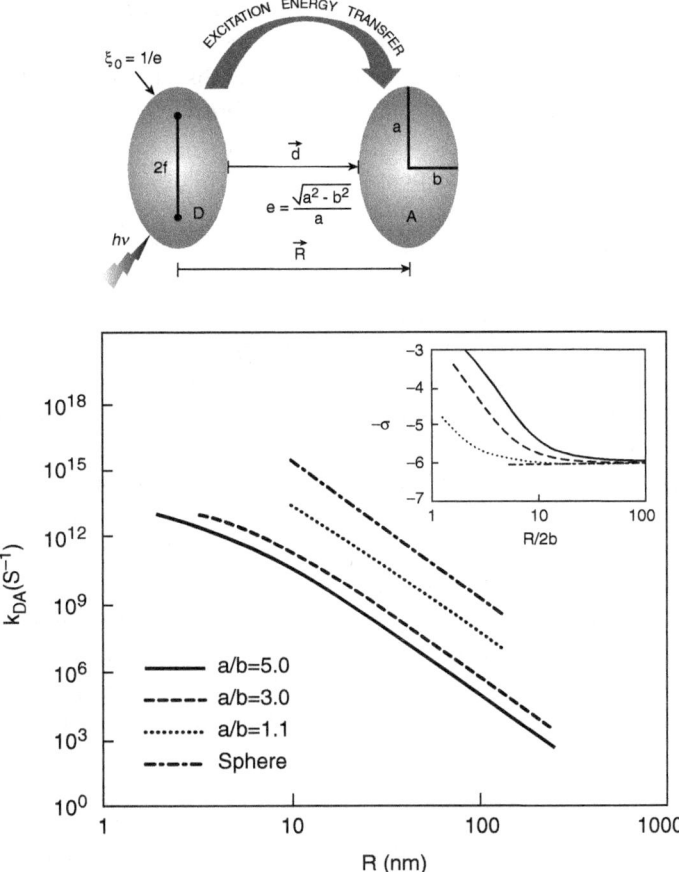

Figure 11.10 Rate of EET as a function of center-to-center distance R for prolate Ag nanoparticles of different aspect ratio (a/b) but with the same value of semimajor axis $(a = 5\text{ nm})$. The inset shows the dependence of distance exponent σ. The deviation of EET rate from Förster distance dependence (R^{-6}) becomes more pronounced with the increase in the asphericity of nanoparticles. Reprinted with permission from *J. Phys. Chem. B* **113**, 1817 (2009). Copyright (2009) American Chemical Society.

11.4.5 FRET and Conjugated Polymers

In the field of materials science, continuous efforts are being made to develop optoelectronic devices based on conjugated polymers, such as substituted polyphenylene vinylene (PPV). The novel optoelectronic properties of conjugated polymers and the availability of simple, cost-effective device-fabricating techniques make these polymers ideal for developing such organic-based devices as light-emitting diodes, photovoltaic cells, and optical sensors [13]. However, performance of these devices depends on the transport of excitation among and across the polymer chains.

But what are the factors that control excitation energy transfer across polymer chains? A typical chain of a conjugated polymer consists of chromophores of varying lengths. The presence of different lengths of chromophores is a result of break in conjugation along the chain. This break in conjugation is commonly referred to as "defects" that are either caused by the conformational disorder present in the polymer chains or are due to the presence of "chemical defects." The latter could be due to the tetrahedral sites in conjugated polymers. The chemical defect concentration in a polymer chain depends on the method of synthesis. Even after synthesis, the chemical defects may result from the photooxidation of unsaturated sites on the polymer chain.

It is believed that upon excitation, the energy in these polymers is photochemically funneled toward the longer chromophores, which are of low electronic energy and have larger oscillator strengths. Using a relatively simple model for conjugated polymers, the study of excitation-energy dynamics based on FRET can capture the qualitative features of this photochemical funnel. Such a description uses a simple quantum-mechanical description of a "particle in a box" for the excitation energy and wave function that are needed to evaluate the Coulomb interaction energy between different segments. One then uses a statistical description based on a master equation to describe energy transfer between different segments. If detailed balance is satisfied for the occupational probability of excitation at different chromophore segments, then an initial random population shall move toward the longer segments, which can act as a funnel due to larger oscillator strengths of these segments.

However, despite the conceptual simplicity of the problem, a complete microscopic study of this excitation dynamics is a highly nontrivial problem. The complexity arises because the distance dependence of the excitation energy transfer between different chromophores depends not only on the length of the chromophore segments but also on their spatial arrangement, i.e., relative separation distance and orientation. Figure 11.11 shows the excitation energy migration in a polymer chain containing chromophores which are 2–12 monomers long and with a mean conjugation length of $k = 7$. k denotes the number of monomers present in a given chromophore. The initial excitation is located on a chromophore with $k = 7$. The rise (build up) and fall (decay) in the population of singlet exciton at chromophores of different lengths as a function of time (in picoseconds) confirms the hopping of excitation energy from smaller to longer chromophoric units.

Studies have shown that the excitation energy may not always migrate toward the longest chromophore segments in the polymer chain because the efficiency of nonradiative energy transfer depends strongly on the separation distance between the donor and acceptor chromophores and also on their relative orientation. It is likely that a polymer consists of different domains containing chromophores of varying conjugation lengths and the excitation-energy transfer from shorter to longer segments is efficient only within these domains. This viewpoint is further supported by the presence of more than one emissive species contributing to the emission spectra of the polymer.

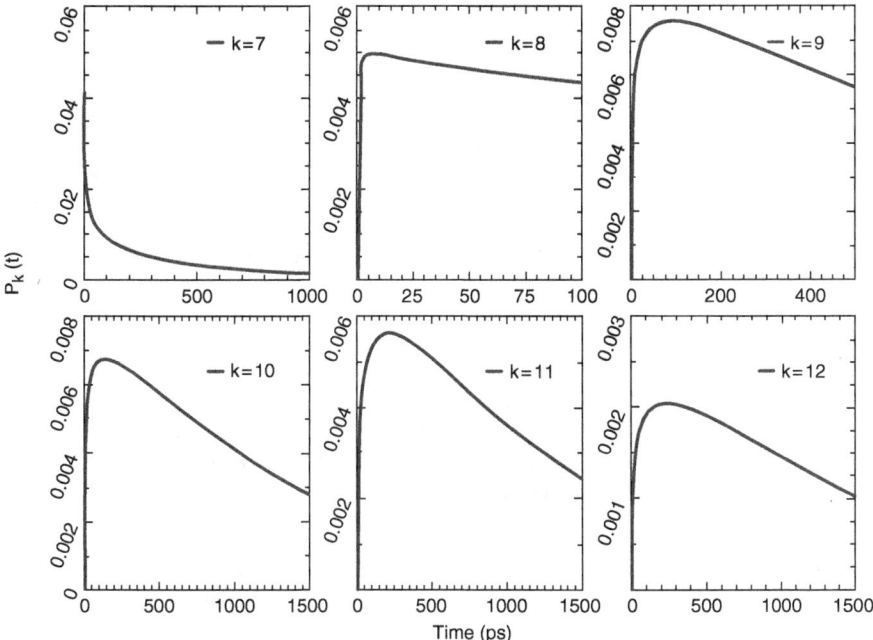

Figure 11.11 The exciton population dynamics in PPV chains of mean conjugation length, $l_0 = 7a$. The rise and fall of the population is shown for chromophores of different lengths, $l = 8a - 12a$, where a is the length of a monomeric unit. For PPV, $a = 0.64$ nm. *Phys. Chem. Chem. Phys.* **12**, 7427 (2010). Reproduced by permission of the PCCP Owner Societies.

As has been mentioned, the optical properties of conjugated polymers strongly depend on the degree of disorder and the conformations adopted by the polymer chain. The information regarding the defects and disorder in a polymer chain and hence the conformation of a polymer chain can generally be obtained by studying the spectroscopic properties of these polymers. Spectroscopic studies, both single molecular and ensemble [14], on poly[2-methoxy-5-(2′-ethylhexyl)oxy-1,4-phenylenevinylene] (MEH-PPV) polymer, along with the computer simulations results showed the existence of defect-coil (noncollapsed) and defect-cylinder (collapsed) conformations (Fig. 11.12) as the dominant species. Photochemical properties of the two types are quite different.

Such ordered structures arise due to the interplay between rigidity toward bending due to conjugation, the presence of defects that break the conjugation, and the attractive interaction between monomers. Note that such structures would entail different conformational characteristics from what would be given by the standard Flory random-walk model of polymers. The latter would predict a collapse chain without structures.

From Monte Carlo simulations, it was found that an equal superposition of the anisotropy distributions of defect-coil and defect-cylinder conformations resulted in a reasonable agreement with the anisotropy distribution data obtained from the polarization spectroscopy measurements. Figure 11.12 shows that both these

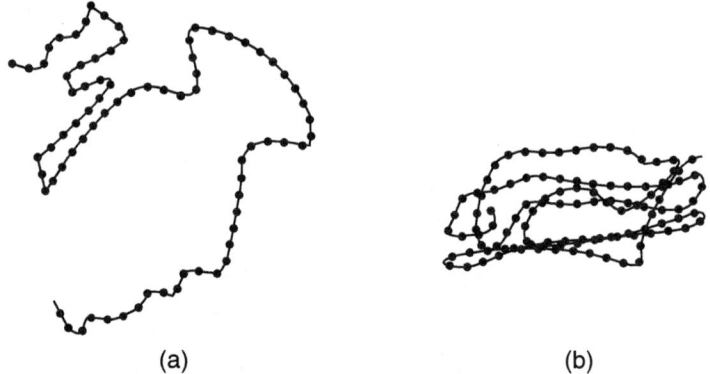

Figure 11.12 The conformations obtained from Monte Carlo simulations: (a) Defect-coil; (b) Defect-cylinder. Reprinted with permission from Macmillan Publishers Ltd: *Nature*, **405**, p. 1030 (2000). Copyright (2000).

conformations are characterized by quasi-straight-chain segments that support the chromophoric picture of the conjugated polymers. The highly ordered defect cylinder conformation of MEH-PPV may be the answer to various puzzling properties of thin films of such polymers. Note that for Monte Carlo simulations the defects in a polymer chain were assumed to be the tetrahedral sites. A separate study found these defects to be thermodynamically stable.

Although the conjugated polymers are being considered as an alternative to silicon in designing the solar cells, the efficiency of these is far less than the already available silicon-based solar cells. A recent study has suggested the use of nanoparticles in developing the solar cells can be an efficient alternative [15].

11.5 BEYOND FÖRSTER FORMALISM

We now discuss several studies that have examined different aspects of Förster final rate expression.

11.5.1 Orientation Factor

Let us rewrite the Förster expression for the rate of energy transfer from D and A as follows:

$$k_{DA} = k_{rad} \frac{3}{2} \kappa^2 \left(\frac{R_F}{R}\right)^6. \tag{11.46}$$

The effect of orientational dynamics of the chromophores enters into the above equation through the orientation parameter κ^2.

While analyzing the experimental results, the exact value of orientation factor given by Eq. (11.10) is usually replaced by a pre-averaged value of κ^2 equal to 2/3. The pre-averaging involves unrestricted averaging over all the three angles of Eq. (11.10). The validity of this pre-averaging is not clear because in principle *one should obtain the dynamics of energy transfer at various angles and then perform the averaging.* As a result, the distance estimated from Förster theory using a pre-averaged value of orientation factor may lead to erroneous conclusion.

Figure 11.13 shows the effect of rotational dynamics of D and A chromophores on the rate of FRET. The rotational diffusion coefficient D_R determines rotational rate of both D and A. The horizontal line in the figure represents the value of $\tau_{FRET}(= k_{DA}^{-1})$ by using $\kappa^2 = 2/3$. On going from the static limit ($D_R = 0$) to the dynamic limit ($D_R \to \infty$) of the rotational diffusion coefficient, the EET rate increases (τ_{FRET} decreases). In the limit of slow rotation (small D_R), the rate of EET is overestimated by the conventional scheme, i.e., $\kappa^2 = 2/3$, by about 20% for a mobile D-A pair. However, exactly the opposite happens for fast rotation. The pre-averaged rate is about 20% smaller, that is, it underestimates the actual rate. It is easy to understand this result. For fast rotation, transfer takes place from the high transfer zone and vice versa. This analysis shows that it is more appropriate to obtain the dynamics of energy transfer at various angles and then perform the averaging instead of using a pre-averaged value of orientational factor.

11.5.2 Point-Dipole Approximation

Clearly, the point-dipole approximation should work at separations larger than the molecular dimensions of D and A. The following cases exemplify the necessity of considering the full columbic interactions between donor and acceptor molecules at smaller distances. Let us first consider a hypothetical case in which the energy transfer takes place from a donor molecule to a dimer acceptor. Figure 11.14

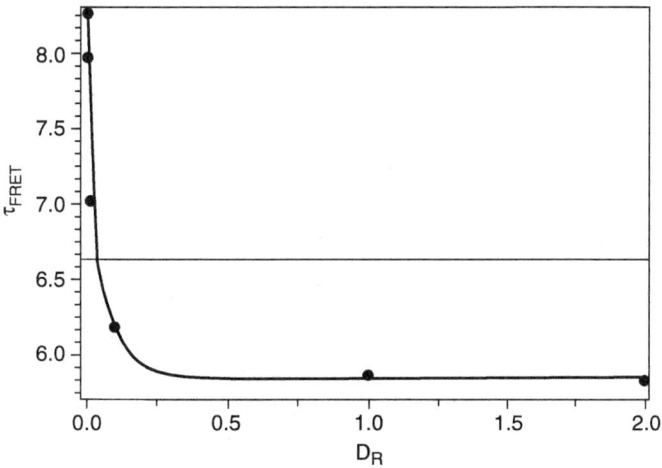

Figure 11.13 The survival time, k_{DA}^{-1}, obtained by Brownian dynamics simulations (filled circles) is plotted as a function of rotational diffusion coefficient D_R for a pair of mobile donor and mobile acceptor molecules. The horizontal line corresponds to τ_{FRET} in dynamically averaged limit, i.e., $\kappa^2 = 2/3$. Reprinted with permission from *J. Phys. Chem. B* **105**, 9370 (2001). Copyright (2001) American Chemical Society.

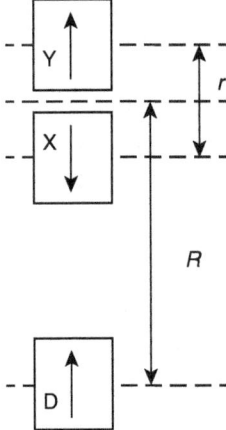

Figure 11.14 The geometrical arrangement of a hypothetical case where an acceptor is a dimer formed by arranging molecules X and Y in a tail-to-tail orientation. $2r$ is the distance between the centers of the X and Y molecules. The donor molecule (D) is positioned at a distance R from the center of X-Y dimer. Reprinted with permission from *J. Phys. Chem. B* **105**, 1640 (2001). Copyright (2001) American Chemical Society.

shows the geometric arrangement of the D-A pair under consideration. For this arrangement, the use the point-dipole approximation will result in no coupling interaction between D and A (Eq. (11.47)). But if we represent the individual molecules of a dimer (X and Y) by transition dipoles placed at the center of these molecules, we will obtain nonzero coupling interaction (Eq. (11.48)):

$$V = \mu_D \mu_+ / R^3 = 0, \qquad (11.47)$$

$$V = \frac{1}{\sqrt{2}} \left[\mu_D \mu_X / (R - r)^3 - \mu_D \mu_X / (R + r)^3 \right] \neq 0, \qquad (11.48)$$

where μ_+ is a net transition dipole of a dimer while $\mu_{X/Y}$ is a transition-dipole moment of individual X/Y molecule.

The distance dependence of rate of energy transfer from a six-unit oligomer of polyfluorene (PF$_6$) to a tetraphenylporphyrin (TPP) molecule was calculated using full Coulombic interactions. This calculation showed a significant deviation from Förster theory (see Fig. 11.15) at shorter distances [16]. The experimentally observed time scale of the EET rate was of the order of few tens of picoseconds while Förster theory predicted the time scale to be in femtosecond range, at small separations. The full Columbic rates calculated using PPP Hamiltonian (a semiempirical quantum-chemical method) were in good agreement with the experimental results.

11.5.3 Contribution of Optically Dark States

The Förster expression expresses rate in terms of fluorescence and absorption spectra of the donor–acceptor system. This result is a consequence of the

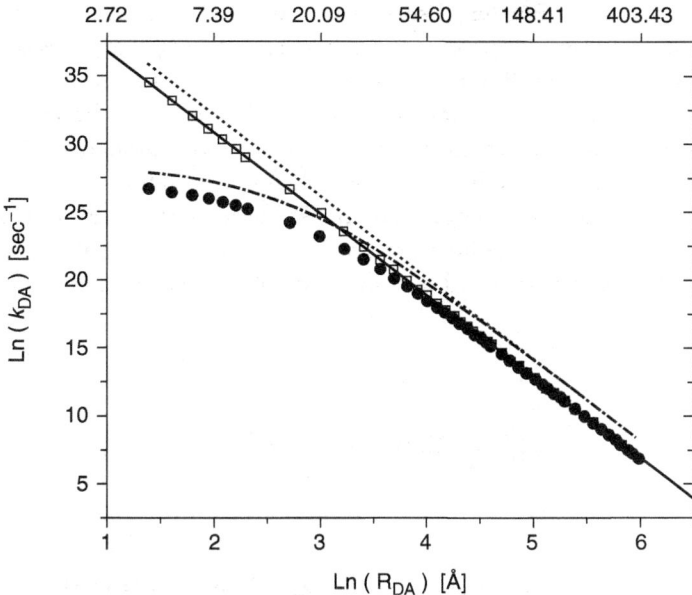

Figure 11.15 Distance dependence of the rate of EET between donor chromophore PF_6 and the acceptor TPP is plotted. The transition-dipole moments of D and A are aligned parallel to each other and are orthogonal to the D-A intermolecular axis. The filled circle (●) denotes the rate calculated using full Coulombic interaction Hamiltonian, while the rate calculated by Förster theory is represented by unfilled square (□). The acceptor states have mid-range oscillator strength of 0.70. The Förster distance dependence (R^{-6}) is shown by solid line. The total Förster (dotted line) and resonance-Coulomb rates (dashed line) are summed over all optically bright and dark states that mediate EET at particular excitation wavelength. Reprinted with permission from *J. Phys. Chem. A* **108**, 5752 (2004). Copyright (2004) American Chemical Society.

point-dipole approximation. However, this result is not a direct consequence of Coulomb interaction. Quantum calculations of the rate from Fermi golden rule using the full interaction potential shows that significant transfer occurs between donor and acceptor states that have low oscillator strength and are, therefore, optically dark states.

Results of such calculations (shown in Fig. 11.15) demonstrate that these dark states indeed make a significant contribution to the rate of energy transfer. The dark states do not contribute toward optical absorption or emission spectra as transitions to and from these states are either spectroscopically forbidden or are of low intensity. As discussed earlier, in Förster theory the strength of coupling interactions among D and A is interpreted in terms of absorption and emission spectra. As a result, Förster theory completely neglects the contribution from these optically dark states.

To summarize, Förster theory works well at separations larger than the dimensions of D and A molecules. At these separations, molecules see an average picture of each other. But at separations smaller than the molecular dimensions, the local interactions predominate as a result, the coupling between D and A is no longer proportional to the optical transition dipole moments of the molecules under consideration, and hence in principle the dark states can contribute toward nonradiative mode of energy transfer. At these separations, the rate may be calculated by using a modified form of Förster theory where the spectral overlap between D and A is weighed by a local coupling interaction element given as in Eq. (11.49):

$$k_{DA} = \frac{2\pi}{\hbar} \left| V_{DA}^{Coul} \right|^2 \int_0^\infty d\omega F_D(\omega) A_A(\omega) \tag{11.49}$$

where $F_D(\omega)$ and $A_A(\omega)$ are area-normalized emission and absorption spectra of donor and acceptor molecules on a energy scale. V_{DA}^{Coul} is a matrix element of Coulomb interaction potential between initial and final electronic states, $V_{DA}^{Coul} = \langle \varphi_{M_D}^g; \varphi_{N_A}^e | H_I | \varphi_{N_D}^e; \varphi_{M_A}^g \rangle$ with interaction Hamiltonian given by

$$H_I = \frac{1}{4\pi\varepsilon} \left(\frac{1}{2} \sum_{j,k} \frac{e^2}{|\vec{R} + \vec{r}_A(j) - \vec{r}_D(k)|} \right), \tag{11.50}$$

where $\vec{R}$ is a distance between D and A, $\vec{r}_A(j)$ and $\vec{r}_D(k)$ respectively denote the position coordinate of jth electron in an acceptor molecule and kth electron in a donor molecule. Figure 11.15 shows the contribution of dark states toward EET as a function of separation distance between D and A calculated using this modified form of Förster theory.

The other examples where the contribution of optically dark states toward the rate of EET has been found to be significant include the bacterial photosynthetic unit [17]. A photosynthetic unit consists of an antenna system and the reaction center (RC). In the case of purple bacteria, the antenna system consists of LH1 and LH2 ring systems. While the LH1 ring system encircles the reaction center, LH2 is located at the periphery (see Fig. 11.16). The solar radiation is first absorbed by the B800 bacteriochlorophyll (BChl) molecules and the excitation energy is nonradiatively transferred to B850 BChls within the LH2 system and subsequently from LH2 to LH1 system then from LH1 to RC. The fast rate of energy transfer (a) from B800 to B850 bacteriochlorophyll molecules in the LH2 ring system, (b) from LH2 to LH1, and (c) in the reaction center from the monomeric BChl unit to a special pair of BChls (P_A or P_B) can be explained only if the contributions of transfer via the optically dark states are taken into consideration.

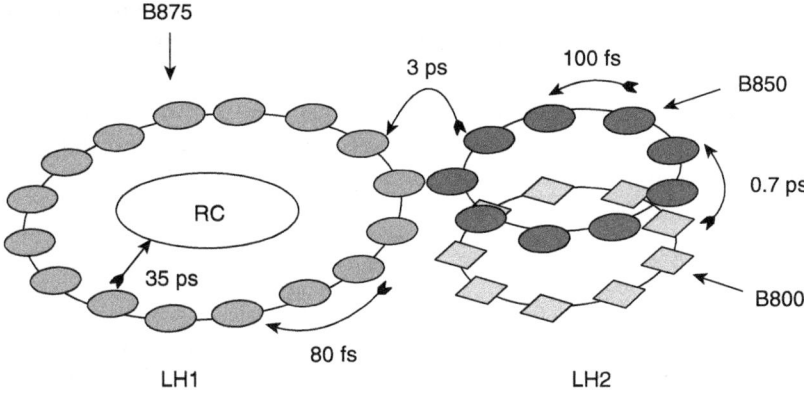

Figure 11.16 The schematic representation of bacterial photosynthetic unit. LH2 contains two types of bacteriochlorophyll (BChl) molecules, one absorbing at 800 nm and the other at 850 nm. These BChl molecules are commonly referred to as B800 and B850. BChls in LH1 absorb at 875 nm and are labeled as B875. LH1 encircles the reaction center (RC). Reprinted from *Curr. Opin. Struc. Biol.* **7**, Graham R. Fleming, Rienk van Grondelle, Femtosecond spectroscopy of photosynthetic light-harvesting systems, p. 738 (1997). Copyright (1997) with permission from Elsevier.

11.6 SUMMARY

FRET has now become a highly popular and widely used technique, finding increasing use as a diagnostic tool in physical, chemical, materials, and life sciences. Very few techniques have such wide applicability. The important point to remember is that as long as the assumptions made by Förster are valid, Förster theory has been highly successful in interpreting the experimental results and has provided a significant insight into various biochemical processes both *in vitro* and *in vivo*. However, uncertainties in accurate estimation of the value of the orientational factor do present some difficulties and have limited reliable applications of FRET.

With the advent of nano science and technology, there is now increasing demand to study distances comparable to the sizes of donor and acceptor. However, the FRET technique finds difficulties at such small distances. The main reason for the failure of Förster theory is the point-dipole approximation. The contribution of optically dark states toward rate of energy transfer is also in a way the manifestation of the breakdown of the point-dipole approximation. In such cases, it becomes necessary to modify the Förster theory and describe the coupling between donor and acceptor using a Hamiltonian that takes into account the full Coulombic interactions among donor and acceptor molecules. An important issue that might find increasing attention is the role of nonequilibrium processes (such as solvation dynamics of the excited donor/acceptor species, vibrational energy relaxation (VER)) in the observed rate of FRET.

FRET is continuously undergoing improvements in technical aspects, improving its applicability, for example, by using better donor–acceptor pairs. Most recent applications include use of metal nano particles and quantum dots in donor–acceptor systems. This improvement has helped increasing Förster distance R to over 20 nm.

In this chapter we have discussed many aspects of FRET with emphasis on fundamental aspects. We have also addressed the limitations of Förster expression. We have discussed several interesting applications of FRET, although many remained untouched.

REFERENCES

1. Förster, Th.; *Ann. Phys.* (Leipzig) 2, 55 (1948), and Förster, Th.; *"Delocalization and excitation transfer"*, in Modern Quantum Chemistry, Istanbul Lectures, Part III: Action of Light and Organic Crystals, edited by Sinanoglu, O., Academic Press: New York (1995).
2. J. R. Lakowicz, *Principles of Fluorescence Spectroscopy*, Plenum, New York (1983).
3. The history of FRET: From conception through the labors of birth" by R. M. Clegg, *Reviews in Fluorescence*, Volume 2006, 1 (2006).
4. D. L. Dexter, *J. Chem. Phys.* 21, 836 (1953).
5. V. May and O. Kühn, *Charge and Energy Transfer Dynamics in Molecular Systems*, Wiley-VCH (2000).
6. G. Wilemski and M. Fixman, *J. Chem. Phys.* 60, 866 (1974); *J. Chem. Phys.* 60, 878 (1974).
7. J. J. Portman and P. G. Wolynes, *J. Phys. Chem. A* 103, 10602 (1999).
8. G. Srinivas and B. Bagchi, *J. Phys. Chem. B* 105, 2475 (2001).
9. L. Stryer and R. P. Haugland, *Proc. Natl. Acad. Sci. U.S.A.* 58, 719 (1967).
10. S. Weiss, *Science* 283, 1676 (1999).
11. P. R. Selvin, *Nature* 7, 730 (2000).
12. S. Saini, G. Srinivas, and B. Bagchi, *J. Phys. Chem. B* 113 1817 (2009).
13. S. R. Forrest and M. E. Thompson, *Chem. Rev.* 107, 923 (2007).
14. D. Hu, J. Yu, K. Wong, B. Bagchi, P. J. Rossky, and P. F. Barbara, *Nature* 405, 1030 (2000).
15. S. Lu and A. Madhukar *Nano Lett.* 7, 3443 (2007).
16. K. F. Wong, B. Bagchi, and P. J. Rossky, *J. Phys. Chem. A* 108, 5752 (2004).
17. (a) G. D. Scholes, X. J. Jordanides, and G. R. Fleming, *J. Phys. Chem. B* 105, 1640 (2001); (b) H. Sumi, *J. Phys. Chem. B* 103, 252 (1999); (c) G. D. Scholes and G. R. Fleming, *J. Phys. Chem. B* 104, 1854 (2000); (d) B. P. Krueger, G. D. Scholes, and G. R. Fleming, *J. Phys. Chem. B* 102, 5378 (1998).

12

Vibrational-Energy Relaxation

12.1 INTRODUCTION

The vibrational state of a given mode is fully determined by the quantum number and the phase of the vibrational wave function. If the vibrational mode is initially in a nonequilibrium excited energy state with a definite initial phase, then with time the system returns to the vibrational ground state with a phase that is uncorrelated with the initial phase. Thus, vibrational relaxation consists of two different processes. One is the relaxation of the energy of the mode. The second is the relaxation of the phase. These two are fundamentally different processes. Here we discuss the energy relaxation and in the next chapter we shall discuss the phase relaxation.

Vibrational-energy relaxation (VER) of molecules in condensed phases is an important dynamical process that is considered as an essential part of physical chemistry/chemical physics. VER plays a critical role in chemical reactions. After all, breaking or formation of a chemical bond is largely determined by the efficiency with which energy can be transferred to a bond or removed from the broken fragments, which are often left in a vibrationally excited state after the bond-breaking event. A simple example of such relaxation is shown in Fig. 12.1.

In contrast to rotational and translational motions of molecules, it is not easy to obtain information about the rates of vibrational-energy relaxation in a bond. The reason is that vibrations entail rapid and periodic changes in the bond length. So, one needs to follow the decay of an envelope. Classical methods like ultrasonic attenuation and infrared line shapes all have had mixed success in finding the rates of vibrational-energy relaxation. Infrared absorption line shape gives some measure of the vibrational-energy relaxation rate, but it is mixed with rotational

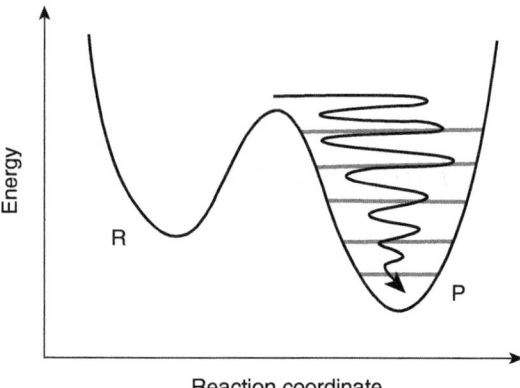

Figure 12.1 Schematic representation of VER after breaking of a chemical bond.

contributions and also inhomogeneous contributions. The latter is common in liquid because molecules in different environments within the same liquid can have slightly different vibrational frequencies. However, taken together these add to the width of the line shape. This is termed inhomogeneous broadening. Inhomogeneous broadening makes study of vibrational relaxation a bit difficult. This situation has improved considerably recently due to the development of infrared lasers, so that vibrational-energy relaxation can be studied directly in the time domain with combination of pulses.

The time scales associated with VER of different modes in different molecules in solution range from subpicoseconds to milliseconds or even longer. This large variation of time scales arises from the fact that different vibrational modes in different molecules (or, maybe in the same molecule) can have different frequencies and relax by different mechanisms. For example, the frequency of vibration of the symmetric stretch mode in water is vastly different from its bending mode, and so is the rate of VER.

As mentioned above, the initial (imperfect) information about VER was obtained from the frequency-domain experiments. One such technique was the dispersion of acoustic attenuation in a liquid. The idea there was that propagating sound waves can excite a chemical bond and the sound energy gets partly dissipated through vibrational-energy relaxation. However, this does not provide a direct access to VER for most systems. Both infrared absorption and Raman scattering techniques were also used to obtain VER rates. The study of VER gained momentum after laser spectroscopy allowed the measurement of population dynamics of vibrational levels. Even then the field had to wait until the development of infrared lasers in the mid-1990s for reliable quantitative information on rates of vibrational-energy relaxation. For the first time, infrared lasers permitted the study of vibrational relaxation in the ground electronic state of the molecules, thus allowing us to avoid the complications due to electronic excitation.

Despite considerable effort, our progress in theoretical understanding of VER has been rather slow mainly because it is an extremely difficult problem involving many different aspects. These include not only the solute–solvent interactions, but also the anharmonic and nonlinear couplings between different vibrational modes of the same and also different molecules. VER provides a very sensitive probe of the intramolecular energy flow.

Theoretical studies on vibrational energy relaxation have mostly been carried out by invoking two basic models—the isolated binary collision (IBC) model in which the collision frequency (a concept that has its origin in the kinetic theory of gases) is modified by the liquid structure, and the weak-coupling model, where the vibrational motion of the molecule weakly couples to the remaining (mainly translational and rotational) degrees of freedom so that a perturbative technique can be used [1].

12.2 ISOLATED BINARY COLLISION (IBC) MODEL

The IBC model of VER was developed by Herzfeld, Litovitz, and Madigosky [2, 3]. This model is based on the assumption that VER takes place via isolated and uncorrelated collisions with solvent molecules. In this model, the VER rate for two-level systems is assumed to be given by

$$\tau_{ij}^{-1} = P_{ij}\tau_c^{-1}, \tag{12.1}$$

where τ_c^{-1} is the collision frequency and P_{ij} is the probability per collision that a transition from level "i" to level "j" will take place per collision. P_{ij} is independent of density but does depend on temperature, whereas τ_c depends on both these state parameters. In the gas phase, τ_c can be obtained from the kinetic theory of gases, but it is difficult to obtain in the condensed phase. If the colliding molecules are approximated by effective hard spheres, then an expression for the collision frequency can be obtained from the Enskog theory:

$$\tau_c^{-1} \propto g(\sigma), \tag{12.2}$$

where $g(\sigma)$ is the value of the radial distribution function at contact, with σ as the molecular diameter. τ_c is proportional to the friction and hence to the viscosity of the gas. Thus, the IBC model with Enskog collision frequency predicts a rate proportional to the viscosity of the gas.

A measure of the collision frequency can also be obtained from the cell model of the liquid. In this model, the vibrating molecule moves in a cell created by its neighbors. The molecule moves with the average thermal velocity given by

$$\bar{v} = \left(\frac{8k_BT}{\pi m}\right)^{1/2}. \tag{12.3}$$

Actually, several different versions of cell model exist. The cell model of Madigosky and Litovitz [3] provides a simple expression of the collision frequency

$$\tau_c = \frac{\rho^{-1/3} - \sigma}{\bar{v}}, \tag{12.4}$$

where ρ is the density of the liquid and the velocity in the denominator of the above expression is given by Eq. (12.3).

The calculation of the second ingredient of the binary collision model, the transition probability P_{ij}, is difficult because it requires the use of quantum-scattering theory, which is highly nontrivial even for molecules of moderate size. To circumvent this difficulty, one often uses semiclassical approximations. One such popular expression is given by

$$P_{k-l}^{i-j}(a,b) = F_{tr}\{P_0(1)P_0(2)[V^{i-j}(a)]^2[V^{k-\ell}(b)]^2\}, \qquad (12.5)$$

where

$$F_{tr} = \left(\frac{8\mu}{2k_BT}\right)\left(\frac{8\pi^3\mu\Delta E}{h^2}\right)^2 \exp\left(-\frac{\phi_0}{k_BT}\right)$$
$$\times \int_0^\infty dv \frac{v}{\alpha^4}\left(\frac{r_c}{r_0}\right)^2 \frac{\exp(q-q')}{(1-\exp(q-q'))^2} \exp\left(-\frac{\mu v^2}{2k_BT}\right), \qquad (12.6)$$

which is self-explanatory in many ways. Here $P_{k-l}^{i-j}(a,b)$ is the probability per collision that the mode Q_a in one molecule will change its state from i to j, while the mode Q_b in the same or a different molecule changes its state from k to ℓ. q, q', and ΔE are defined by

$$q = \frac{4\pi^2\mu v}{\alpha h}, \qquad (12.7)$$

$$q' = \frac{4\pi^2\mu v'}{\alpha L}, \qquad (12.8)$$

$$\Delta E = h\nu_a(i-j) + h\nu_b(k-l), \qquad (12.9)$$

where v and v' are the initial and the final relative velocities of colliding pair, μ is the reduced mass, α is the exponential repulsion parameter in the potential, ϕ_0 is the well depth, r_c is the distance of closest approach, and r_0 is the separation at zero potential energy. The vibrational factors $V^{i-j}(a)$ and $V^{k-l}(b)$ are often calculated in the breathing-sphere approximation, which in effect averages the vibrational motion out over the surface of a sphere.

The application of the IBC model (proposed originally for the gas phase) to study the vibrational-energy relaxation in the liquid phase has an interesting history. Soon after it was proposed, some authors criticized it on several grounds. Fixman [4] argued that the transition probability should be density dependent and the dynamics of the relaxation would be modified by three-body and higher-body interactions. He considered the total force acting on the vibrating molecule at any time as a sum of hard binary collision and a Brownian random force. This treatment led him to suggest that the interactions higher than the two-body-collision interactions are important in determining the VER rate in liquids.

Zwanzig [5] criticized the fundamental assumption of the IBC model that the rate can be given by the product of the collision frequency and the transition probability. He used the weak-coupling perturbative technique to obtain the VER rate, and showed that both the two-body binary collision and three-body interactions are equally crucial in determining the rate of energy relaxation.

Herzfeld [6] responded to the above criticisms by pointing out that the models applied by Fixman and Zwanzig were internally inconsistent. He showed that the IBC model could be used to obtain the VER rate in dense liquids as well. The only condition here is that one should not use the zero-frequency Enskog friction to calculate the rate because it can vastly overestimate the high-frequency value of the collisional frequency. Therefore, a systematic approach is needed to obtain the proper high-frequency spectrum of the collisional friction. This can be significantly different for molecules interacting with continuous potential, like Lennard-Jones, from those interacting with hard-sphere potentials. A detailed analytic treatment of the dynamic friction based on the generalized Langevin equation (GLE) was first provided by Berne and coworkers [7] who tested the validity of IBC approach by molecular-dynamics-simulation studies. The simulation studies on a diatomic molecule dissolved in Lennard-Jones argon (with rigid-bond approximation) indicates that the IBC model is accurate in describing the vibrational-energy relaxation at moderately high frequency.

It is worthwhile to reflect on the arguments mentioned above. It is only the high-frequency or the short-time part of the friction that is relevant in VER. The story is similar to the one we faced in the study of isomerization reaction with sharp activation barrier, discussed in Chapter 7. In that case also one makes a mistake if the zero-frequency friction is used, and the hard-sphere model is used to estimate this friction. For continuous potentials, the short-time friction could be vastly different from the zero-frequency friction.

These interesting discussions also bring out the difference between hard-sphere potential and continuous potential. Hard-sphere model of interaction gives a nonzero value of friction on the bond even at large frequency. This value is the Enskog friction. For a continuous potential, on the other hand, the friction goes to zero at high frequency, which is the correct behavior. Therefore, hard-sphere model can vastly overestimate the rate of vibrational-energy relaxation if used in the IBC model.

12.3 LANDAU–TELLER EXPRESSION: THE CLASSICAL LIMIT

This approach considers the vibrational-energy relaxation as a classical process where energy is dissipated to the medium by the usual frictional mechanism. Then, one can adopt a stochastic approach. For VER, it is sometimes reasonable to assume that the vibration is harmonic. Under this condition, one can write a generalized Langevin equation of motion for the normal coordinate Q

$$\mu \ddot{Q}(t) = -\mu \omega^2 Q(t) - \int_{-\infty}^{t} dt' \zeta(t-t') \dot{Q}(t') + R_Q(t), \qquad (12.10)$$

where $R_Q(t)$ is the normal force on the chemical bond that is related to the time-dependent friction $\zeta(t)$ by the fluctuation–dissipation theorem (see **Chapter 1**). The following important observation is made at this point. Even at as low a bond frequency as 100 cm^{-1}, the solvent frictional force on the bond is very small. Thus, while the zero-frequency friction can be large, the relaxation of Q probed at large frequency can be in the underdamped limit.

The rate of the VER of a classical oscillator is given by the simple Landau–Teller expression:

$$\frac{1}{T_1} = \frac{\zeta_{real}^{bond}(\omega_v)}{\mu}, \tag{12.11}$$

where μ is the reduced mass of the oscillator (the vibrating bond), ω_v is the harmonic vibrational frequency of the bond, and $\zeta_{real}^{bond}(\omega_v)$ is the real part of the friction acting on the vibrational coordinate. The friction at the bond frequency is the cosine integral of the force–force time-correlation function (FFTCF) acting on the bond. This friction is responsible for population redistribution in vibrational levels because energy dissipates through friction.

We now present a simple derivation of the Landau–Teller expression from the generalized Langevin equation of the reaction coordinate. This derivation is based on the assumption that the average of the normal coordinate time-correlation function can be written in the following form:

$$\frac{\langle Q(0)Q(t)\rangle}{\langle Q(0)Q(0)\rangle} = \exp(i\omega_0 t)\exp(-(iS + 1/T_1)). \tag{12.12}$$

This expression gives a Lorentzian line shape, where S gives the shift from the resonance vibration frequency ω_0 and $1/T_1$ gives the width. Both the shift and the width of the Lorentzian form are due to the frequency-dependent friction of the assumed Langevin equation.

The vibrational line shape due to population relaxation is given by the Fourier transform of $\langle Q(0)Q(t)\rangle$. Therefore, this line shape can be obtained directly from the generalized Langevin equation of Q given above. That is, we obtain $\langle Q(0)Q(t)\rangle$ from the generalized Langevin equation and take the cosine transformation to obtain the Raman line shape. This line shape is given approximately by

$$I(\omega) = \frac{\mathrm{Re}\,\zeta(\omega_0)}{[\omega - \omega_0 + \mathrm{Im}\,\zeta(\omega_0)]^2 + [\mathrm{Re}\,\zeta(\omega_0)/\mu]^2}. \tag{12.13}$$

A comparison between this and the one given by the assumed form immediately gives the Landau–Teller expression.

For high-frequency modes, the classical Landau–Teller description provides an inaccurate description because the quantum effects become significant in this case.

The molecular-dynamics-simulation studies of Berne and coworkers [7] have shown that if the cross correlation between the solvent forces on each atom of the diatom is neglected and the bond is held rigid, then $\zeta_{real}^{bond}(\omega_v)$ could be approximated as

$$\zeta_{real}^{bond}(\omega_v) = \frac{\zeta_{real}(\omega_v)}{2}, \qquad (12.14)$$

where $\zeta_{real}(\omega_v)$ is the friction experienced by one of the atoms of the vibrating homonuclear diatomic molecule. Equation (12.14) is obtained by approximating the solvent force on each atom in the bond by one-half of the force that it would experience in the free state as an isolated atom. Next, the solvent force on the bond is then the sum of the forces on each atom. When cross correlation between forces from each atom is neglected, then we have two additive pure terms but each with a prefactor of 1/4, which leads to the factor of 1/2 in Eq. (12.14).

12.4 WEAK-COUPLING MODEL: TIME-CORRELATION-FUNCTION REPRESENTATION OF TRANSITION PROBABILITY

We would like to remind the reader of the analyses presented in **Chapter 4** where we discussed the relationship between theory and experiment. We pointed out that establishment of such a relationship involves three steps, one of which employs, for absorption or scattering experiments, an expression for transition probability or rate of transition between two levels. The most popular expression is the Fermi golden rule (FGR). Here we shall provide one more demonstration of the scheme outlined in **Chapter 4**.

If one assumes that the coupling between the vibrational degrees of freedom and the bath is weak, then one can use the first-order time-dependent quantum perturbation theory (Fermi golden rule) to express the relaxation rate from one vibrational state to another, by using the theoretical methods described in **Chapter 4**.

Suppose the Hamiltonian has the form

$$H = H_0 + H_B + V, \qquad (12.15)$$

where H_0 is the Hamiltonian for the subsystem (vibrational degrees of freedom, assumed to be *harmonic* in the present discussion) whose eigenstates are denoted by i, j. H_B is the *bath* Hamiltonian with eigenstates α, β, which includes the rotational and the translational degrees of freedom of the vibrationally excited molecule (solute). V couples the system to the bath. We have further assumed that the relaxation of the bath is very fast (on the time scale of relaxation of vibrational degrees of freedom) so that the initial state of the bath can be given by the thermal equilibrium distribution. The transition rate (k_{ij}) from level i to j is then given by the Fermi golden rule as

$$k_{ij} = \frac{2\pi}{\hbar} \sum_\alpha P_\alpha \sum_\beta |\langle i, \alpha | \hat{V} | j, \beta \rangle|^2 \delta(E_{i\alpha} - E_{j\beta}), \qquad (12.16)$$

where $P_\alpha = \dfrac{e^{-E_\alpha/kT}}{\sum_\alpha e^{-E_\alpha/kT}}$, $E_{i\alpha} = E_i + E_\alpha$, and $E_{j\beta} = E_j + E_\beta$. E_i, E_j are the energies of subsystem eigenstates $|i\rangle$, $|j\rangle$ and E_α, E_β are the energies of bath eigenstates $|\alpha\rangle$, $|\beta\rangle$.

We now use the integral representation of the Dirac-δ function

$$\delta\left(E_{i\alpha} - E_{j\beta}\right) = \frac{1}{2\pi\hbar} \int_{-\infty}^{\infty} dt\, e^{\frac{i\left(E_{i\alpha}-E_{j\beta}\right)t}{\hbar}}. \tag{12.17}$$

We use this representation in Eq. (12.16) to rewrite this equation as

$$k_{ij} = \frac{1}{\hbar^2} \int_{-\infty}^{\infty} dt\, e^{\frac{i(E_i-E_j)t}{\hbar}} \sum_\alpha \sum_\beta P_\alpha e^{\frac{iE_\alpha t}{\hbar}} \langle i, \alpha | \hat{V} | j, \beta \rangle e^{\frac{-iE_\beta t}{\hbar}} \langle j, \beta | \hat{V} | i, \alpha \rangle. \tag{12.18}$$

Equation (12.18) is still quite complex. We next use the following facts. First, E_α and E_β are the eigenstates of the bath Hamiltonian. The other parts of the total Hamiltonian do not act in the Heisenberg representation

$$\begin{aligned} e^{\frac{iE_\alpha t}{\hbar}} \langle i, \alpha | \hat{V} | j, \beta \rangle e^{\frac{-iE_\beta t}{\hbar}} &= \langle i, \alpha | e^{\frac{iH_B t}{\hbar}} V e^{\frac{-iH_B t}{\hbar}} | j, \beta \rangle \\ &= \langle i, \alpha | \hat{V}(t) | j, \beta \rangle \\ &= V_{i\alpha, j\beta}(t). \end{aligned} \tag{12.19}$$

Here

$$\langle i, \alpha | \hat{V} | j, \beta \rangle = V_{i\alpha, j\beta}(0). \tag{12.20}$$

And the term

$$\sum_\alpha \sum_\beta P_\alpha V_{i\alpha, j\beta}(t) V_{j\beta, i\alpha}(0), \tag{12.21}$$

is the ensemble average $\langle V_{ij}(t) V_{ji}(0) \rangle$ over the bath states. Hence, final expression can be written as

$$k_{ij} = \frac{1}{\hbar^2} \int_{-\infty}^{\infty} dt\, e^{i\omega_{ij} t} \langle V_{ij}(t) V_{ji}(0) \rangle_q, \tag{12.22}$$

where $\omega_{ij} = (E_i - E_j)/\hbar$.

In the case when $V_{ij} = F \cdot q_{ij}$ where F is an operator in the bath subspace and q is the coordinate operator for the oscillator, then the above equation reduces to

$$k_{ij} = \frac{1}{\hbar^2} |q_{ij}|^2 \int_{-\infty}^{\infty} dt\, e^{i\omega_{ij}t} \langle F(t)F(0) \rangle_q. \qquad (12.23)$$

From Eq. (12.23) it is clear that the VER occurs as a result of fluctuating forces exerted by the environment on the chemical bond. The Fourier transform of this quantum-mechanical force–force correlation function at ω_{ij} is quantum-mechanical frequency-dependent friction exerted on the subsystem by the bath. For a harmonic solute, only transitions between eigenstates $|n\rangle$ and $|n \pm 1\rangle$ are allowed. Using the harmonic oscillator matrix element, $q_{n,n-1} = \sqrt{n\hbar/2\mu\omega}$, Eq. (12.23) becomes

$$k_{n-1 \leftarrow n} = \frac{n}{2\mu\hbar\omega} \int_{-\infty}^{\infty} dt\, e^{i\omega t} \langle F(t)F(0) \rangle_q. \qquad (12.24)$$

In Eq. (12.22) because the average is over the quantum bath and the evaluation of the quantum-mechanical correlation function is *prohibitively* difficult, many attempts have been made to replace this correlation by a classical one because the bath degrees of freedom (rotations and translations) may be treated classically.

A direct classical approximation in Eq. (12.22) gives $k_{ij} = k_{ji}$, which is in contrast with the thermodynamic equilibrium where detailed balance $k_{ij} = k_{ji} e^{\beta\hbar\omega_{ij}}$ must satisfy. Oxtoby [1] proposed that before making a classical approximation the above equation should be rewritten in terms of the time-symmetrized anticommutator as

$$k_{ij} = \frac{2\hbar^{-2}}{1 + \exp[-\beta\hbar\omega_{ij}]} \int_{-\infty}^{\infty} dt\, \exp[i\omega_{ij}t] \left\langle \frac{1}{2}[V_{ij}(t), V_{ji}(0)]_+ \right\rangle, \qquad (12.25)$$

where $[A, B]_+$ is the symmetrized anticommutator defined by

$$[A, B]_+ = AB + BA. \qquad (12.26)$$

The symmetrized time-correlation functions of the form $[A, B]_+$ are insensitive to the order of A and B operators, suggesting that the quantum-correlation function in Eq. (12.25) can be approximated by its classical counterpart, that is, $\left\langle \frac{1}{2}[V_{ij}(t), V_{ji}(0)]_+ \right\rangle$ can be replaced by $\left\langle [V_{ij}^{class}(t), V_{ji}^{class}(0)] \right\rangle$. This leads to the following approximated semiclassical expression for the rate constant:

$$k_{ij}^{SC} = \frac{2\hbar^{-2}}{1 + \exp[-\beta\hbar\omega_{ij}]} \int_{-\infty}^{\infty} dt\, e^{i\omega_{ij}t} \times \left\langle V_{ij}^{class}(t) V_{ij}^{class}(0) \right\rangle. \qquad (12.27)$$

To calculate the VER rate using Eq. (12.25), the interaction Hamiltonian V is usually expanded in a Taylor series with respect to the normal-mode coordinate $\{Q_\alpha\}$ of the excited molecule (solute):

$$V = V_0 + \sum_\alpha F_1^\alpha Q^\alpha + \frac{1}{2}\sum_\alpha \sum_\beta F_2^{\alpha\beta} Q^\alpha Q^\beta + \ldots, \qquad (12.28)$$

where $F_1^\alpha \equiv \left.\frac{\partial V}{\partial Q^\alpha}\right|_{Q=0}$ and $F_2^{\alpha\beta} \equiv \left.\frac{\partial^2 V}{\partial Q^\alpha \partial Q^\beta}\right|_{Q=0}$ are forces on the normal-mode and the Hessian matrices, respectively.

If one keeps only the first term (that is, V is linear in Q), then one recovers the *Landau–Teller* expression in the classical limit.

The vibrational-transition rate can be rewritten as

$$k_{ij}^{SC} = \frac{2\hbar^{-2}}{1+\exp[-\beta\hbar\omega_{ij}]} \int_{-\infty}^{\infty} dt\, \exp[i\omega_{ij}t]$$

$$\times \left\langle \left[\sum_\alpha (Q^\alpha)_{ij} F_1^\alpha(t) + \frac{1}{2}\sum_\alpha\sum_\beta (Q^\alpha Q^\beta)_{ij} F_2^{\alpha\beta}(t)\right] \right.$$

$$\left. \times \left[\sum_\alpha (Q^\alpha)_{ij} F_1^\alpha(0) + \frac{1}{2}\sum_\alpha\sum_\beta (Q^\alpha Q^\beta)_{ij} F_2^{\alpha\beta}(0)\right] \right\rangle. \qquad (12.29)$$

For the calculation of relaxation rate one needs to know $F_1^\alpha(t)$ and $F_2^{\alpha\beta}(t)$.

12.5 VIBRATIONAL RELAXATION AT HIGH FREQUENCY: QUANTUM EFFECTS

The calculation of the VER rate of the high-frequency modes is considerably more difficult than that of the low-frequency modes. Unfortunately, there is yet no fully reliable method to treat the VER of a high-frequency mode. There are several reasons for that. First, for high-frequency modes, the system cannot be treated classically and the quantum effects play important roles in determining the rate of VER. The coupling of the vibrating system with the bath also becomes nonlinear when the probing frequency is large. At high frequency, multiphonon processes affect the VER rate significantly. Second, the solvent density of states at high frequency is often very small, leading to very small rates from the classical Landau–Teller expression. Thus, the relaxation becomes too slow to be calculated directly from the nonequilibrium molecular-dynamics simulations. Third, one needs to Fourier transform the FFTCF at a frequency much higher than the characteristic frequencies of the correlation function. This in turn means measuring a *weak* signal in presence of a large signal-to-noise ratio. These difficulties have made theoretical study of VER involving high frequencies nontrivial. Recently, some useful attempts have been made in this direction where a semiclassical approach has been undertaken. In this approach, the vibrating mode is treated quantum mechanically whereas the solvent molecules (bath) are treated classically. In this semiclassical scheme, the quantum effect is often introduced through a *quantum-correction factor*. However, there is no general prescription for relating a classical-correlation function to a quantum-correlation function. The precise relationship depends on the underlying Hamiltonian generating the classical and quantum dynamics.

Bader and Berne [7] have developed a theoretical framework to extract relaxation times for quantum solutes in a quantum bath from classical simulation studies. Bader and Berne's approach is briefly discussed here. If one assumes that the coupling term V is linear in the solute coordinate and then uses the harmonic oscillator matrix elements, then one can write Eq. (12.25) as

$$k^{qm}_{n-1 \leftarrow n} = \frac{n}{\mu \hbar \omega [1 + \exp(-\beta \hbar \omega)]} \int_{-\infty}^{\infty} dt \, e^{i\omega t} \left\langle \frac{1}{2} [F(t), F(0)]_+ \right\rangle_{qm}. \tag{12.30}$$

Let us say, $\zeta_{qm}(t) = \frac{1}{2k_B T} \left\langle [F(t), F(0)]_+ \right\rangle_b^{qm}$, then the Laplace transform of $\zeta_{qm}(t)$ is

$$\tilde{\zeta}_{qm}(i\omega) = \int_0^{\infty} dt \, e^{-i\omega t} \zeta_{qm}(t) = \tilde{\zeta}'_{qm}(\omega) - i \tilde{\zeta}''_{qm}(\omega). \tag{12.31}$$

Equation (12.30) can be written as

$$k^{qm}_{n-1 \leftarrow n} = \frac{2n}{\mu \beta \hbar \omega [1 + \exp(-\beta \hbar \omega)]} \tilde{\zeta}^i_{qm}(\omega). \tag{12.32}$$

Let us now represent the overall relaxation process by a master equation for the probability P_n to be in energy state n of the oscillator,

$$\frac{dP_n}{dt} = k_{n \leftarrow n-1} P_{n-1} + k_{n \leftarrow n+1} P_{n+1} - k_{n-1 \leftarrow n} P_n - k_{n+1 \leftarrow n} P_n. \tag{12.33}$$

Solving the above master equation and noting that $(\hbar \omega / 2) \coth(\beta \hbar \omega / 2)$ is the thermal-equilibrium average energy $\langle E \rangle_{eq}^{qm}$ ($\langle n \rangle \hbar \omega$), one obtains the following rate for VER:

$$k^{qm}_{VER} = \frac{\tanh(\beta \hbar \omega / 2)}{\mu \hbar \omega} \int_{-\infty}^{\infty} dt \, e^{i\omega t} \left\langle \frac{1}{2} [F(t), F(0)]_+ \right\rangle_{qm} = \frac{\tanh(\beta \hbar \omega / 2)}{\beta \hbar \omega / 2} \frac{\tilde{\zeta}'_{qm}(\omega)}{\mu}, \tag{12.34}$$

where the VER rate is defined as

$$\frac{d\langle E\rangle}{dt} = k_{VER}^{qm}\left[\langle E\rangle - \langle E\rangle_{eq}\right]. \tag{12.35}$$

In the classical limit, $\hbar\omega \approx k_B T$, we obtain

$$k_{VER}^{cl} = \frac{\tilde{\zeta}'_{qm}(\omega)}{\mu}. \tag{12.36}$$

For the case of harmonic bath Hamiltonian and of linear coupling of system to the bath, where $\zeta_{qm}(t)$ can be computed exactly, both quantum and classical mechanically, the expression for $\tilde{\zeta}'_{qm}(\omega)$ in terms of $\tilde{\zeta}'_{cl}(\omega)$ is

$$\tilde{\zeta}'_{qm}(\omega) = \left(\frac{\beta\hbar\omega}{2}\right)\coth\left(\frac{\beta\hbar\omega}{2}\right)\tilde{\zeta}'_{cl}(\omega). \tag{12.37}$$

From Eqs. (12.34) and (12.37) the vibrational-energy-relaxation rate of a quantum solute in a quantum bath in terms of the classical bath correlation function is given by

$$k_{VER}^{qm} = \frac{\tilde{\zeta}'_{cl}(\omega)}{\mu}. \tag{12.38}$$

This equation indicates that the rate of VER for a quantum solute in the quantum bath is exactly the same as the VER rate for the classical solute in the classical bath.

The exact expression for vibrational transition of a quantum solute in a quantum bath written in terms of a classical bath correlation function (measured directly from molecular-dynamics simulations) is given by

$$k^{qm}_{n-1\leftarrow n} = \frac{n}{\mu\hbar\omega\left[1+\exp\left(-\beta\hbar\omega_{ij}\right)\right]}\frac{\hbar\omega}{2k_B T}$$
$$\times \coth\left(\frac{\hbar\omega}{2k_B T}\right)\int_{-\infty}^{\infty} dt\, e^{i\omega t}\,\langle F(t)F(0)\rangle_c. \tag{12.39}$$

Comparing Eqs. (12.30) and (12.39) it is obvious that to obtain exact rate constant for a quantum subsystem one must multiply the semiclassical rate constant by a *quantum-correction factor* that depends on the magnitude of the energy difference between the two states. The quantization of solvent modes enhances the rate of vibrational transitions and VER rates by a factor of $\frac{\beta\hbar\omega}{2}\coth\left(\frac{\beta\hbar\omega}{2}\right)$.

12.6 EXPERIMENTAL STUDIES OF VIBRATIONAL-ENERGY RELAXATION

Early vibrational-energy-relaxation measurements in condensed phases were based on ultrasonic attenuation measurements in liquids, which provide an overall relaxation time for slow relaxing modes in small molecules. Although this method provides the evidence for the large effect of impurities having low-frequency modes on the VER rate and the effect of temperature and pressure on VER rate, this technique has certain severe limitations. First of all this method does not provide the relaxation of individual vibrational levels and the second limitation is that it allows measurement of relaxation life times only if they lie in the nanosecond time domain. Thus, slowly relaxing modes (as in liquid O_2 or N_2) or rapidly relaxing modes (as in H_2O) could not be studied.

It is probably fair to say that the modern understanding of VER began with the well-known picosecond spectroscopy experiments of Eisenthal and coworkers [8] on the recombination of photodissociated, nascent iodine atoms. In this study the recombination was studied by following the ground-state recovery of I_2 molecules. The latter was again determined by measuring the population of the ground vibrational level of the I_2 molecule. It came as a surprise that the recombination was very slow—longer than 10 ps or so. It was later realized that the recombination was not slow, rather *it was the population relaxation to the ground vibrational state of the newly recombined I_2 molecule* that was very slow.

It was later found that VER in diatomics like N_2 was even slower. Historically, the surprise at the slow rates of VER for small molecules came from the then prevailing disbelief that the large rates deduced from the isotropic Raman linewidth were due to VER. It was realized only several years later that the large rates observed in isotropic Raman linewidth should be attributed to vibrational dephasing. Thus, the study of VER experienced several interesting turns.

Lasers revolutionized the VER rate measurements by providing better time resolution and specific state resolution. The major breakthroughs in early ultrafast VER experiments were made by Kaiser and coworkers [9]. Their studies were mostly concerned with the relaxation of high-frequency modes of polyatomic molecules in solution. Quantitative interpretation of vibrational-relaxation rates in polyatomic liquids is difficult because of the involvement of many processes, which implies a correspondingly large number of relaxation rate constants. But it is possible to measure first the fast relaxation (redistribution of energy between different vibrational modes) and then the slow relaxation of an ensemble of different vibrational modes. One typical example is the VER of C–H stretching mode of CH_3CH_2OH in liquid phases (using stimulated Raman scattering to excite the vibrations, and spontaneous anti-Stokes-Raman scattering to probe the subsequent relaxation) where one first observes the fast decay with relaxation time $\tau = 20 \pm 5$ ps followed by a remarkable slow decay with relaxation time $\tau = 0.25$ ns [1].

Recently, Hochstrasser and coworkers [10] measured the VER of a cyanide ion in H_2O and D_2O by IR-pump-probe experiments. They varied the isotopic composition of the ion in order to vary the oscillation frequency of the CN^-

vibrational mode. In D_2O, they observed a decrease in the vibrational relaxation time from 120 to 71 ps on increasing the vibrational frequency from 2004 to 2079 cm^{-1} and in H_2O, the time constant between 31 and 28 ps. A significant correlation between the VER of the solute and the IR absorption cross section of the solvent was found, providing experimental evidences for a dominating contribution to the vibrational relaxation of Coulomb interactions and the importance of coupling to internal solvent modes.

Other experimental techniques have also been used to study fast vibrational energy relaxation of complex molecules in solution and we direct the reader to Ref. 11 for details. An excellent history of early VER studies is given by Oxtoby [1].

12.7 COMPUTER-SIMULATION STUDIES OF VIBRATIONAL-ENERGY RELAXATION

12.7.1 Vibrational-Energy Relaxation of Water

Vibrational-energy relaxation of water provides a microscopic understanding of the motions of water molecules. In the liquid state one finds that VER in water is surprisingly fast. A water molecule is characterized by three vibrational modes—symmetric O–H stretch, antisymmetric O–H stretch, and the symmetric H–O–H bend. The frequencies of these three modes for liquid water are 3656 cm^{-1}, 3755 cm^{-1}, and 1594 cm^{-1}, respectively.

The O–H stretching frequency of water depends on the OH...O hydrogen bond. The stronger the hydrogen bond, the weaker is the O–H covalent bond and the smaller is the corresponding O–H stretching frequency. In liquid water, intramolecular vibrational modes get shifted and mixed in a spectrum.

Because of hydrogen bonding, several low-frequency intermolecular vibrational modes appear in liquid water. This is unique to liquid water and these low-frequency modes have far-reaching consequences in determining the dynamical response of liquid water. Among the low-frequency modes, the mode at 685 cm^{-1} is assigned to the librational mode, the one at 200 cm^{-1} to the intermolecular vibrational mode, and the one near 50 cm^{-1} to the hindered translational motion due to the caging of the water molecules. In order to understand chemical reactions in water, an in-depth understanding of the vibrational-energy relaxation of the various modes of water is required. Unfortunately, very little is still known about the energy relaxation and energy exchange of the low-frequency vibrational modes, such as libration, and these remain challenging problems for future studies. However, considerable progress has recently been made in understanding energy relaxation of the high-frequency modes because of the development of femtosecond spectroscopy in the mid-infrared region.

Rey and Hynes [12] studied the VER of the O–H stretch of an HOD molecule immersed in liquid D_2O by molecular-dynamics simulation. They assumed that the Hamiltonian of the system can be written

$$H = H_{HOD} + H_{coupling} + H_{bath}, \qquad (12.40)$$

where H_{HOD} is the quantum mechanical (anharmonic) subsystem (vibrational motion of HOD molecule) Hamiltonian. H_{bath} represents the classical Hamiltonian for D_2O solvent, which also includes the translational and rotational contributions of the HOD molecule. $H_{coupling}$ is the coupling of the subsystem to rest of degrees of freedom (bath). $H_{coupling}$ has three important contributions:

$$H_{coupling} = H_{V-B} + H_{Cor} + H_{Cen}, \quad (12.41)$$

where H_{V-B} represents the coupling of the molecular vibrational coordinate to the surrounding solvent molecules. H_{cor} represents the Coriolis coupling between normal modes and H_{Cen} represents the centrifugal coupling (the effect that the variation of moment of inertia with the vibrational motion may have on the relaxation).

The V–B coupling Hamiltonian, to the first order in three HOD dimensionless normal coordinates $\{q_i\}$, is given by

$$H_{V-B} = -\sum_i q_i F_i, \quad (12.42)$$

where F_i is the intermolecular force on the harmonic normal-mode coordinate q_i due to the bath. The Coriolis contribution to the Hamiltonian has the form

$$H_{Cor} = -\vec{\Omega} \cdot \vec{J}, \quad (12.43)$$

where J is the total rotational angular momentum of the solute and $\vec{\Omega}$ is a bilinear function of the normal-mode coordinates and conjugate momenta. For the case of planar HOD one can write

$$H_{Cor} = -\Omega_z J_z, \quad (12.44)$$

where z denotes the principal axis of inertia perpendicular to the plane of molecule.

Finally, the centrifugal-coupling Hamiltonian is of the form

$$H_{cen} = -\frac{\hbar^{1/2}}{2} \sum_{\alpha\beta} \sum_s \frac{J_\alpha J_\beta}{I_\alpha I_\beta} \frac{a_s^{(\alpha\beta)}}{\omega_s^{1/2}} q_s, \quad (12.45)$$

where the subscripts α, β indicate the principal axes of inertia of the HOD molecule, $\{I_\alpha\}$ represents the associated moments of inertia, and $\{J_\alpha\}$ the corresponding angular momenta. $\{a_s^{(\alpha\beta)}\}$ depend on the geometry and the intramolecular potential of the molecule.

Because we now know the Hamiltonians corresponding to different modes of coupling, using Eq. (12.25), it is easy to find the rate of VER of O–H stretch and different contributions (Vibration-Bath, Coriolis, and centrifugal coupling) to the relaxation rate.

Table 12-1. $k_{1\to 0}$ AT 70 K AND 80 K, THEORY AND EXPERIMENT. ALL NUMBERS LISTED IN UNIT OF s^{-1} [13].

	70 K	80 K
No quantum correction	0.006 ± 0.002	0.035 ± 0.005
Bader-Berne	0.19 ± 0.04	0.99 ± 0.16
Egelstaff	1200 ± 200	1900 ± 300
Experiment	360 ± 30	405 ± 25

Simulation studies based on the above theoretical formalism revealed that the relaxation of $\nu = 1$ state of O–H stretch mode for HOD in D_2O involves the following sequential steps:

(a) A transition to the second excited state of the HOD bending mode with a characteristic time of 7.2 ps, which is remarkably closer to the pump-probe result (8 ± 2 ps). *The major contribution comes from the effects of the anharmonic part of the intramolecular potential on the V–B coupling, with a small enhancement due to Coriolis coupling.*
(b) The above transition is followed by a fast relaxation (≈ 1 ps) to the first excited state of the bending mode. This transition is also introduced by V–B coupling but internal anharmonicity does not play any role.
(c) Finally, the (010) bend state relaxes to the ground state, with a characteristic time of 1.6 ps via V–B coupling.
(d) Centrifugal coupling has a negligible effect in all transitions.

Understanding of the VER of O–H stretch of water constitutes a landmark achievement in physics of liquids.

12.7.2 Vibrational-Energy Relaxation in Liquid Oxygen and Nitrogen

The lifetimes of the first-excited vibrational states in liquid oxygen and nitrogen are very long, 2.5 ms and 56 s, respectively. The theoretical studies of these long lifetimes is challenging because in these cases the vibrational-energy gap is much higher than $k_B T$ and than characteristic translational and rotational frequencies in the liquid. Skinner and coworkers [13, 14] performed molecular-dynamics simulations of rigid molecules of both oxygen at 70 K and 80 K and for nitrogen at 77 K. The theoretical results for oxygen are shown in Table 12.1.

From the wide variation of numbers in Table 12.1 it is immediately apparent that an accurate quantum factor is very important for obtaining the correct rate constant. The corrected results of Bader-Berne [7] are not in good agreement with experiment. On the other hand, the results obtained by Egelstaff are within a factor of 5. The poor agreement between theoretical and experimental predictions is mainly due to the fact that quantum corrections are not accurate and errors in the potential model, can easily account for an order of magnitude discrepancy.

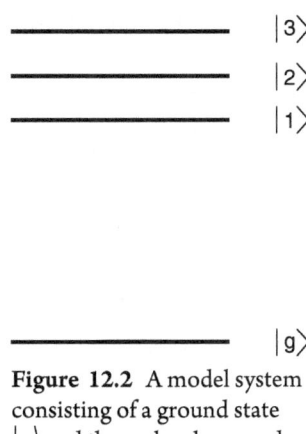

Figure 12.2 A model system consisting of a ground state $|g\rangle$ and three closely spaced excited states $|1\rangle$, $|2\rangle$, and $|3\rangle$.

The VER study of nitrogen is much more challenging than oxygen because the energy gap is significantly larger and the VER rate is slower. Skinner and coworkers concluded on the basis of their computer-simulation studies that in spite of nitrogen having a substantial quadrupole moment, insertion of quadrupolar interaction does not significantly change the VER rate. The theoretical result for the VER rate is about factor of 35 slower than experiments. In fact, it is believed that VER in liquid nitrogen is actually radiative, i.e., the nonradiative rate is significantly slower than experimental rate.

12.8 QUANTUM INTERFERENCE EFFECTS ON VIBRATIONAL-ENERGY RELAXATION IN A THREE-LEVEL SYSTEM: BREAKDOWN OF THE RATE EQUATION DESCRIPTION

Starting from IBC, all the theoretical models have assumed the validity of a rate law description when discussing energy relaxation through vibrational levels. Such a description assumes that we can write a master equation for population relaxation of individual levels with transition rates only between the adjacent energy levels. This description has been extensively used in our discussions in Chapter 7.

However, when quantum effects are important, such a simple rate equation description might break down due to *interference between adjacent pathways*. Such quantum interference effects on VER were studied within a model system consisting of ground state $|g\rangle$ and three closely spaced excited states $|1\rangle$, $|2\rangle$, and $|3\rangle$, as shown in Fig. 12.2 [15].

Two interesting new features appear in the three-level system. The first new feature is that the stochastic forces that couple the different pair of levels may differ. There can obviously be two limiting situations: (a) all the stochastic forces are the same due to the fact that they arise from the same bath, or (b) they all are completely different of each other, arising from uncorrelated baths.

The rate-law approach predicts that relaxation will be identical in both these situations. According to the rate law, the energy exchange between any pair of levels depends only on the stochastic forces that couple these two levels. In a multilevel system, like the one considered here it implies that the rate of energy exchange between any two levels is invariant to a change in the statistical properties of the coupling between any other two levels. The rate equation description for the population changes can be obtained by solving the following differential equations:

$$\dot{N}_1 = -(k_{12} + k_{13})N_1 + k_{21}N_2 + k_{31}N_3,$$
$$\dot{N}_2 = -(k_{21} + k_{23})N_2 + k_{12}N_1 + k_{33}N_3, \quad (12.46)$$
$$\dot{N}_3 = -(k_{32} + k_{31})N_3 + k_{13}N_1 + k_{23}N_2.$$

Because the energy levels are equally spaced, the following relations between the rate constants:

$$k_{12} = k_{21} = k_{23} = k_{32} = K, \quad (12.47)$$
$$k_{13} = k_{31} = K'. \quad (12.48)$$

The populations $N_i(t)$ are then given by

$$N_1(t) - N_1(\infty) = \frac{1}{2}[N_1(0) - N_3(0)]\exp[-(K + 2K')t] \\ - \frac{1}{2}[N_2(0) - N_2(\infty)]\exp[-3Kt], \quad (12.49)$$

$$N_2(t) - N_2(\infty) = [N_2(0) - N_2(\infty)]\exp[-3Kt], \quad (12.50)$$

$$N_3(t) - N_3(\infty) = \frac{1}{2}[N_3(0) - N_1(0)]\exp[-(K + 2K')t] \\ - \frac{1}{2}[N_2(0) - N_2(\infty)]\exp[-3Kt], \quad (12.51)$$

The rate constants K and K' can be obtained in weak-coupling limit using time-dependent perturbation theory. The time-dependent perturbation theory gives the following expression for the transition rate between level i and j

$$k_{ij} = \frac{8\pi^2}{h^2[1 + \exp(-\beta\hbar\omega_{ij}/2\pi)]}\int_0^\infty dt\,\exp(i\omega_{ij}t)\langle V_{ij}(t).V_{ji}(0)\rangle, \quad (12.52)$$

where ω_{ij} is the frequency difference between levels i and j, $\beta = 1/k_B T$, and $\langle\ldots\rangle$ is over the stochastic forces. The stochastic forces V_{ij}, which couple levels i and j, can either be Gaussian or Poissonian and the correlation function of stochastic forces can be given by

$$\langle V_{ij}(t).V_{ji}(0)\rangle = \langle V_{ji}(0)^2\rangle \exp(-bt). \qquad (12.53)$$

Equation (12.51) gives the following expressions for K and K':

$$K = \frac{2\langle V_{ij}^2(0)\rangle b}{1+b^2} \text{ and } K' = \frac{2\langle V_{ij}^2(0)\rangle b}{4+b^2}. \qquad (12.54)$$

In the Markovian limit ($b \to \infty$), K and K' become equal and the rate law predicts single exponential decay for all the levels. In the non-Markovian limit ($b \to 0$), $K' \approx K/4$.

The rate law predictions imply that the rate of population relaxation between any two levels depend only on the stochastic forces that couple these two levels. That is, the rate of relaxation is invariant to a change in the stochastic properties of the coupling between any other two levels. However, this is not true. There is a significant effect of the change in the statistical properties of the coupling between the *any* two levels in a multilevel system on the population transfer between two levels, in both Markovian and non-Markovian limits.

According to the rate-law prediction, the relaxation dynamics for level 1 are identical to those for level 3, but those for level 2 are different. The decay level 2 does not depend on the channel 3 $\to$ 1, i.e., independent of the interaction between level 1 and 3 (V_{13}).

However, in quantum-mechanical calculations carried out by turning off channel 3 $\to$ 1 and comparing the results with the resulting rate with the channel 3 $\to$ 1 kept open, a remarkable effect of channel 3 $\to$ 1 on the population relaxation is observed, which is shown in Fig. 12.3. The relaxation rate of level 2 changes for both the same bath and the independent bath cases. The change is more pronounced for the same bath case than the independent. However, in the Markovian limit and independent bath case the relaxation of level 2 does not change significantly, i.e., the interference between pathways is absent and behaves according to the rate law.

12.9 VIBRATIONAL LIFE TIME DYNAMICS IN SUPERCRITICAL FLUIDS

Because of the strong influence of the surrounding solvent molecules on vibrational-energy and phase-relaxation rates of a chemical bond, it can be anticipated that lifetime of an excited vibrational mode can exhibit interesting density dependence across the gas–liquid critical point. Figure 12.4 shows the observed vibrational lifetime of the T_{1u} asymmetric CO stretching mode of $W(CO)_6$ in ethane [16, 17].

Note that the broad density range near the critical point where the lifetime of vibrational relaxation is insensitive to the change in density. In contrast, the vibrational lifetime changes smoothly with density along the 50 °C isotherm.

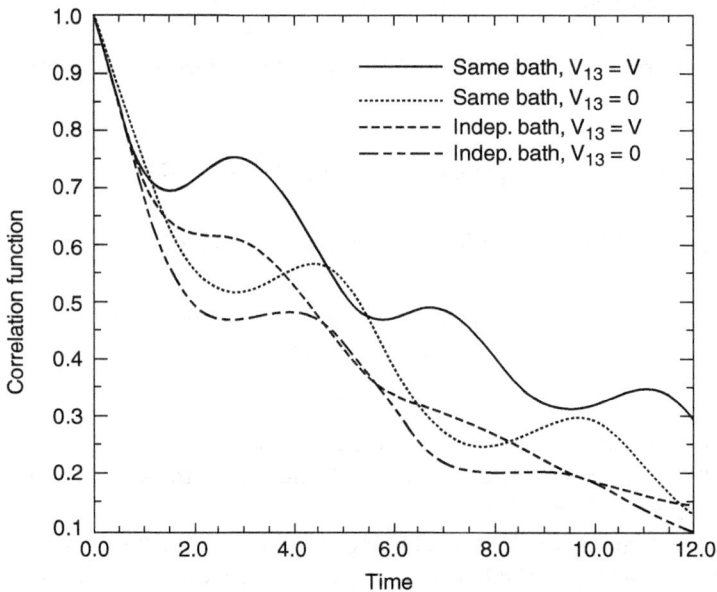

Figure 12.3 Figure showing the interference between pathways for a Poisson bath in the non-Markovian limit. Reprinted with permission from *J. Chem. Phys.* **77**, 1391 (1982). Copyright (1982) American Institute of Physics.

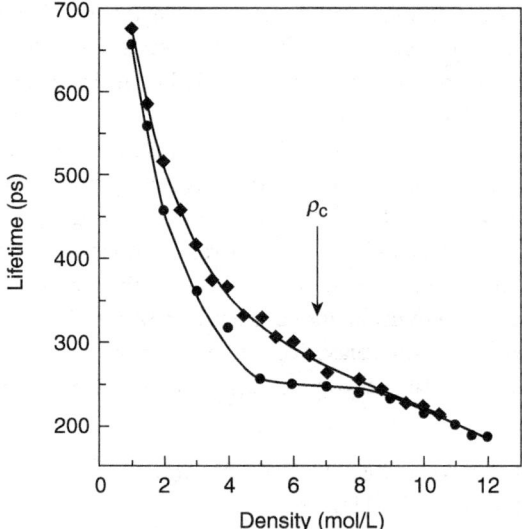

Figure 12.4 Vibrational lifetime of the T_{1u} asymmetric CO stretching mode of $W(CO)_6$ versus density of C_2H_6 at 34 °C (circles; $T_r \approx 1.01$) and 50 °C (diamonds; $T_r \approx 1.06$). A plateau region centered near ρ_c is the most prominent feature of the 34 °C data. The critical temperature for ethane is 32.2 °C and the critical density (marked with an arrow) is 6.87 mol/L. Reprinted with permission from *J. Chem. Phys.* **107**, 3747 (1997). Copyright (1997) American Institute of Physics.

12.10 SUMMARY

Understanding energy loss of an excited vibrational mode to its own bath modes (rotation and translation) or to other vibrational modes of the liquid remains a difficult problem because of the quantum nature of the vibrational mode. As described above, one models such a system as a quantum mode in contact with classical modes. Except in the case where the classical bath can be modeled as a collection of harmonic oscillators linearly coupled to the vibrational coordinate, we do not have any exact solution of the problem. Although many approximations have been put forward, the situation is still not satisfactory. On the whole, however, understanding has improved considerably in the last few decades.

On the experimental front the situation has improved considerably, as discussed above. Particular note is to be taken of the recently emerged understanding of VER of the O–H stretch in liquid water. Subquadratic quantum number dependence of the overtone is another interesting recent development which has been discussed here.

REFERENCES

1. D. W. Oxtoby, *Adv. Chem. Phys.* 47, 487 (1981).
2. K. F. Herzfeld, *J. Chem. Phys.* 20, 288 (1952).
3. W. M. Madigosky and T. A. Litovitz, *J. Chem. Phys.* 34, 489 (1961).
4. M. Fixman, *J. Chem. Phys.* 34, 369 (1961).
5. R. Zwanzig, *J. Chem. Phys.* 34, 1931 (1961).
6. K. F. Herzfeld, *J. Chem. Phys.* 36, 3305 (1962).
7. J. S. Bader and B. J. Berne, *J. Chem. Phys.* 100, 8359 (1994).
8. T. J. Chuang, G. W. Hoffman, and K. B. Eisenthal, *Chem. Phys. Lett.* 25, 201 (1974).
9. A. Laubereau and W. Kaiser, *Rev. Mod. Phys.* 50, 607 (1978).
10. P. Hamm, M. Lim, and R. M. Hochstrasser, *J. Chem. Phys.* 107, 10523 (1997).
11. J. C. Owrutsky, D. Raffery, and R. M. Hochstrasser, *Annu. Rev. Phys. Chem.* 45, 519 (1994).
12. R. Rey and J. T. Hynes, *J. Chem. Phys.* 108, 142 (1998).
13. K. F. Everitt, S. A. Egorov, and J. L. Skinner, *Chem. Phys.* 235, 115 (1998).
14. K. F. Everitt and J. L. Skinner, *J. Chem. Phys.* 116, 179 (2002).
15. B. Bagchi and D. W. Oxtoby, *J. Chem. Phys.* 77, 1391 (1982).
16. R. S. Urdahl, D. J. Myers, K. D. Rector, P. H. Davis, B. J. Cherayil, and M. D. Fayer, *J. Chem. Phys.* 107, 3747 (1997).
17. S. C. Tucker, *Chem. Rev.* 99, 391 (1999).

13

Vibrational-Phase Relaxation

13.1 INTRODUCTION

Vibrational-phase relaxation (VPR) is quite distinct from vibrational-energy relaxation (VER). The factors that are responsible for the vibrational-phase relaxation are also quite distinct from those involved in the energy relaxation. VPR is a vital probe to study the "soft" many-body interaction of a chemical bond with the surrounding solvent molecules while VER is often sensitive to the "hard" binary interactions, as the success of the isolated binary collision (IBC) model demonstrates [1, 2]. Vibrational-phase relaxation is also a sensitive probe of the dynamics of the surrounding solvent as the phase is sensitive to small perturbations of the energy. The study of vibrational-phase relaxation can also provide information about some important *molecular* properties, such as the anharmonic coupling between different vibrational degrees of freedom (such as bending and stretching). Generally, such information is not easily available from other sources.

Although the study of vibrational-phase relaxation has always been an important area of study, the development of *infrared lasers* in the mid-1990s has given a new impetus to this area. This is because one now could study vibrational-phase relaxation in the ground state directly. Additionally, new time-domain techniques have been developed, with infrared lasers, to study the evolution of phase of a collection of molecules. Many of these techniques use multiple laser pulses. By varying the time lag between the pulses and the frequencies, one can construct multidimensional infrared and Raman spectroscopy of liquids. Many theoretical and experimental groups have demonstrated that such measurements conducted using sequences of femtosecond pulses may be used to probe intramolecular and intermolecular nuclear motions directly. These techniques provide an additional insight into the nature of the nuclear spectral density observed in optical Kerr measurements [3].

The easiest access to vibrational-phase relaxation is through isotropic Raman linewidth. This is given by the one-sided Fourier transform (or the cosine transform) of the normal coordinate–normal coordinate time correlation function, $\langle Q(0)Q(t) \rangle$. We have discussed this relation in **Chapter 4**.

Raman line shapes of molecules in liquids get shifted and broadened compared to the same in the gas phase. This arises from a delicate balance between several interactions that exist within the liquid. The two main origins in the evolution of the liquid structure and its dynamics are of opposite nature; the long-range attractive forces that try to bring the molecules closer together so as to maintain the liquid at high densities and ambient pressure, and the relatively short-range harsh repulsive forces between the nearest-neighbor molecules that tend to force them apart. Vibrational-phase relaxation is sensitive to *both* these forces. While the microscopic molecular arrangement of most liquids can be rather well accounted for in terms of repulsive forces only, the attractive forces are mainly responsible for the shifts of vibrational/electronic frequency between the vapor and the liquid phases. Therefore, the measurement of frequency shift and vibrational linewidth together is important to obtain information on the structure and the dynamics of liquids.

Physically, the process of vibrational phase relaxation can be imagined in the following way. For simplicity, let us assume that the vibrational motion can be approximated by that of a harmonic oscillator. The time-dependent vibrational wave function of the molecule is determined by the occupation number of various levels of the oscillator and the phase of the oscillation. Let us assume that at time $t = 0$ we expose a set of such oscillators to radiation, which establishes a phase relation between the zeroth and the first vibrational level of the oscillators, that is, excite the fundamental. Thus, the optical pulse not only populates both the levels $\nu = 0$ and $\nu = 1$, but also creates a definite phase of the wave function in each of the oscillators. That is, the excitation creates a macroscopic vibrational amplitude. As time evolves, the phases of the individual oscillators become uncorrelated. This can happen by elastic collisions with surrounding molecules in the gas phase or by interactions with the surrounding liquid molecules. As has already been discussed in Chapter 4, this phase relaxation is studied most easily by isotropic Raman linewidth. The broadening of an isotropic Raman line shape of a vibrational mode mainly arises from the following three processes.

(a) The phase relaxation or pure vibrational dephasing due to elastic processes. This arises from fluctuations of the energy gap between the two vibrational levels whose dephasing is being probed. However, the amplitude and the rate of this fluctuation of the energy gap depend on several factors, both molecular (such as, anharmonicity) and surrounding solvent properties, such as collision frequency, which again depends on temperature and density:

(b) Resonant energy transfer between different molecules that leads to the loss of phase coherence. This mechanism can be important at high density.

(c) Vibrational-energy relaxation due to hard inelastic transitions.

The mechanism (a), the pure dephasing, involves the continuous, stochastic fluctuations of the energy levels of a molecule interacting with its surroundings. It is usually the dominant dephasing mechanism in liquid and is also often much faster than the other two dephasing processes mentioned above. Isotopic dilution studies have shown that the effect of resonant transfer on the linewidth can be at most about 10–20% and is often much less; population relaxation studies have shown that it is often much slower than dephasing by factors ranging from 5 to 10^{12}. Therefore, it is generally sufficient to consider the effect of pure dephasing on line shapes. However, in a few cases, the timescales of the above processes could be comparable and their interplay could result in strong cross-correlations between them.

Even when the pure dephasing is the primary mechanism for phase relaxation, the situation is quite complex. To understand this, let us consider the following simple case, the dephasing of fundamental of the stretching of N_2. In this case, there are three different sources of fluctuations of the frequency gap: (i) anharmonicity-mediated linear (in the vibrational coordinate) coupling to the surrounding molecules, (ii) nonlinear (at least second-order or quadratic) coupling of the normal mode with surrounding molecules, and (iii) vibration–rotation coupling.

Off-diagonal anharmonicity, present in more-complicated molecules, is absent in diatomic molecules. One needs to consider contributions not only of the pure anharmonic terms, but also from cross-correlations, as explained below. Therefore, understanding of vibrational-phase relaxation is nontrivial and proper care is required to extract information from experimental results.

The time scales of molecular dynamics in liquids span a wide range from the ultrafast (collisional interactions) to the almost static (long-range attractive-force interactions). While it has been known for a long time that VER is sensitive only to the high-frequency (or short-time) response of the liquid, it was believed that such short-time dynamics is not very relevant for vibrational dephasing. Recent studies with ultrafast laser spectroscopy seem to portray a different picture.

Fortunately, despite the inherent complexity, most aspects of vibrational-phase relaxation can be understood from a beautiful theory, which we describe below.

13.2 KUBO–OXTOBY THEORY OF VIBRATIONAL LINE SHAPES

Our understanding of vibrational dephasing is based primarily on Kubo's stochastic theory of line shape. Kubo's theory is a general theory applicable to any transition, electronic, vibrational, rotational, or NMR. For vibrational dephasing, an insightful implementation of the Kubo theory was made by Oxtoby in an important work in 1979 and we shall refer to this theory as the Kubo–Oxtoby theory [1, 2]. This theory is rather simple and consists of the following straightforward steps.

The first step is to identify the appropriate dynamical variable that is responsible for the linewidth. For electronic and vibrational line shapes, it is the energy gap between the two quantum levels whose line shape is being probed. If we denote the dynamical variable by $x(t)$, then it is assumed that $x(t)$ obeys the following equation of motion:

$$\frac{dx(t)}{dt} = i\omega(t)x(t). \tag{13.1}$$

The above equation is solved to obtain the following formal solution:

$$x(t) = x(0) \exp\left(i \int_0^\infty dt'\, \omega(t')\right). \tag{13.2}$$

The line shape is given by one-sided Fourier transform of the time autocorrelation function of $x(t)$:

$$I_{lineshape}(\omega) = \int_0^\infty dt\, \exp(i\omega t) \langle x(0)x(t)\rangle, \tag{13.3}$$

where $\langle x(0)x(t)\rangle$ is the autotime-correlation function of the dynamic variable $x(t)$.

Since the relevant dynamical variable $x(t)$ in this case is the energy gap between two vibrational levels, the theoretical analysis of line shape is somewhat tricky. This is because the primary source of energy gap fluctuation is not obvious *a priori* and it has created much confusion in the past. In order to understand the problem, let us write the Hamiltonian of a vibrating system with normal coordinate Q as

$$H(Q) = H_v + V_B(Q), \tag{13.4}$$

where H_v is the Hamiltonian for the vibrational degree of freedom in question and $V_B(Q)$ is the coupling of the normal coordinate with the bath. The bath consists not only of the solvent molecules but also of other vibrational and rotational modes of the molecule in question. This coupling term is evaluated by expanding $V_B(Q)$ in powers of Q:

$$V_B(Q) = V_B(Q=0) + \left[\frac{\partial V_B(Q)}{\partial Q}\right]_0 Q + \frac{1}{2}\left[\frac{\partial^2 V_B(Q)}{\partial Q^2}\right]_0 Q^2 + \cdots\cdots . \tag{13.5}$$

Now, the energy difference between two levels, say i and j, involve in the first order in Q, $\langle i|Q|i\rangle - \langle j|Q|j\rangle$. If H_v is fully harmonic, then this first-order term is zero and the contribution starts with the second-order term. This has sometimes been referred to as nonlinear coupling between the system and the bath. A paradoxical situation arose because the second-order term gave rates of dephasing that, for most cases, were three to four orders of magnitude smaller than experimental values.

In his pioneering work, Oxtoby applied Kubo's theory of vibrational dephasing and demonstrated that anharmonicity could play an important role in enhancing dephasing rates by many orders of magnitude because now the contribution of the first-order term (in Q) is nonzero. The final expression of Oxtoby involves a force–force time-correlation function, which acts on the normal coordinate. This force is coming from the surrounding solvent molecules. Oxtoby related the dephasing rate to the solvent viscosity by a hydrodynamic argument.

The broadened isotropic Raman line shape $(I(\omega))$ is the Fourier transform of the normal coordinate (Q) time-correlation function by

$$I(\omega) = \int_0^\infty \exp(i\omega t) \langle Q(t)Q(0) \rangle, \qquad (13.6)$$

where ω is the Laplace frequency conjugate to time, t. The experimental observables are either the line shape function, $I(\omega)$, or the normal coordinate time-correlation function, $\langle Q(0)Q(t) \rangle$. From Eq. (13.2), where we now take $x(t) = Q(t)$, the normal coordinate time correlation is related to frequency-modulation time-correlation function by

$$\langle Q(t)Q(0) \rangle = \operatorname{Re}\left[\exp(i\omega_0 t) \left\langle \exp\left[i \int_0^t dt' \Delta\omega_{mn}(t') \right] \right\rangle \right], \qquad (13.7)$$

where ω_0 is the vibrational frequency and $\hbar\Delta\omega_{mn}(t) = V_{nn}(t) - V_{mm}(t)$ is the fluctuation in energy between vibrational levels of n and m, where n and m represent vibrational quantum numbers. V_{nn} is the Hamiltonian matrix element of the coupling of the vibrational mode to the solvent bath. $\Delta\omega_{mn}(t)$ is, therefore, the instantaneous shift in the vibrational frequency due to interactions with the solvent molecules.

A cumulant expansion, with truncation at the second order, of Eq. (13.7) gives the following well-known expression for the normal coordinate time-correlation function:

$$\langle Q(t)Q(0) \rangle = \operatorname{Re}\left[\exp\left(i\omega_0 t + i \langle \Delta\omega \rangle_{mn} t \right) \right.$$

$$\left. \times \exp\left[-\int_0^\infty dt' \int_0^{t'} dt'' \langle \Delta\omega_{mn}(t') \Delta\omega_{mn}(t'') \rangle \right] \right]. \qquad (13.8)$$

The double integration in Eq. (13.8) is now rewritten in the following way:

$$\langle Q(t)Q(0)\rangle = \text{Re}\left[\exp\left(i\omega_0 t + i\langle\Delta\omega\rangle_{mn} t\right)\right.$$

$$\left.\times \exp\left[-\int_0^t dt'\,(t-t')\langle\Delta\omega_{mn}(t')\Delta\omega_{mn}(0)\rangle\right]\right], \quad (13.9)$$

where $\langle\Delta\omega_{mn}(t')\Delta\omega_{mn}(0)\rangle$ is the frequency-modulation time-correlation function. If the frequency–frequency modulation time-correlation function decays fast, the upper limit of integration can be extended to infinity and $(t - t')$ can be replaced by t. As a result one obtains the following exponential decay of $\langle Q(t)Q(0)\rangle$

$$\langle Q(t)Q(0)\rangle \approx \exp(-t/\tau_v), \quad (13.10)$$

where τ_v is the average dephasing time defined by

$$\tau_v^{-1} = \int_0^\infty dt'\,\langle\Delta\omega_{mn}(t')\Delta\omega_{mn}(0)\rangle. \quad (13.11)$$

The exponential decay described in Eq. (13.10) gives rise to a Lorentzian line width with a half-width at half-maximum (HWHM) of τ_v^{-1}. Thus, assumption of the exponential decay (as in Eq. (13.10)) allows us to determine τ_v directly from Eq. (13.11). However, $\langle Q(t)Q(0)\rangle$ may be nonexponential (which is actually observed in the majority of cases) and as discussed later, may give rise to an overall subquadratic quantum number dependence of the rate. However, the above description of phase relaxation provides a relation between dephasing time and frequency-modulation time-correlation function in a succinct way.

As mentioned earlier, the bath Hamiltonian H_B is the sum of the Hamiltonian for the rotational and translational degrees of freedom. These two degrees of freedom are treated classically. V represents the anharmonic oscillator–medium interaction. Oxtoby used the following Hamiltonian for the vibrating mode:

$$H_{vib} = K_{11}Q^2 + K_{111}Q^3, \quad (13.12)$$

where K_{111} is the coefficient that gives rise to the anharmonicity in the vibration.

If V is the anharmonic oscillator–medium interaction, the energy fluctuation between vibrational levels 0 and n can be obtained by expanding V in the vibrational coordinate Q using Taylor's series as follows:

$$\hbar\Delta\omega_n(t) = (Q_{nn} - Q_{00})\left(\frac{\partial V}{\partial Q}\right)_{Q=0}(t)$$
$$+ \frac{1}{2}(Q_{nn}^2 - Q_{00}^2)\left(\frac{\partial^2 V}{\partial Q^2}\right)_{Q=0}(t) + \ldots. \quad (13.13)$$

In the estimation of $\Delta\omega_n(t)$, only the first two terms are considered, neglecting the higher-order term. $(Q_{nn} - Q_{00})$ and $(Q_{nn}^2 - Q_{00}^2)$ are the quantum-mechanical expectation values for an anharmonic oscillator. They are easily calculated using perturbation theory to be

$$Q_{nn} - Q_{00} = \frac{3n\hbar(-K_{111})}{\omega_0^3}, \quad (13.14)$$

$$Q_{nn}^2 - Q_{00}^2 = \frac{n\hbar}{\omega_0}. \quad (13.15)$$

The derivatives appearing in Eq. (13.13) can be rewritten in terms of the atomic displacement ε_i, if the potentials are taken to be atom-additive functions of relative atomic separations $V(|r_i - r_j|)$. The derivatives are given as

$$\frac{\partial V}{\partial Q} = \sum_i \left(\frac{\partial V}{\partial \varepsilon_i} \cdot \frac{\partial \varepsilon_i}{\partial Q}\right) = -\sum_i F_i \left(\frac{\partial \varepsilon_i}{\partial Q}\right), \quad (13.16)$$

where F_i represents the force on the atom i. Also, if the potential has a strong repulsive part, $\left(\frac{\partial^2 V}{\partial Q^2}\right)$ can be approximated as

$$\frac{\partial^2 V}{\partial Q^2} \approx -\frac{1}{L}\sum_i F_i \left(\frac{\partial \varepsilon_i}{\partial Q}\right)\left|\left(\frac{\partial \varepsilon_i}{\partial Q}\right)\right|, \quad (13.17)$$

where L defines the characteristic potential range.

The atomic displacement ε_i is related to the normal-coordinate Q through

$$\varepsilon_i = \frac{l_{ik}}{Q_k} m_i^{1/2}, \quad (13.18)$$

where m_i is the mass of the ith atom and l_{ik} is related to a characteristic vector, $\mathbf{l}_{ik}$, along the normal mode Q_k as $\mathbf{l}_{ik} = l_{ik}u_{ik}$. For a diatomic molecule, $\mathbf{l}_{ik} = \sqrt{(m_i/\mu)}\gamma_i$, where $\gamma_i = m_i/(m_i + m_j)$. μ represents the reduced mass of the system.

Using the simplifying approximation that the forces on the different atoms of the diatom are uncorrelated and that the area of contact of each atom with the solvent is a half-sphere, Oxtoby derived the following expression for the time-correlation function of the frequency modulation between two quantum levels n and m in terms of the dynamic friction on the atom involved:

$$\langle \Delta\omega_{mn}(t)\Delta\omega_{mn}(0)\rangle = \frac{(m^2 - n^2)}{2} \sum_i \left[\frac{3(-K_{111})l_{ik}}{\omega_0^3 m_i^{1/2}} + \frac{l_{ik}^2}{2\omega_0 L m_i}\right]^2 \langle \mathbf{F}_i(t)\mathbf{F}_i(0)\rangle, \quad (13.19)$$

where L is the characteristic potential range and $\langle \mathbf{F}_i(t)\mathbf{F}_i(0)\rangle$ represents the force–force correlation function on the atom i moving along the direction of vibration.

We next need to find the value of the anharmonicity parameter, K_{111}, and this is obtained as follows. First, it is assumed the vibrational-bond energy to be given by the Morse potential of the following form:

$$V(r) = D_e \left[1 - \exp\left[-\beta(r - r_e)\right]\right]^2, \quad (13.20)$$

where β and D_e are the Morse potential fit parameters and r_e is the equilibrium bond length. Then a Maclaurin series expansion of the potential about the equilibrium position of the vibration and a term by term comparison with Eq. (13.13) produces the expression for K_{111}.

Equation (13.19) is the expression used in many studies of vibrational dephasing and even in computer simulations. Note that it does not include the vibrational–rotational contribution to dephasing or the resonant energy transfer between different molecules.

The presence of n^2 in Eq. (13.19) is the reason why the vibrational dephasing rate is usually assumed to exhibit the *quadratic* quantum-number dependence for overtones and hot bands. However, while $\langle \Delta\omega_{mn}(t)\Delta\omega(0)\rangle$ may have an n^2 dependence, the average dephasing rate (τ_v^{-1}) can show subquadratic dependence when $\langle Q(t)Q(0)\rangle$ follows a Gaussian decay at short times with the form

$$\langle Q(t)Q(0)\rangle = \exp\left[-n^2 t^2/\tau^2\right]. \quad (13.21)$$

However, the decay of the vibrational-correlation function in dense liquids is expected to be more complex because the frictional response of dense liquid is strongly biphasic, containing both an ultrafast and a slow component. Therefore, a switchover from quadratic to subquadratic quantum-number dependence will be largely controlled by the respective amplitude of each of the components (Gaussian and exponential) of the bimodal frictional response.

13.3 HOMOGENEOUS VS. INHOMOGENEOUS LINEWIDTHS

Raman line shape (or, for that matter any transition line shape) is found to be quite broadened in the liquid phase. The width of this broadened line shape derives contributions primarily from two sources, namely, the existence of adjacent transition frequencies and the interaction with the surrounding solvent molecules. Here we concentrate on the latter reason as it gives information of the structure and dynamics of the liquid phase. Also, in many cases of interest, the adjacent transition frequencies are absent naturally or can be made to disappear by methods such as dilution.

Depending on the relative time scales of solvent relaxation and dephasing (which again depends on solvent), we often classify a line shape as homogeneous or inhomogeneous. These terms are often confusing and we attempt to demystify them below.

Let us assume that a given vibrational mode (with harmonic frequency (ω_0)) of all the molecules in a liquid are prepared at a given phase initially by the application of an ultrafast laser pulse. This phase coherence between different molecules is destroyed by two independent solvent mechanisms. The first one involves nearly elastic collisions with the surrounding solvent molecules. This interaction leads to small frequency shifts ($\Delta\omega(t)$) from the average frequency ($\bar{\omega}$) of the solvent molecules in liquid. Thus, the instantaneous frequency of a vibration of a given mode of a particular molecule is given by

$$\omega(t) = \bar{\omega} + \Delta\omega(t), \qquad (13.22)$$

where $\Delta\omega(t)$ represents the stochastic modulation of $\omega(t)$ due to interactions with the environment. Note that $\bar{\omega}$ contains the shift from the gas-phase frequency, ω_0. Pure dephasing is controlled by the fluctuation in $\Delta\omega(t)$.

When $\Delta\omega(t)$ is characterized by statistical properties *common* to all the vibrating molecules involved, the mechanism is called the *homogeneous mechanism of dephasing* and the linewidth is called the *homogeneous linewidth*.

In dense liquids there are, however, motions that are very slow compared to the time of vibrational dephasing. For example, exchange of different species in a binary mixture is a slow process. Therefore, a different statistical distribution of a particular species around the vibrating solute may lead to different frequencies in different molecules. *That is, $\bar{\omega}$ itself would be different for different molecules!* This inhomogeneity is further accentuated if the fluctuations in $\Delta\omega(t)$ are large. This will also lead to the loss of vibrational coherence among different molecules and then, according to the Kubo–Oxtoby theory, the rate of dephasing will also be large! This is called the *inhomogeneous mechanism of vibrational dephasing* or line broadening.

However, inhomogeneous broadening gives erroneous information about the rate of dephasing. It is the homogeneous rate that is the true rate. The extra broadening caused by inhomogeneity is an artifact due to long-lived density or structure fluctuations and has nothing to do with true dephasing.

Understanding homogeneous and inhomogeneous mechanisms of vibrational dephasing has remained an active area of research for several decades. One would certainly like to know the homogeneous linewidth of a given line shape, as this is usually considered as the true dephasing mechanism. In the ideal limit, only the homogeneous mechanism contributes to the vibrational linewidth. This limit is satisfied when the inhomogeneous part is frozen or decays extremely slowly.

In liquids, however, such a clear separation may not exist, as the different structures relax with time.

13.4 RELATIVE ROLE OF THE ATTRACTIVE AND REPULSIVE FORCES

A unified molecular theory which allowed an understanding of the relative role of attractive and repulsive forces on solvent-induced vibrational dephasing was developed in an elegant work by Schweizer and Chandler [4]. Based on the assumption of separation of time scale between the attractive and repulsive forces, they could analyze the respective roles of repulsive and attractive contributions and pointed out that the vibrational dephasing is largely controlled by the attractive forces.

A rough idea of relative roles of attractive and repulsive forces can also be obtained by using mode-coupling theory (MCT), which is based on the separation of time scales between the binary collision and the repeated re-collisions. For the Lennard-Jones potential, the contribution to the binary part $(\zeta_B(t))$ arises primarily from the attractive part of the potential. On the other hand, the density part $(R_{\rho\rho}(t))$ is determined by both the attractive and repulsive contributions, although it is the latter that renders it the dominant contribution. The important role of the attractive part implied by MCT in dephasing is in complete agreement with the conclusions arrived at in Schweizer and Chandler work.

13.5 VIBRATION–ROTATION COUPLING

The effective potential for the vibration–rotation (VR) centrifugal coupling is the rotational kinetic energy, which is given by

$$V_{VR} = \frac{J^2}{2I}, \qquad (13.23)$$

where J is the angular momentum, $I = \mu r^2$ is the moment of inertia, and μ is the reduced mass. Expanding this as a Taylor series around the equilibrium bond length, r_e, results in

$$V_{VR} = \frac{J^2}{2I_m r_e}\left[\frac{r_e}{2} - \Delta r + \frac{3(\Delta r)^2}{2r_e} - \ldots\ldots\right], \qquad (13.24)$$

where I_m is now the moment of inertia value at r_e, and Δr gives the displacement from equilibrium.

The centrifugal (VR) contribution to the broadening of the line shape is then given by

$$\langle \Delta\omega_{VR}(t)\Delta\omega_{VR}(0)\rangle_{n0} = \frac{\langle[V_{VR,nn}(t) - V_{VR,00}(0)][V_{VR,nn}(t) - V_{VR,00}(0)]\rangle}{\hbar^2}$$

$$= \left(\frac{\Delta R_{n0}}{\hbar I_m r_e}\right)^2 \langle \Delta J^2(t)\Delta J^2(0)\rangle, \qquad (13.25)$$

where

$$\Delta R_{n0} = (Q_{nn} - Q_{00}) - 3(Q_{nn}^2 - Q_{00}^2)/2r_e. \tag{13.26}$$

$Q_{nn} - Q_{00}$ and $(Q_{nn}^2 - Q_{00}^2)$ are the quantum-mechanical expectation values of the bond-length displacement and its square, $(r - r_e)$ and $(r - r_e)^2$, for the nth vibrational level of the isolated molecule. These can be easily calculated using quantum perturbation theory, as mentioned before and are given by Eqs. (13.15) and (13.16). On writing the correlation function $\langle \Delta J^2(t) \Delta J^2(0) \rangle$ as

$$\langle \Delta J^2(t) \Delta J^2(0) \rangle = \langle J^2(t) J^2(0) \rangle - \langle J^2 \rangle^2, \tag{13.27}$$

and then rewriting the correlation function $\langle J^2(t) J^2(0) \rangle$

$$\langle J^2(t) J^2(0) \rangle = 0.5 \langle J^4 \rangle \left[1 + \langle \mathbf{J} \cdot \mathbf{J}(t) \rangle^2 / \langle J^2 \rangle^2 \right], \tag{13.28}$$

and finally, using $\langle J^4 \rangle = 2 \langle J^2 \rangle^2$ and $\langle J^2 \rangle = 2Ik_BT$, one obtains

$$\langle \Delta \omega_{VR}(t) \Delta \omega_{VR}(0) \rangle_{n0} = (2k_B T \Delta R_{n0}/\hbar r_e)^2 \left[J_{corr}(t) \right]^2, \tag{13.29}$$

where J_{corr} is the normalized angular-momentum autocorrelation function, $\langle \mathbf{J} \cdot \mathbf{J}(t) \rangle^2 / \langle J^2 \rangle^2$. This is available experimentally from anisotropic Raman line shape data for some systems.

Note that $\langle \Delta \omega(t) \Delta \omega(0) \rangle_{n,VR}$ also has a quadratic n dependence as both $Q_{nm} - Q_{00}$ and $Q_{nm}^2 - Q_{00}^2$, which are involved in determining ΔR_{n0}, have an n dependence (Eqs. (13.15) and (13.16)). As the angular momentum–momentum correlation function related to the vibrational–rotation friction too can be highly nonexponential, this again could significantly alter the n^2 dependence of the dephasing rate for reasons described earlier.

13.6 EXPERIMENTAL RESULTS OF VIBRATIONAL-PHASE RELAXATION

Many experimental methods (IR, Raman line shapes, laser spectroscopy) have been developed over the years and many studies have been carried out on the several systems to observe vibrational-energy and phase-relaxation of molecular solutes in condensed medium and its dependence on temperature, pressure, and concentration. In the following we discuss several aspects with emphasis on semiquantitative and general features.

13.6.1 Semiquantitative Aspects of Dephasing Rates in Solution

The rate of dephasing can be obtained by combining Eqs. (13.8) and (13.19), and making a Markovian (or exponential) approximation for the decay of the force–force time-correlation function. For a homo–diatomic molecule (like N_2 and O_2), the predominant contribution to the dephasing rate then comes from the combination $K_{111}^2/(\omega^4 \zeta^{bond})$, where ζ^{bond} is the friction on the chemical bond of the diatomic. As discussed in Chapter 12, if we neglect the cross-correlation between the solvent forces on the two atoms, then this bond friction is equal to one-half of the friction on each atom of the homo–diatom.

Therefore, vibrational-dephasing rate depends both on the properties of the oscillator, through K_{111}^2/ω^4, and on the solvent properties through friction, ζ. The latter is proportional to T/η. Thus, the vibrational-dephasing rate is faster for systems with larger anharmonicity and smaller frequency.

Experimental results on the vibrational-dephasing rate are usually discussed in terms of the Raman linewidth (the full width at half-maximum) or vibrational lifetime, τ_v. For an assumed Lorentzian line shape the full width at half-height is $2/\tau_v$.

The quantitative agreement of the dephasing rate between the Kubo–Oxtoby prediction and experimental results has been quite good for most cases. For example, for liquid nitrogen, both give values of lifetime $\tau_v \approx 150\,\text{ps}$ [1]. The full width at half-maximum of the Lorentzian lines has been found to be 0.067 cm^{-1} for N_2 and 0.125 cm^{-1} for O_2. For the C–I stretch of methyl iodide, the value of the frequency is much less and the value of the anharmonicity parameter K_{111} is more. Because of these opposing effects, the value of the lifetime is near, again in good agreement with Kubo–Oxtoby theory.

13.6.2 Subquadratic Quantum Number Dependence

Experiments involving echo techniques on dephasing have revealed interesting subquadratic observations of the dephasing rate (directly proportional to the linewidth) as a function of quantum number n[5] as against the theoretically predicted quadratic dependence by earlier theoretical models such as the Kubo–Oxtoby stochastic theory of line shapes. For example, in systems like $CDCl_3$ and CD_3I, the ratio of the dephasing rates between the overtone and the fundamental of the C–D stretching mode are 2.1 and 2.4, respectively, as against the theoretically expected value of 4 in the both cases. This is somewhat surprising because in liquids the frequency time-correlation function is expected to be short and in the motional narrowed limit. Overtone transition studies that have been attempted earlier [6] have also reported similar subquadratic quantum-number dependence, although these observations involve considerable amount of uncertainty because of the signal-to-noise factors of the bandwidth measurements.

Recent theoretical and experimental studies of overtone dephasing of C–H stretching in CH_3I seem to have revealed the following picture of vibrational dephasing:

(a) The subquadratic quantum-number dependence due to Gaussian decay of the force–force time correlation function can occur only when the time scale of decay of the frequency time-correlation function and the normal coordinate time-correlation function are comparable and these two functions overlap. A Gaussian time decay of the normal coordinate correlation function can occur due to small frictional forces at very short times.

(b) This overlap is possible only when (1) the harmonic frequency is not so large, (2) the anharmonicity is significant, and (3) the mean square fluctuation $(\langle\Delta\omega^2\rangle)$ is large.

Recent experimental study of vibrational dephasing of I_3^- in ethanol and methyl tetrahydrofurane solutions from 300 to 100 K, for the vibrational fundamental and its first overtone, suggests that dephasing is homogeneous throughout the temperature range studied [7]. Vibrational population relaxation has negligible contribution to the dephasing rate and shows subquadratic number dependence (ratio between 2 and 2.7) at the temperature range studied.

The explanation of the subquadratic quantum number dependence in terms of a dominant Gaussian time decay of the $\langle Q(0)Q(t)\rangle$ time-correlation function is still an unsettled issue.

13.7 VIBRATIONAL DEPHASING NEAR THE GAS–LIQUID CRITICAL POINT

Because the phase-relaxation rate is sensitive to the density of the surrounding solvent, this can be a sensitive probe of density fluctuations. In a novel study of vibrational-phase relaxation near the gas–liquid critical point, interesting insights into the time scales of density fluctuations [8, 9] were obtained. In a series of experiments, Musso et al. [8] measured the Raman linewidth in nitrogen across the phase boundary and also along the phase boundary up to the critical point. The results of this study are quite interesting and the results are shown in Fig. 13.1. Note the lambda-type rise in the line width. A major contribution to this rise is the large fluctuation in the frequency due to large density inhomogeneity in the system near the critical point. It turns out that this inhomogeneity or static density fluctuation is more important than dynamic effects.

13.8 MULTIDIMENSIONAL IR SPECTROSCOPY

A common problem with conventional experimental probes of vibrational dephasing is their inability to distinguish between homogeneous and inhomogeneous contributions. Sometimes the inhomogeneous contribution fluctuates at such a slow time scale that it is termed static inhomogeneity. In such cases, one can burn a hole in the inhomogeneously broadened spectrum to separately study the homogeneous broadening because by exciting a subpopulation from an inhomogeneously broadened spectrum, one can study a homogeneous contribution. However, although such spectral hole-burning experiments have

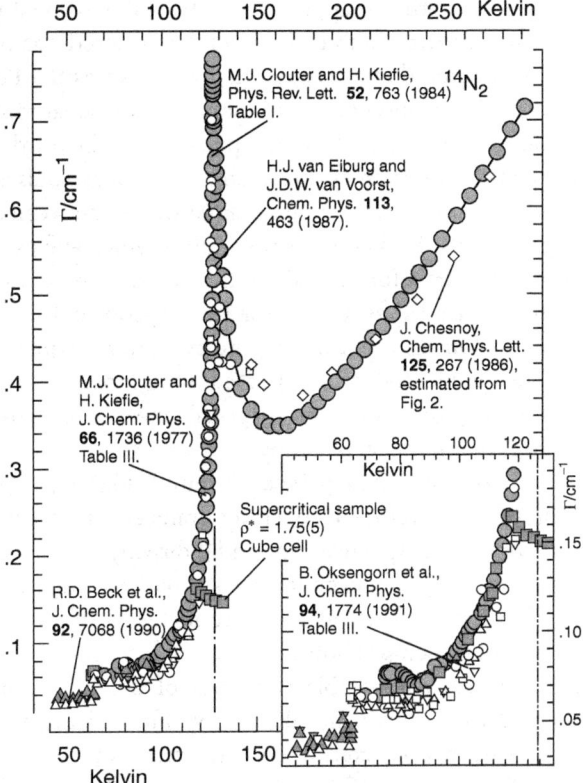

Figure 13.1 Full width at half-height Γ of the Raman band of $^{14}N_2$ along the solid–gas coexistence line and the liquid–gas coexistence line, and in the isochoric state for the critical sample, $\rho^* = 1.00$ (gray circles), and the supercritical sample, $\rho^* = 1.75$ (gray squares), and a comparison with literature data (open symbols). Reprinted with permission from *J. Chem. Phys.* **116**, 8015 (2002). Copyright (2002) American Institute of Physics.

been successful in the study of glasses and also in distinguishing protein conformational substates, they have not been used for liquids where the environment is not static.

Thus, one requires a different set of tools to study inhomogeneously broadened spectrum in liquids. Inspired by the success of the two-dimensional NMR techniques, the 2D-IR spectroscopy is being developed to address this issue [10–13]. This is a method still under progress but already significant new results have been obtained. In this spectroscopy, the inhomogeneously broadened spectrum is interrogated to obtain information about the structure and dynamics of the system. *That is, the cause of the inhomogeneity is turned into an asset!*

In standard 2D-IR spectroscopy, one adopts one of the two approaches, depending on whether the experiments are carried out in the time or in the frequency domain. In both the approaches a two-dimensional plot is generated where the intensity of the emitted light (ω_3) is plotted against the excitation frequency ω_1. In the time-domain experiments, known as the Fourier Transform Method, three short laser pulses are sent in sequences at separations τ_1 and τ_2. The first two pulses are referred to as the pump pulse, while the third pulse is the probe pulse. Usually, the time duration τ_1 (between the two pump pulses) is referred to as the coherence time, while the second time duration τ_2 (between the second pump pulse and the probe pulse) is known as the waiting time. The excitation frequency is obtained by Fourier transforming along the τ_1 axis. The emission frequency is that of the photon emitted after stimulation by the probe pulse.

In the frequency-domain experiments, the excitation frequency is scanned and emission frequency is measured, as expected. One looks for cross correlation in the two-dimensional plot. Along the diagonal one obtains the absorption and emission spectra as in linear spectroscopy.

As mentioned, the two-dimensional infrared (2D-IR) spectroscopy is a promising tool being developed to study transient molecular structure and dynamics. It is a vibrational spectroscopy and it directly interrogates the vibrations of chemical bonds. The off-diagonal peaks give the amplitude of coupling between two vibrations and reveal structure in terms of connectivity, distance, or orientation between chemical bonds.

There have been several notable successes of the 2D-IR spectroscopy. For example, it helped to discover the ultrafast vibrational dephasing of the O–H stretch of water with a time constant of the order of 100 fs, which has been explained successfully by using Kubo–Oxtoby theory discussed above.

2D-IR spectroscopy has been used to determine the lifetime of the hydrogen bond in solution, dynamics of molecular complexes, secondary structure in proteins, to name a few examples. The main difference from 2D-NMR is that the IR spectroscopy provides a time resolution that is a picosecond or faster, so it can capture transient dynamics. The limitation of course is that the coupling between the different modes is not as precisely known as in NMR, hindering precise structure determination.

We have provided a few relevant references [10–14] for further reading on non-linear spectroscopy.

13.9 SUMMARY

Vibrational-phase relaxation remains a highly active area of research. We have discussed here only the basic concepts and some of the recent results and issues. Particularly, we have discussed the elegant Kubo–Oxtoby theory of vibrational-phase relaxation which emphasized the important role of anharmonicity in dephasing. Among the new experimental results, we have discussed the anomalous increase in the rate of vibrational dephasing of nitrogen–nitrogen bond in neat liquid nitrogen near the gas–liquid critical region in the pressure–temperature plane, ultrafast vibrational phase relaxation of O–D stretch in liquid water, and subquadratic quantum number dependence of dephasing. While some of these results can be understood from Kubo–Oxtoby theory, there remains some that are not yet fully understood, but they underline the power of this relaxation method to study various physicochemical processes.

REFERENCES

1. D. W. Oxtoby, *Adv. Chem. Phys.* 40, 1 (1979).
2. D. W. Oxtoby, *Anuu. Rev. Phys. Chem.* 32, 77 (1981).
3. S. Mukamel, *Principles of Nonlinear Optics and Spectroscopy*, Oxford University Press (1995).
4. K. S. Schweizer and D. Chandler, *J. Chem. Phys.* 76, 2296 (1982).
5. K. Tominaga and K. Yoshihara, *Phys. Rev. Lett.* 74, 3061 (1995).
6. M. R. Battaglia and P. A. Madden, *Mol. Phys.* 36, 1601 (1978).
7. E. Gershgoren, Z. Wang, S. Ruhman, J. Vala, and R. Kosloff, *J. Chem. Phys.* 118, 3660 (2003).
8. M. Musso, F. Matthai, D. Keutel, and K. L. Oehme, *J. Chem. Phys.* 116, 8015 (2002).
9. S. Roychowdhury and B. Bagchi, *Phys. Rev. Lett.* 90, 075701 (2003).
10. *Ultrafast Infrared and Raman Spectroscopy* (Practical spectroscopy series, volume 26), Edited by M. D. Fayer Dekker (2001).
11. S. Mukamel, *Ann. Rev. Phys. Chem.* 51, 691 (2000).
12. R. M. Hochstrasser, *Proc. Nat. Ac. Sc.* 104, 14190 (2007).
13. M. Khalil, N. Demirdoven, and A. Tokmakoff, *J. Phys. Chem A* 107, 5258 (2003).
14. E. T. J. Nibbering and T. Elsaesser, *Chemical Reviews* 104, 1887 (2004).

14

Epilogue

In this book we have brought together several different relaxation phenomena in liquids under a common umbrella. Our attempt throughout the book has been to correlate recent experimental results obtained with relatively modern techniques with recent theoretical developments, such as molecular hydrodynamics and mode-coupling theory.

Such close interaction between experiment and theory in this area goes back to the works of Einstein, Smoluchowski, and Kramers. The seminal works of van Hove and de Gennes in describing neutron-scattering data further underlie this closeness in a remarkable fashion. This collaboration between experiment and theory adds additional attraction to the problems in this area.

The focus on the experimental side has been in the use of ultrafast laser pulses to study relaxation directly in the time domain. This has allowed study, initiated in the early 1980s, of relaxation in the picosecond and subpicosecond time scales. This was a remarkable advance because many of the fundamental chemical processes occur precisely in this range and were inaccessible before the 1980s. From that time until recently, an enormous wealth of information has been generated by many groups around the world, discovering many interesting phenomena that fuelled further growth in this field.

Some of the studies successfully undertaken and completed are quite impressive. For example, the problem of unraveling the mechanism of vibrational-energy relaxation of the O–H (or, the O–D) bond of water in liquid water is a case of point. We should also mention the study of ultrafast solvation dynamics in liquid water and acetonitrile. No one really anticipated the existence of such a large component with subhundred femtosecond time constant in the relaxation of solvation energy in these liquids. Furthermore, these results helped in explaining the relative insensitivity of the electron-transfer rate on solvent dynamical properties even for an adiabatic electron-transfer reaction where earlier theories based on overdamped

Smoluchowski-type equation predicted a strong dependence on solvent dynamical properties. Similarly, the explanation of the observed weak dependence of a high barrier chemical reaction on solvent frictional forces necessitated the generalization of Kramers' theory to include the non-Markovian or memory effects of solvent frictional forces.

The seemingly different phenomena studied in the area are often closely related at a fundamental level. This explains why relatively small, although fairly sophisticated, theoretical tools, have been successful in explaining a wealth of phenomena at a semiphenomenological level.

Perhaps the most important understanding that has emerged from all these studies is the importance of intermolecular correlations at molecular- (that is, short-) length scales. By this we mean that dynamical processes are determined largely by the spatial and orientational correlations present at a length scale comparable to molecular diameter. Thus, the friction experienced by a solute molecule in a solvent is determined by the molecules responsible for the first and the second peak in solute–solvent radial distribution function. Thus, for many chemical relaxation processes, the hydrodynamic approach to solvent dynamic influence turned out to be inadequate.

As a result, the theoretical approach to relaxation processes in liquids has taken a mixed approach. While we still use generalized Langevin equation with frequency-dependent friction or, non-Markovian Smoluchowski equation, but we use the frequency- or time-dependent friction obtained from mode-coupling theory. The latter does justice both to the short-time collisional forces and the long-time hydrodynamic forces. The friction so obtained has a biphasic character, when plotted against frequency. It is the high-frequency component of this friction that determines many of the chemical-relaxation phenomena, like activated barrier-crossing dynamics, adiabatic electron-transfer reactions, vibrational energy, and phase relaxation. However, the viscosity of the medium is coupled largely to the low-frequency component of the friction. Interestingly, when viscosity of the medium is varied by changing the temperature and/or the pressure, the high-frequency component of the friction changes along with the low-frequency component, and they both change in the same way. Thus, one can be led to believe that the hydrodynamic description provides a valid description of the solvent effects on relaxation phenomenon, which is of course not correct.

The time scales of the processes considered here vary over a wide range, from subnanosecond to a few tens of femtoseconds, that is, they vary over almost six orders of magnitude. It is indeed gratifying that quantitative description of such processes now seem to be in place, although it took more than 50 years of efforts of the scientific community to arrive at the present rather satisfactory situation. Many laboratories and groups around the world participated in the journey. We hope we have been able to capture a glimpse of the exciting developments that gave rise to much new science in this area of research. The field of course continues to evolve. The new frontiers opened are the dynamics in complex systems, such as protein folding, protein–DNA interaction, and water in various natural and biological environments. The lessons learned in the studies of simple systems are proving to be extremely useful in this new frontier.

Index

Page numbers in *italics* indicate figures and tables.

2D IR spectroscopy. See
    Two-dimensional infrared
    spectroscopy

Absorption spectrum, expressions for, 237–238
Acetonitrile, 103–104, 106, 220–221, *221*
Activation barrier-crossing dynamics
    Grote-Hynes theory and, 132–143, *137, 140, 142*
    Kramers' theory and, 127–132, *130*
    memory effects in chemical reactions and, 132–143, *137, 140, 142*
    microscopic aspects of, 119–126, *120, 121, 126*
    multidimensional Kramers' theory and, 150–153
    multidimensional reaction surfaces and, 144–146, *145*
    overview of, 117–119, *118*, 149–150
    quantum transition-state theory and, 148–149
    transition path sampling and, 146–148, *147*
    transition state theory and, 126–127
    variational transition-state theory and, 143–144
Activation energy, negative, *173*, 173–174
Activationless reactions. See Barrierless reactions
Adiabatic potential energy surfaces, *197*, 197–198
Anharmonic coupling, 280, 282, 285–287
Anharmonicity parameter, 287
Anisotropy rotational, 39, 44
Arcsin law, 163
Attractive forces, vibrational phase relaxation and, 289
Average rate constant, defined, 158
Average solvation time, 93

Bacteriochlorophylls, 256–257
Bader-Berne theoretical framework, 269–270
Bagchi, Fleming, Oxtoby (BFO) model, 167, *167, 172*
Barrier crossing. See Activation barrier-crossing dynamics

Barrierless reactions
    experimental results analysis, 174–177, *176*
    inertial effects in, 166–170, *167, 168, 169*
    main features of, 172–173, *173*
    memory effects in, 170–172, *172*
    multidimensional reaction potential energy surface and, 174
    standard model of barrierless reactions, 158–166
BDL model. See Brownian-dipolar-lattice model
Berne treatment of Markovian theory of collective orientational relaxation, 59–68
Betaine-30, 210, 213
BFO model. See Bagchi, Fleming, Oxtoby model
Bimolecular reactions, diffusion-controlled, 190–193
Binary collision model, 262
Boot strapping, different dynamical quantities and, 30
Born-Oppenheimer approximate wave functions, 197–198, 232
Bottlenecks
    diffusion-controlled bimolecular reactions, 190–193
    diffusion in random Lorentz gas, *183*, 183–184
    diffusion in two-dimensional periodic channel, 181–183
    diffusion over rugged energy landscape, 186–189, *187, 188*
    dynamical disorder, 184–186
    as entropic free-energy barriers, 180
Brillouin doublet, 23
Broadening, line shapes 45, 230, 260, 287–289, 292–293
Brownian-dipolar-lattice (BDL) model, 100–102, *101*
Brownian motion, 73, 192
Brownian oscillator model, 88
Budo formula, 88

Canonical variational transition-state theory (CVTST), 143–144
CARS. See Coherent anti-Stokes Raman scattering
CDX cavity. See Cyclodextrin cavity
Cell models of liquid, 261–262
Centrifugal coupling, 274, 289–290
Charge state, Marcus theory and, 199
Chemical defects, 250–251
Chemical reactions
    diffusion-controlled bimolecular, 190–193
    memory effects in, 132–143, *137, 140, 142*
    nonequilibrium solvation effects in, 107–110, *109*
    stochastic models of, 120
    vibrational-energy relaxation and, 259–260
    vibrational-phase relaxation and, 280
Chlorophylls, 226–227, 256–257
Coherent anti-Stokes Raman scattering (CARS), 45
Cole-Cole plots, 66, *68*
Collective density relaxation, 36
Collective orientational density, 53
Collective orientational relaxation, Markovian theory of, 59–68
Collective relaxation, 38–39
Collision frequency, 261
Colloid coagulation, 190–191

Complex systems, solvation dynamics in, 95–97
Condon approximation, 232
Conjugated polymers, FRET and, 249–252, 252, 253
Constitutive relations, hydrodynamic 21
Continuity equations, hydrodynamic 21
Continuum limit, orientational relaxation and, 67–68
Continuum model theories, 54, 79, 86–93
Coumarin 343, solvation dynamics in water, 103, 104
Critical points, gas-liquid, vibrational phasing near, 292
Crystal violet isomerization, 177, 178
CVTST. See Canonical variational transition-state theory
Cyanide, vibrational-energy relaxation of, 271–272
Cyclodextrin (CDX) cavity, solvation dynamics in, 96

Dark states, FRET and, 255–257, 257
Debye model ,dielectric relaxation 39
Debye relaxation time, 64–65, 87
Decay rate, friction parameter and, 169–170
Defect-coil conformation, 251–252, 252
Defect-cylinder conformation, 251–252, 252
Defects, 250–251
De Gennes narrowing, 25–27, 66–67
Delta function sink, 164–165
Density, hydrodynamics approach and, 20–22

Density relaxation, time scales of, 20
Dephasing. See Vibrational phase-relaxation
Depolarization, fluorescence, relationship between theory and experiments, 39–40
Depolarized (VH) Rayleigh scattering, 36
Dexter mechanism, 230–231, 231
Dextran, 239, 240
Diabatic potential energy surfaces, 197, 197–198
Dielectric constant, fluctuations in, dynamic light scattering and, 34–36
Dielectric friction, as time or frequency dependent, 68–69
Dielectric models
  homogeneous, 86–88
  inhomogeneous, 89, 89–91, 90
Dielectric relaxation
  molecular theory of, 64–65, 67
  orientational relaxation of water and, 72–73
  of pure liquid, 98
  relationship between theory and experiments, 38–39
Diffusion coefficients, 5, 8–9, 218
Dimethyl sulfoxide (DMSO), 75, 76
Dipolar correlation function, 64
Dipole-dipole interactions, 229
Dissipation. See Fluctuation-dissipation theorem
Distance-dependent dielectric function, 89
DNA (deoxyribonucleic acid), 97, 115, 187
Dynamical disorder, overview of, 184–186

Dynamic disorder models, 180, 191
Dynamic exchange model, 91–93, *92*
Dynamic response functions, overview of, 5
Dynamic solvent effects, on electron transfer reactions, 208–210, 220–221
Dynamic structure factor, 22–24, *24*, 36

Echo signals, 45–46, 80, 83–85, *95*
Effective relaxation time, 218
Eigenvalue analysis, 119–122, *120*, *121*
Einstein relation, generalized, 9
Elastic neutron-scattering experiments, 42
Electrolyte solutions, 107, *109*, 111
Electromagnetic field, energy flux density for, 239
Electronic matrix element, 204
Electron solvation, dynamics of, 111–112
Electron transfer reactions. See also Barrierless reactions; Marcus theory
  classical intramolecular vibrational modes and, 210–213, *211*
  classification of, 196–197, *197*
  collective solvent polarization coordinate and, 180
  dynamical solvent effects and, 208–210
  high-frequency vibration modes and, 213–215, *214*
  hybrid model of, 215–216
  multidimensional, theoretical formulation of, 216–220
  overview of, 195, 221–222
  ultrafast solvation in, 107
  in water, acetonitrile, and methanol, 220–221, *221*
Ellipsoid in a sea of spheres (EISS) model, 51–53, *52*
Emission spectrum expressions, 234–237
Enskog friction, 141, 142, *142*, 263
Enskog kinetic theory, 30, 261
Enskog theory, generalized, 27, 29
Entropic bottlenecks. See Bottlenecks
Enzyme kinetics, diffusion over rugged energy landscape and, 186
Euler's equations, 201
Excitation energy migration, 250–251, *251*
Excitation wavelength, barrierless reactions and, 172, *173*, 177
Experiment-theory relationship. See Theory-experiment relationship
Eyring prefactor, 126–127

FDT. See Fluctuation-dissipation theorem
Fermi Golden Rule (FGR), 33–35, *34*, 205, 237, 265
Fluctuation-dissipation theorem (FDT)
  activation barrier-crossing dynamics and, 133–134, 135
  overview of, 8
  solvation dynamics and, 81–82
  vibrational-energy relaxation and, 264
Fluctuations, response functions and, 4–6
Fluid mechanics, basic equations of, 21–22

Fluorescence depolarization, relationship between theory and experiments, 39–40
Fluorescence quantum yield, 157–158
Fluorescence quenching, 47–48, 190
Fluorescence resonance energy transfer. See Förster resonance energy transfer
Fluorescence spectrum expressions, 234–237
Fluorescence Stokes shift, time dependent, 40–41, 45–46, 55–57, 79–80
Fluorescence up-conversion method, 85
Fokker-Planck equations, 14–15, 17, 123, 128, 132, 166–167, 186
Force-force time-correlation function (FFTCF), 264
Förster equation, 230–239, *231*, 255–256
Förster resonance energy transfer (FRET)
  conjugated polymers and, 249–252, *252*, *253*
  derivation of rate equation for, 230–239, *231*
  glucose sensor based on, 239
  historical perspective on, 229–230
  macromolecular dynamics and, 239–243, *240*
  nonorganic dyes as donor-acceptor labels and, 247–249, *248*, *249*
  optically dark states and, 255–257, *257*
  orientation factor and, 252–254, *253*
  overview of, 226–229, *228*, 257–258
  point-dipole approximation and, 254–255, *255*

  single-molecule spectroscopy and, 243–247, *244*, *245*, *246*, *247*
Fourier Transform Method, 294
Franck-Condon transitions, 80, *81*, 174, 175, 198
Free energy surfaces, calculation of, 200–203, *203*
Frequency, viscosity and diffusion constant as function of, 5
Frequency-dependent friction, 220–221, *221*
Frequency domain experiments, 260
Frequency-fluctuation time-correlation function, 45–46, 80, 84–85, 95
Frequency shift, vibrational-phase relaxation and, 281
Friction
  barrier crossing-rate in solution and, 127–132
  frequency dependence of, 138–143
  frequency-dependent diffusion and, 9–10

Gas-liquid critical point, vibrational phase relaxation near, 292
Gaussian sink, 165–166, *166*, 168, 168–169, *169*
Generalized Langevin equations (GLE), 133, 170–171, 263–264
Generalized Smoluchowski equation (GSE), 60–62, 170–171
Geometric bottlenecks. See Bottlenecks
Glucose sensor, FRET-based, 239
Green's function technique, 16, 164–165, 218–219, 222–224, 241
Grote-Hynes generalization of Kramers' theory, 132–143, *137*, *140*, *142*

INDEX

GSE. See Generalized Smoluchowski equation

Hamiltonian dynamics, dynamic disorder models and, 180
Hard-sphere model, 263
Harmonic bath, activation barrier-crossing dynamics and, 134–136
Harmonic expansion, spherical, 57, 62–63
Harmonic potential, as special case of phenomenological approach, 14, 15, 16–17
Heat, hydrodynamics approach and, 20–22
Heaviside functions, 125, 126
Hexane, isomerization of stilbene in, 12–13
Homegoneous broadening, 45, 287–289, 292–293
Homogenous mechanism of dephasing, 287–288
Hybrid model of electron transfer reactions, 215–216
Hydration layers, 91–93, 92, 96–97, 114–115
Hydrodynamics
 for estimating frequency dependence of friction in Grote-Hynes formula, 138–140, 140
 hydrodynamics between self-diffusion coefficient and viscosity and, 24–25
 at intermediate length scales, 27–29
 at large length scales, 20–22
 at large wavenumbers, 25–27

mode-coupling theory and, 29–30
molecular descriptions, 97–98
relations between self-diffusion coefficient and viscosity, 24–25

IBC model. See Isolated binary collision model
Idealized models, of solvation dynamics, 100–102
Inelastic neutron scattering experiments, 42
Inhomogeneous broadening, 45, 260, 287–289, 292–293
Interference, quantum, 275–277
Inverse snow ball effect, 79, 111–112
Ionic solutes, dielectric models and, 89–91
Ions, solvation dynamics of, 105
Isolated binary collision (IBC) model, 261–263
Isomerization, 175–177, 176, 178. See also Barrierless reactions
Isorhodopsin, 175–177, 176
Isotropic Raman linewidth, 43–45, 48, 281–282, 287–289
Isoviscosity plots, 131

Jablonski diagram, 228
Jortner-Bixon model, 213–216

Kerr relaxation, relationship between theory and experiments, 38
Kirkwood's g factor, 71
Kohlrausch-Williams-Watts function, 93
Kolmogorov equation, 16
Kramer's equation. See Fokker-Planck equation
Kramers-Moyal expansion, 123

Kramers' theory
  Grote-Hynes generalization of, 132–143, *137, 140, 142*
  multidimensional, 150–153
  overview of, 127–132, *130*
Kubo, R, 5, 6
Kubo-Oxtoby theory of vibrational line shapes, 282–287

Landau-Placzek formula, 23
Landau-Teller expression, 263–265, 268
Langevin equations, 8, 13–14
Lattice models, of orientational relaxation, 74–75
LDS-750, solvation time in methanol and butanol, 93
Lennard-Jones (LJ) interactions, 102, 113, 263, 289
LH1 ring system, 256–257, *257*
LHNC model. See Linearized hypernetted-chain model, 63
LIC. See Limiting ionic conductivity
Ligand participation, classification of electron transfer reactions based on, 196
Light scattering, 34–36
Limiting ionic conductivity (LIC), 107, *109*
Linear response theory (LRT), 5, 6–7, 124. See also Onsager's linear regression theorem
Liouville approach, stochastic, 6, 15, 186
Liouville equations (noise-free), 124
Longitudinal components, in orientational relaxation, 65–66

Long-time rate constant, defined, 158
Lorentz gas, random, diffusion in, *183*, 183–184

Macro-micro relations, 70–72, 73
Macromolecular dynamics, FRET and, 227, 239–243, *240*
Magnetic resonance experiments, 32, 36–38, 72–73
Marcus expression for electron transfer rate in nonadiabatic case, 205
Marcus parabolic dependence, 205–206, *206*
Marcus reorganization energy, 202
Marcus theory
  derivation of electron transfer rate and, 203–206, *204, 206*
  experimental verification of, 206–208, *208*
  free energy surface calculation, 200–203, *203*
  overview of, 197–198
  reaction coordinate, *198*, 198–200
Markovian theory of collective orientational relaxation (Berne treatment), 59–68
Master equations, overview of, 16
Mean first passage time (MFPT), defined, 188–189
Mean-spherical-approximation (MSA) model, 63
Memory effects
  in barrierless reactions, 170–172, *172*
  in chemical reactions, 132–143, *137, 140, 142*
  in orientational relaxation, 68–70, *70*
Methanol

electron transfer reactions in, 220–221, *221*
solvation dynamics of, 104, 106
MFPT. See Mean first passage time
Micelles, solvation dynamics in, 95–96, 114
Microcanonical variational transition-state theory (MVTST), 144
Microemulsions, solvation dynamics in, 95–96
Mode-coupling theory (MCT)
  for attractive and repulsive forces in vibrational-phase relaxation, 289
  density relaxation, friction and, 20
  for estimating frequency dependence of friction in Grote-Hynes formula, 140–142, *142*
  overview of, 29–30
  solvation dynamics and, 111
Molecular approaches
  to activation barrier-crossing dynamics, 119–126, *120*, *121*, *126*
  De Gennes narrowing and, 25–27
  to dielectric relaxation, 64–65
  dynamics at intermediate length scales, 27–29, *29*
  hydrodynamics at large length scales, 20–22
  hydrodynamics between self-diffusion coefficient and viscosity and, 24–25
  mode-coupling theory and, 29–30
  to nonequilibrium solvation dynamics, 97–100
  to orientational correlations, 54–55, 57–59

overview of, 10, 19–20, 30
Rayleigh-Brillouin spectrum and, 22–24, *24*
self-diffusion coefficient and viscosity and, 24–25
Momentum, hydrodynamics approach and, 20–22
MSA model. See Mean-spherical-approximation model
Multidimensional electron transfer reactions, theoretical formulation of, 216–220
Multidimensional Kramers' theory, 150–153
Multidimensional reaction surfaces, 144–146
Multidimensional transition state theory, 144–146
Multiexponential solvation dynamics, discovery of, 93–94, *94*
MVTST. See Microcanonical variational transition-state theory
Myoglobin, 184

Narrowing, de Gennes, 25–27, 66–67
Navier-Stokes equation, 25, 138
Nee-Zwanzig continuum model, *101*
Negative activation energy, barrierless reactions and, *173*, 173–174
Neutron scattering experiments, relationship with theory, 41–42
Nile Blue system, 210
Nitrogen (liquid), vibrational-energy relaxation in, *274*, 274–275
NMR experiments. See Nuclear magnetic resonance experiments
Noble metal nanoparticles, 247–248, *248*

Nonassociated liquids, orientational relaxation of, 75–76
Nonequilibrium solvation dynamics, 96–97, 114–115
Nonorganic dyes, as donor-acceptor labels in energy transfer, 247–249, *248*, *249*
Nonpolar solvents, solvation dynamics in, 112–113
Nonradiative energy transfer, 227. See also Förster resonance energy transfer
Nuclear magnetic resonance (NMR) experiments, 32, 36–38, 72–73

Occupation number, 281
OD-stretch of water H-O-D, 73–74, *74*
Onsager-Glarum expression, 71, 73
Onsager's inverse snowball effect, 79, 111–112
Onsager's linear regression theorem, 6
Orientational relaxation
 equilibrium and time-dependent correlation functions, 55–57
 experimental observables and, 54, 57
 lattice models of, 74–75
 macroscopic vs. microscopic, 70–72, 73
 Markovian theory of collective (Berne treatment), 59–68
 memory effects in, 68–70, *70*
 as molecular hydrodynamic description of orientational motion, 57–59
 of nonassociated liquids, 75–76
 overview of, 51–55, *52*, 76
 of water (special case), 72–74, *74*
Orientation factor, 252–254, *253*

Ornstein-Zernike relations (molecular/generalized), 55, 66
Oscillator strength, 250
Oster-Nishijima model of barrierless reactions, *160*, 160–161
Oxygen (liquid), vibrational-energy relaxation in, *274*, 274–275

3PEPS. See Three-pulse photon echo peak shift measurements
Perrin, F. and J., 229–230
Phenomenological approaches
 continuum-model descriptions of solvation dynamics, 86–93
 Fokker-Planck equation, 14–15
 harmonic potential as special case, 16–17
 Langevin equation, 13–14
 master equations, 16
 overview of, 3–4, 10, 12–13, 17
 Smoluchowski equation, 15
Photoassisted isomerization, 155, *156*
Photochemical funnels, 250
Photochemical reactions, exactly solvable models for, 159–163, *160*, *162*, *164*
Photoisomerization, ground-state potential energy surface and, 174–175
Photon echo spectroscopy, 80, 83–85, 95
Photosynthesis, 226–227, 256–257, *257*
Picosecond spectroscopy experiments, 271
Pinhole sink model of barrierless reactions, 162–163, *164*, 167, *167*, 212–213

INDEX

Point-dipole approximation, 232, 254–255, *255*
Polarizability
  electron transfer reaction rate and, 208–210
  Kerr relaxation experiments and, 38
  of pure liquid, 98
  Raman linewidth and, 42–43
Polarization function, 201
Polarization relaxation, 97–98, *99*
Polar solutes, dielectric models and, 89–91
Pollak's approach to Grote-Hynes generalization, 134–135
Polyfluorene, 254–255, *255*
Polymer folding, 243
Polymers, conjugated, FRET and, 249–252, *252*, *253*
Polyphenylene vinylene (PPV), 249–250
Population distribution, 16
Potential energy surface reactions, *197*, 197–198, 216–220
Poynting vector, 239
PPV. See Polyphenylene vinylene
Protein folding, 144, 146, 186, *187*
Protein hydration layers, 96–97, 114–115

Quantum correction factors, 269, *270*, 274–275
Quantum effects, 268–270
Quantum interference in vibrational energy relaxation, 275–277
Quantum-mechanical frequency-dependent friction, 267
Quantum-scattering theory, 262
Quantum theory, theory-experiment relationships and, 33–34
Quantum transition-state theory, 148–149
Quenching, fluorescence, 47–48, 190

Radial distribution function, 55
Radiative lifetime, 236
Radiative rate of excited electronic state, 236
Raman line-shape measurements, 43–45, 48, 281–282, 287–289
Raman scattering, coherent anti-Stokes, 45
Random Lorentz gas, diffusion in, *183*, 183–184
Rate law descriptions
  in activation barrier-crossing dynamics, 119, 122–124
  separation of transient behavior from, 124–126, *126*
  vibrational-energy relaxation and, 276–277
Rayleigh-Brillouin spectroscopy (RBS), 22–24, *24*, 36
Rayleigh lines, 23
Rehm-Weller behavior, 206, *207*
Relationships between theory and experiments. See Theory-experiment relationships
Renormalized kinetic theory, 30
Repulsive forces, vibrational phase relaxation and, 289
Response functions, overview of, 4–6
Reverse micelles, solvation dynamics in, 95–96, 114
Rhodopsin, 155, 156, *156*, 175–177, *176*

Ribozyme molecule folding process, 244, 246, 247
Rotational dynamics, FRET and, 253, 253
Rotational friction, defined, 10
Rouse model, 241–242
Rugged energy landscape, diffusion over, 186–190, *187*, *188*

Self-consistency, importance of, 30
Self-diffusion coefficients, 9, 24–25
Semiclassical approximations, 262
Sensitized fluorescence, 229
SET. See Surface energy transfer
Shape, dependence of rotational relaxation on, 52
Shear viscosity, frequency dependence of hydrodynamic friction and, 139, *140*
Simulations
   Grote-Hynes generalization of Kramers' theory and, 142–143
   Kramers' theory vs., 132
   limitations of, 12–13
   of solvation dynamics, 113–115
   of vibrational-energy relaxation, 272–275, *274*, *275*
Single-molecule spectroscopy FRET (SMS-FRET), 49, 243–247, *244*, *245*, *246*, *247*
Single-particle relaxation, 36–37, 56, 57
Slip boundary condition, 52
Slow time constant, 92–93
Smoluchowski equation, 15, 17, 187, 189, 192–193, 212. See also Generalized Smoluchowski equation
Smoluchowski limit, 129, 132, 193
Smoluchowski operator, 217–218

Smoluchowski rate, 129, 191
Smoluchowski-Vlasov-type equations, 26
SMS-FRET. See Single-molecule spectroscopy FRET
Snow ball effect, inverse, 79, 111–112
Solute probe, solvation process of, 80, *81*
Solution phase, effects of solvation on chemical processes in, 106–110
Solvation dynamics
   in aqueous electrolyte solutions, 111
   chronological overview of experimental results, 93–97
   computer simulation studies of, 113–115
   continuum-model descriptions of, 86–93
   effects of solvation on chemical processes in solution phase, 106–110
   of electron solvation, 111–112
   microscopic theories of, 97–100
   nonpolar, 112–113
   overview of, 78–79, *81*
   physical concepts and measurement, 79–85
   relationship between theory and experiments, 40–41, 46
   simple idealized models of, 100–102
   in supercritical fluids, 112
   in water, acetonitrile, and methanol, 102–106
Solvation free energy, nonequilibrium solvation effects and, 107–110
Solvation time correlation function (STCF)
   for coumarin 343 in water, 103
   history of study of, 93, *94*

for important solvents, 95, *96*
overview of, 80–82, *82*
Solvent effects, on electron transfer reactions, 208–210
Solvent reaction time-correlation function, 218
Solvent reorganization energy, 202
Space, response functions and, 5
Spatial averages, 188–189
Spectral broadening, 230
Spectral density, 135
Spectral hole-burning experiments, 292–293
Spectral overlap integral, 227
Spectroscopic ruler, FRET as, 227, 243
Spectroscopy
    Förster resonance energy transfer (FRET), 243–24t
    photon echo, 80, 83–85, 95
    picosecond experiments, 271
    Rayleigh-Brillouin (RBS), 22–24, *24*, 36
    single-molecule FRET (SMS-FRET), 49, 243–247
    two-dimensional infrared (2D IR), 48–49, *293*, 293–294
    ultrafast infrared pump-probe, 73
    vibrational-phase relaxation and, 48–49, 292–294
Spherical harmonic expansion, 57, 62–63
Staircase model of barrierless reactions, 161–162, *162*
Standard model of barrierless reactions, 159, 164, 179
Static disorder, rate processes with, 184–185
Static response functions, overview of, 4–5

STCF. See Solvation time correlation function
Stick boundary condition, orientational relaxation and, 52
Stilbene, 12–13, 47, *47*
Stochastic models, 119–122, *120*, *121*, 275–277
Stockmayer liquid model, 100, 102
Stokes-Einstein relation, 10, 24–25
Stokes shift. See Time-dependent fluorescence Stokes shift
Subquadratic quantum numbers, dephasing rate and, 291–292
Sumi-Marcus model, 210–213, *211*, 215–216
Supercooled liquids, 13, 186
Supercritical fluids, 112, 277, 278
Surface energy transfer (SET), 248
Survival probability, 218–219, 243, 244
Symmetrized anticommutators, 267

TCF. See Time correlation functions
TDFSS. See Time-dependent fluorescence Stokes shift
TDGL approach. See Time-dependent Ginzburg-Landau approach
Tetrahymena thermophila, 244, 246, 247
Tetraphenylporphyrin, 254–255, *255*
Thallium, 229
Theory-experiment relationships
    coherent anti-Stokes Raman scattering, 45
    dielectric relaxation, 38–39
    dynamic light scattering, 34–36
    echo techniques, 45–46
    fluorescence depolarization, 39–40
    fluorescence quenching, 47–48

Theory-experiment relationships (*Contd.*)
  Grote-Hynes generalization of Kramers' theory, 142–143
  Kerr relaxation, 38
  Kramers' theory, 131–132
  magnetic resonance experiments, 32, 36–38
  Marcus theory, 206–208, *207, 208*
  neutron scattering, 41–42
  orientational relaxation and, 54, 57
  Raman line-shape measurements, 43–44
  single-molecule spectroscopy, 49
  solvent dynamics, 40–41
  two-dimensional infrared (2D IR) spectroscopy, 48–49
  ultrafast chemical reactions, 47, *47*
  vibrational-energy relaxation and, 265, 271–272
  vibrational phase relaxation and, 290–292
Three-pulse photon echo peak shift (3PEPS) measurements, 80, 83–85, *95*
Time, response functions and, 5
Time correlation functions (TCF)
  to activation barrier-crossing dynamics, 124–126, *126*
  defined, 6
  diffusion, friction, viscosity and, 8–10
  fluctuation-dissipation theorem and, 8
  linear response theory and, 6–7
  response functions, fluctuations and, 4–6
  theory-experiment relationships and, 34, *34*

Time-dependent fluorescence Stokes shift (TDFSS), 40–41, 45–46, 55–57, 79–80
Time-dependent Ginzburg-Landau (TDGL) approach, 59
Time-dependent interaction Hamiltonian, 234–235
Time-dependent polarization vector, 98–99
Time-dependent vibrational wave function, occupation number and, 281
Time scales
  of molecular relaxation processes, 33, *33*
  relaxations in dense liquids and, 20
  of solvation dynamics, 82
  of vibrational-energy relaxation, 260
  of vibrational-phase relaxation, 282
Time-symmetrized anticommutator, 267
Torque-torque time-correlation function (TTTCF), *101*, 101
TPM dyes. See Triphenyl methane dyes
Transient behavior, separation from rate law descriptions, 124–126, *126*
Transition dipoles, 254
Transition path sampling, 146–148, *147*
Transition-state theory (TST), 126–127, 143–144, 144–146
Translational diffusion, solvation dynamics and, 98–99, *99*
Translational motions, in orientational relaxation, 52, 64, 65–66

INDEX

Transverse components, in orientational relaxation, 65–66, 67, 68
Triphenyl methane (TPM) dyes, 3, 157, 174–175
TST. See Transition-state theory
TTTCF. See Torque-torque time-correlation function
Tunneling effects, quantum transition-state theory and, 148–149
Two-dimensional infrared (2D IR) spectroscopy, 48–49, 96, 293, 293–294
Two-dimensional periodic channel, diffusion in, 181, 181–183

Ultrafast, sub-100 fs decay, 83–85, 104–106
Ultrafast chemical reactions, relationship between theory and experiments, 47, 47
Ultrafast infrared pump-probe spectroscopy, 73
Ultrafast solvation dynamics
  electron transfer reactions and, 220–222
  subpicosecond, discovery of, 94–95, 95, 96
  vibrational-energy relaxation and, 271–272

Variational approximation method, reaction rate 241
Vibrational-energy relaxation (VER)
  computer simulation studies of, 272–275, 274, 275
  electron transfer reactions and, 210–216
  experimental studies of, 271–272
  of high frequency modes, 268–270
  isolated binary collision model of, 261–263
  Landau-Teller expression and, 263–265
  overview of, 259–261, 260, 279
  quantum interference effects in three-level system, 275–277, 278
  supercritical fluids and, 277, 278
  weak-coupling model and, 265–268
Vibrational lifetime, 291
Vibrational-phase relaxation (VPR)
  attractive and repulsive forces and, 289
  coherent anti-Stokes Raman spectroscopy and, 48
  experimental verification of, 290–292
  homogenous vs. inhomegoneous linewidths and, 45, 287–289, 292–293
  IR absorption spectroscopy and, 45–47, 48
  Kubo-Oxtoby theory of vibrational line shapes and, 282–287
  multidimensional IR spectroscopy and, 292–294, 293
  overview of, 280–282, 294–295
  Raman linewidth and, 42–43, 48
  two-dimensional infrared spectroscopy and, 48–49
  vibrational dephasing near gas-liquid critical point and, 292
  vibration-rotation coupling and, 289–290
Vibrational-transition rate, 268
Vibration-rotation coupling, 289–290

Viscosity
  barrierless reactions and, 156, 157, 166–170
  defined, 5
  as measure of frictional force on molecule, 10
  rate turnover in barrierless reactions and, 166–170, *167*, *168*, *169*
  and self-diffusion coefficient, hydrodynamics between, 24–25
Vision transduction process, 155, *156*
VPR. See Vibrational-phase relaxation

Walden product, *108*
Water
  electron transfer reactions in, 220–221, *221*
  orientational relaxation as a special case, 72–74, *74*
  solvation dynamics of, 94, 102–103, 105–106, 113
  supercritical, solvation dynamics in, 112
  vibrational-energy relaxation of, 272–274
Wavelength, excitation, barrierless reactions and, 172, *173*, 177
Wave numbers, 5, 25–27
Weak-coupling model, 265–268
WF theory. See Wilemski-Fixman theory
Wilemski-Fixman (WF) theory, 240–243

Zeolite cages, 181
Zero-frequency friction, 171, 263
Zero friction limit, *129*, 129–130
Zusman model, 208–209
Zwanzig derivation, 14
Zwanzig lattice model, 100–102, *101*
Zwanzig, dynamic disorder models, 181–184
Zwanzig, diffusion on a rugged landscape 186